PRINTREADING
for HEAVY COMMERCIAL CONSTRUCTION

Third Edition

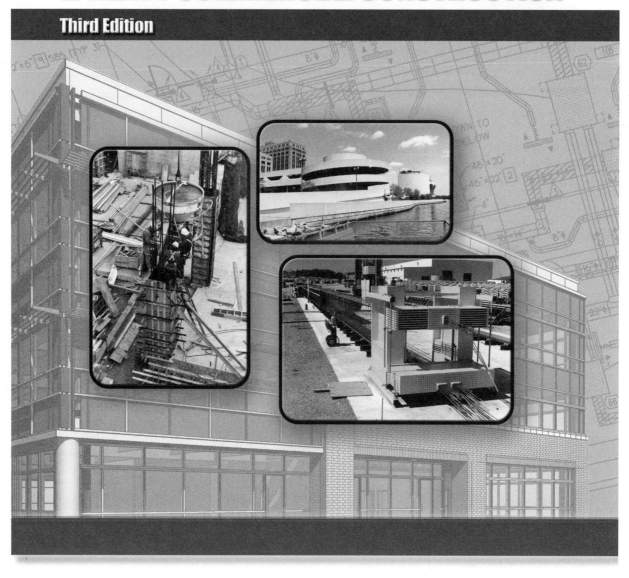

atp AMERICAN TECHNICAL PUBLISHERS, INC.
ORLAND PARK, ILLINOIS 60467-5756

Leonard P. Toenjes

American Technical Publishers, Inc., Editorial Staff

Editor in Chief:
Jonathan F. Gosse
Vice President—Production:
Peter A. Zurlis
Art Manager:
James M. Clarke
Technical Editor:
Charles A. Vescoso Jr.
Copy Editor:
Jeana M. Platz
Cover Design:
Jennifer Hines
Illustration/Layout:
Thomas E. Zabinski
Multimedia Manager:
Carl R. Hansen Jr.
CD-ROM Development:
Robert E. Stickley
Daniel Kundrat

3 4 5 6 7 8 9 – 10 – 9 8 7 6 5 4 3 2 1

Printed in the United States of America

ISBN 978-0-8269-0461-4

 This book is printed on recycled paper.

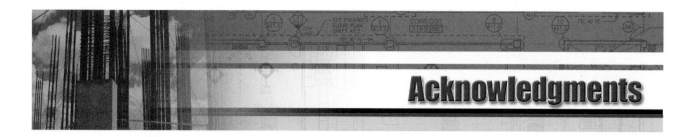

Acknowledgments

The author and publisher are grateful to the following companies and organizations for providing plans, technical information, and assistance:

American Hardwood Export Association

APA—The Engineered Wood Association

Butler Manufacturing Company, Buildings Division

Carlson Software/The Engineering Groupe, Inc.

INTEGRUS Architecture, P.S.

John Deere Construction & Forestry Company

Joint Center for Higher Education

Leica Geosystems

National Wood Flooring Association

Portland Cement Association

Reed Manufacturing Co.

Republic Steel Corporation

Riley Engineering, Inc.

RoseWater Engineering, Inc.

Spokane Intercollegiate Research and Technology Institute

Symons Corporation

Titus

U.S. Green Building Council

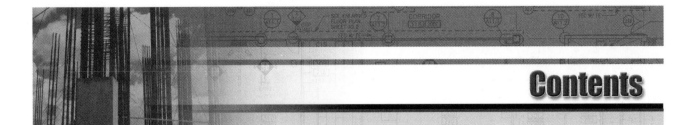

Contents

1 Types of Construction 1

Owners, architects, contractors, suppliers, and tradesworkers play an integral role in commercial building construction. Up-to-date information regarding building methods, materials, standards, and codes is essential to a successful construction project. Prints and specifications must be properly read and interpreted by all members of the building team to ensure that the building design aligns with the needs of the building owner and inhabitants.

2 Specifications 21

Construction of industrial and commercial buildings requires working drawings and written specifications. Specifications and prints describe the entire building process and project. Standard forms, language, and format provide direction for all building team members involved to ensure clear and accurate communication. The specifications must be completely reviewed during the bidding and the construction process to fully understand the complete building project.

3 Sitework 49

Site plans illustrate the existing building site conditions and the layout of the planned final construction. Site plans include topographical, paving, landscape, and detail drawings for drainage and underground piping. Site plans may also indicate locations for traffic control and security features. The planned locations of new construction are determined and shown on site plans after existing conditions are determined. Site plans contain extensive information about sitework that must be completed before building construction begins.

4   Structural Steel Construction

75

Structural steel construction involves the use of steel members to construct the structural framework for a building or other structure. Structural steel is used in the construction of industrial buildings, storage structures, high-rises, and bridges. Four general structural steel construction methods are beam and column, long span, wall bearing, and pre-engineered metal building construction. Joints between adjacent steel members are secured with bolts or by welding. After the structural framework is erected, fastened, and braced, panels are commonly attached to the frame.

Review Questions   99
Trade Competency Test   103

5   Reinforced Concrete Construction

109

Reinforced concrete is concrete containing tensile reinforcement in order to resist forces exerted on a structure. The structural properties of concrete and reinforcement are combined to achieve exceptional overall tensile and compressive strength. Reinforced concrete may be cast-in-place or precast. Cast-in-place concrete is concrete that is deposited in the place where it hardens as part of the structure. Precast concrete is concrete cast at a job site and lifted into place, or formed and cast at a casting yard, transported to the job site, and lifted into place.

Review Questions   135
Trade Competency Test   139

6   Mechanical and Electrical Systems

145

Mechanical prints include information regarding plumbing pipe and fixtures, heating and cooling equipment, air circulation and ventilation equipment, and ductwork. Fire protection systems, designed in accordance with National Fire Protection Association standards and governmental regulations, are also included as part of the mechanical prints. Electrical systems include connections to power supplies, wiring and luminaires, power outlet provisions, electrical equipment attachment and power installation, and wiring and finish details for communications and alarm systems.

Review Questions   171
Trade Competency Test   175

7   Finish Construction

181

Building finishes are applied after foundation systems are placed; the structural skeleton is erected; mechanical, electrical, plumbing, heating, ventilating, and air conditioning systems are roughed in place; and floor and roof structures are built. Finish construction involves the use of exterior materials applied for appearance and weather protection and interior finishes for floor, wall, and ceiling coverings. Architectural prints, including plan views, exterior and interior elevations, sections, and details, contain the majority of finish information.

Review Questions   203
Trade Competency Test   207

8   Plans—SIRTI Building
(Spokane Intercollegiate Research and Technology Institute)

Success in commercial construction requires the ability to accurately read and interpret prints by obtaining information from various portions of the prints and sections of the specifications. Cross-referencing between architectural, structural, electrical, and mechanical prints, as well as their associated details, allows everyone on the building team to completely visualize the intentions of the architect, owner, and other building team members. Reviewing each room and area within a building and studying all elements of the structure provides tradesworkers with the ability to properly plan and construct each building element in the proper sequence.

9   Final Exam

Appendix

Glossary

Index

Interactive CD-ROM Contents

- Quick Quizzes®
- Illustrated Glossary
- Flash Cards
- Print Sets
- Printreading Tests
- Media Clips
- Link to ATPeResources.com

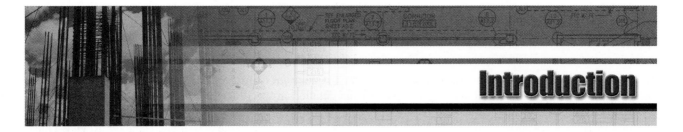

Printreading for Heavy Commercial Construction, 3rd Edition, provides printreading experience relating to commercial construction, including sitework, mechanical and electrical systems, and structural steel, reinforced concrete, and finish construction. This book covers the skills needed to interpret plans and printreading symbols commonly included on prints for large commercial structures. Expanded topics include construction materials and methods, the roles of building-process participants, project delivery methods, and *CSI MasterFormat 2004.* New topics in this edition include:

- The latest LEED® green building requirements and gray water systems
- Solar thermal heating and geothermal systems
- GPS systems and sustainable roofing

Prints from three heavy commercial construction projects—a total of 58 sheets—are included in the storage folder that accompanies *Printreading for Heavy Commercial Construction.* Sheets 1 and 3 to 30 provide printreading experience using the Trade Competency Tests included at the ends of Chapters 2 to 7. Sheets 31 to 46 are an integral part of the walk-through in Chapter 8 and can be referred to when covering related topics. Chapter 9—Final Exam comprehensively assesses mastery of the information presented throughout the book using Sheets 2 and 47 to 58.

Information on using the *Printreading for Heavy Commercial Construction* CD-ROM is included on the last page of the book. To obtain information about related training products, visit the American Technical Publishers web site at www.go2atp.com.

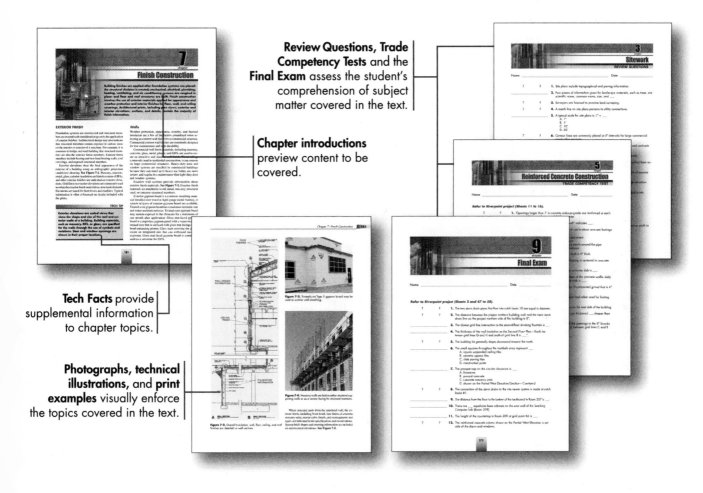

Review Questions, Trade Competency Tests and the **Final Exam** assess the student's comprehension of subject matter covered in the text.

Chapter introductions preview content to be covered.

Tech Facts provide supplemental information to chapter topics.

Photographs, technical illustrations, and **print examples** visually enforce the topics covered in the text.

Using *Printreading for Heavy Commercial Construction*

To obtain maximum benefit from *Printreading for Heavy Commercial Construction,* 3rd Edition, read each chapter carefully, noting the new terms introduced, the elements detailed in the illustrations, and the related information provided in the tech facts and photographs. Identify and review the key concepts presented in the chapter before completing the Review Questions and/or Trade Competency Test included at the end of each chapter. Review Questions are based on the content included in the chapter. Trade Competency Tests are based on prints associated with particular chapters and with reference material included in the Appendix. Each of these learning activities provides an opportunity to reinforce and apply the concepts presented in the chapter.

Interactive CD-ROM Features

The *Printreading for Heavy Commercial Construction,* 3rd Edition, CD-ROM in the back of the book is a self-study aid designed to supplement content and learning activities in the book. The CD-ROM includes Quick Quizzes®, an Illustrated Glossary, Flash Cards, Print Sets, Printreading Tests, Media Clips, and a direct link to ATPeResources.com.

Using This Interactive CD-ROM provides information about components included on the CD-ROM.

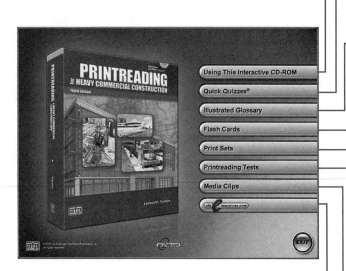

Quick Quizzes® reinforce fundamental concepts, with 10 questions for each chapter.

Illustrated Glossary provides a helpful reference to common printreading and construction terms. Selected terms are linked to interactive illustrations or media clips.

Flash Cards provide a self-study/review tool to match terms and definitions found in the textbook.

Print Sets consists of an electronic version of the prints used in the textbook. The electronic prints can be viewed on screen for easy reference.

Printreading Tests present an opportunity to identify and interpret abbreviations and symbols commonly included on prints.

Media Clips consist of animated illustrations and video clips that reinforce and expand upon textbook content.

ATPeResources.com provides a comprehensive array of instructional resources including Internet links to manufacturers, associations, and related resources.

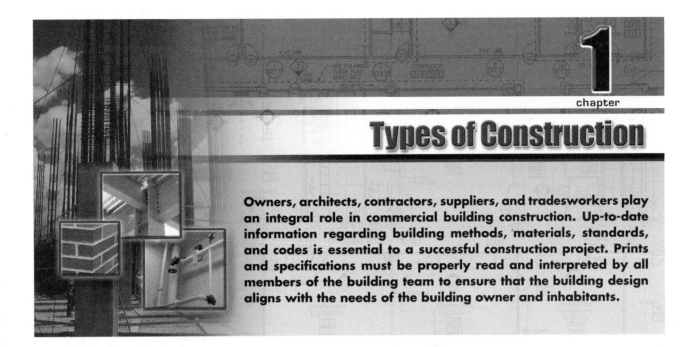

Types of Construction

Owners, architects, contractors, suppliers, and tradesworkers play an integral role in commercial building construction. Up-to-date information regarding building methods, materials, standards, and codes is essential to a successful construction project. Prints and specifications must be properly read and interpreted by all members of the building team to ensure that the building design aligns with the needs of the building owner and inhabitants.

BUILDING PLANNING

A variety of types and styles of buildings are constructed to fulfill many different functional and artistic requirements. Industrial and commercial construction involves the planning and building of many types of structures including roads, bridges, retail establishments, manufacturing facilities, office buildings, schools, places of worship, public gathering places, and many other large structures. **See Figure 1-1.** All components included in a large set of prints must be accurately interpreted for a construction project to be successful.

Building Process Participants

To properly plan and build complex structures, many individuals must communicate with each other in a clear and concise manner. Owners, architects, engineers, contractors, suppliers, and tradesworkers involved in the construction project must all have printreading skills in order to understand the overall size and scope of the building project.

Owners. An *owner* is the purchaser of a construction project. The owner, as an individual or organization, must fully and accurately communicate the building needs to an architect to begin the building process. Building considerations include location, size, placement, appearance, use, and cost. Owners or their representatives can benefit from printreading knowledge to ensure their building needs are met and potential problems can be resolved before actual construction begins.

Portland Cement Association

Figure 1-1. A variety of construction systems, including reinforced concrete, structural steel, and masonry construction, are used to build heavy commercial structures.

Architects. An *architect* is a construction professional who designs and creates plans for a structure. An architect's primary responsibility is to listen to an owner and accurately interpret their needs and requirements into a building design and plan. Architects meet with owners and develop prints and specifications to guide the construction process. **See Figure 1-2.** The prints and specifications are also used by owners and governmental agencies to obtain financing, competitive bids, and building permits. An architect must ensure that the building design conforms to all structural, building, mechanical, electrical, and fire protection codes.

Figure 1-2. Architects and owners define the scope of a building project and communicate their project needs to building team members using prints and specifications.

Engineers. An *engineer* is an individual educated in a specific discipline who is trained to analyze data and solve issues related to a construction project. Structural, mechanical, environmental, soil, and electrical engineers assist an architect in the planning process. Engineers provide accurate information concerning the allowable stresses that various components of a structure can bear including load-bearing capacity of soil, live and dead loads placed on structural members, electrical loads, and the ability of mechanical systems to properly heat, cool, and ventilate an area. **See Figure 1-3.**

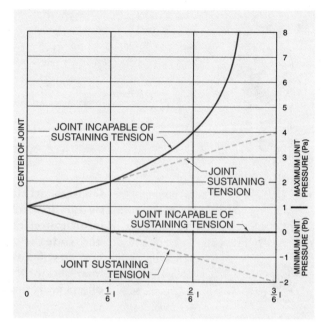

Figure 1-3. Engineering data and charts provide the specific scientific information necessary to ensure the integrity and safety of a structure.

General Contractors/Construction Managers. General contractors and construction managers provide a broad range of services to owners and the construction team. A *general contractor* is an individual or company that agrees to fulfill an entire building agreement with various items or types of work to be completed such as carpentry and electrical and plumbing work. A *construction manager* is an individual or company managing administrative and technical responsibilities of a construction project.

General contractors and construction managers use prints and specifications to generate estimates for quantities of building materials and labor times. Estimators are employed by contractors and construction managers to calculate various types of preliminary and final estimates. An *estimator* is a construction professional who determines overhead expenses, material quantities and costs, and labor needs and costs and assembles bids from others in the construction team to calculate project cost estimates. The prints and specifications are also used to plan and perform the actual construction when a contract is awarded.

Both general contractors and construction mangers utilize subcontractors to perform varying degrees of on-site construction work. General contractors commonly use in-house tradesworkers to perform a significant portion of the on-site construction work but may hire subcontractors for specialized work such as electrical, HVAC, structural steel, and plumbing. Construction managers commonly use a construction system that minimizes or eliminates the need to hire tradesworkers directly and instead contracts subcontractors to do the majority, if not all, of the on-site work.

Subcontractors. Much of the specific trade work performed on a construction project is provided by subcontractors. A *subcontractor* is an individual or company that performs work in one specific trade area on a construction project and works under an agreement with the general contractor. For example, electrical work is installed on a commercial construction project by an electrical subcontractor who has the corporate specialization and tradesworkers dedicated to this one area of construction. Subcontractors dedicate most of their focus to one specific area of the prints when providing a project bid to a contractor or construction manager.

Suppliers. A *supplier* is a company that supplies materials, building components, or equipment to contractors and subcontractors. Architects, engineers, and contractors must rely on suppliers for the latest information concerning available materials. Suppliers rely on architects, engineers, and contractors to specify their materials in the construction project. Specifications often mention a specific material or building component by manufacturer name and product number. **See Figure 1-4.**

MANUFACTURER PRODUCT NUMBER

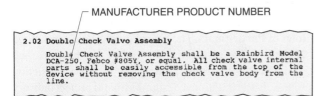

```
2.02 Double Check Valve Assembly

     Double Check Valve Assembly shall be a Rainbird Model
     DCA-250, Febco #805Y, or equal. All check valve internal
     parts shall be easily accessible from the top of the
     device without removing the check valve body from the
     line.
```

Figure 1-4. Specifications may note building products by manufacturer name and product number.

Project Consultants. As construction materials and methods become more complex, additional consultants are becoming common participants on the construction team. Two examples of project consultants are a LEED® Accredited Professional (AP) and a commissioning agent. A *LEED Accredited Professional (AP)* is consultant with extensive knowledge of green building practices and the Leadership in Energy and Environmental Design (LEED) rating system and is employed by an owner, contractor, or construction manager to ensure that environmental processes are followed where specified.

A *commissioning agent* is a project consultant who ensures that all building systems perform according to the design intent and the owner's operational needs. A commissioning agent is employed by an owner, construction manager, or contractor to inspect the completed structure and various systems to insure that they are performing according to the design specifications. For example, a commissioning agent will check the airflow through a particular completed building structure to insure it meets all temperature, flow, and cleanliness requirements for specialized design requirements.

Tradesworkers. It is ultimately the responsibility of the superintendents, foremen, and journeymen on the job site to properly build the structure. Superintendents are responsible for the overall day-to-day operations of job site construction. A *superintendent* is a construction professional who coordinates the many diverse operations on a construction site, such as managing the delivery and placement of project materials, working with various construction team members on labor allocations, and coordinating schedules for subcontractors.

A *foreman* is a skilled trades professional who manages a small trade-specific crew to perform work projects on a construction site. A *journeyman* is a skilled trades professional who accomplishes the actual tasks of construction such as setting concrete forms, installing electrical conduit, or setting plumbing fixtures in place. Knowledge of printreading by all the tradesworkers helps a construction project to be completed efficiently.

TECH TIP

To become a LEED® Accredited Professional (AP), an individual must pass a general knowledge exam, a specialty exam, and provide documented LEED project experience.

Project Delivery Methods and Components

It is in the best interest of the owner and architect to obtain the construction services that are most suitable for their project needs and ensure that the construction team selected to build the project meets certain quality and reliability standards. In some states and municipalities, contractors and subcontractors must be licensed to perform various types of construction work. In other states, architects and owners must make judgments concerning the skill and integrity of contractors.

Some of the factors considered by the owner during the selection of the construction team include price, quality, and reliability standards and the capacity of the construction team to meet other owner requirements related to technology and project delivery system requirements. Owners also consider items such as bonding capacity, association affiliation, quality of previously completed projects, and references from prior and current customers when selecting the construction team.

There are a variety of project delivery systems evolving as the construction marketplace changes. Design-bid-build has been the most common project delivery system for many years for both public and private construction jobs. The new project delivery systems that are emerging include design-build and integrated project delivery. These delivery systems allow contractors and construction managers to provide a broad range of services that bring together all the construction service components, such as architectural services, engineering of the project, oversight of the bidding process, and on-site building services.

Design-Bid-Build. The traditional project delivery system is known as design-bid-build. *Design-bid-build* is a project delivery system where multiple contractors competitively bid against one another on a construction project. With this system, an architectural firm works with the owner to design and engineer a project. Contractors and/or construction managers review the project and solicit bids from subcontractors as appropriate. After all bids are received and compiled, a proposal is submitted to the owner containing all of the elements the owner and architect require. The owner then selects the bid that most closely meets the requirements of the project.

Public owners, typically agencies of federal, state, or local government, are often required by law to select the lowest price bid from among the qualified bidders. Private owners have greater flexibility in construction team selection. They often select the bidder with the lowest price but may choose to select a higher-priced bid that more closely aligns with their overall project requirements.

Design-Build. One alternative project delivery system is design-build. *Design-build* is a project delivery system where an owner negotiates the price of the construction project with a contractor. The owner selects a construction team that includes both design and construction services to assist with putting the project in place. Various design-build construction teams will submit proposals to the owner that describe the design and building services available through their project team. The owner then reviews the qualifications of the various submittals and selects the construction team most aligned with the owners requirements.

Effective communication between construction professionals is essential to ensure a project is completed on time and on budget.

Integrated Project Delivery. *Integrated project delivery (IPD)* is a project delivery method that provides an open information exchange for all those involved in the construction process. Owners, architects, engineers, construction managers, contractors, subcontractors, and suppliers all share information concerning schedules, pricing, and work processes. With this method, everyone on the construction team gains from productivity improvements and looses from substandard performance. The proper formation of an IPD team is an important part in utilizing this delivery system.

Standard Forms. The American Institute of Architects (AIA) series of construction documents and the Consensus-DOCS construction documents are the two most common standard contract forms for bidding on construction projects. **See Figure 1-5.** These contract forms cover a broad range of construction delivery methods, and members of the construction team provide reliable and consistent information for all involved in the bidding and construction process. Governmental agencies also commonly use their own proprietary standardized bidding documents for federal, state, and local agency projects.

Roles of Participants. The type of project delivery method selected by the owner has an impact on the roles and requirements of various participants in the construction process. In the traditional design-bid-build project delivery system, architects distribute plans and specifications to interested contractors, construction managers, and subcontractors with a closing date for bids.

Contractors and subcontractors prepare their bids according to the plans and specifications and submit bids for their work. Subcontractors bid their particular component of a project and submit their costs to a general contractor or construction manager. The general contractor or construction manager combines the subcontractor bids with the costs for general contracting services to develop a total project bid cost.

Alternative project delivery systems such as design-build and integrated project delivery have changed participant roles. Prior to beginning work on a project, it is important for all participants on the project team to fully review the contract documents to determine the project delivery system required by the owner and completely understand the impact this particular delivery system has on the project.

For the traditional project delivery system, after the bid closing date, the architect, owner, or governmental agency opens and reviews the bids and determines if a qualified bid has been received within their projected cost range. If not, the project may be modified and rebid. If the bids are satisfactory, a contract is typically awarded to the lowest-priced qualified bidder. Permits are obtained, materials are ordered, and construction begins.

Figure 1-5. Standard bidding forms provide all participants in the bidding process with a common method for communicating bid information.

In other project delivery systems, choice of the construction team can be more subjective. A team may be chosen based on the quality of the overall design, the schedule suggested, the members of the construction team, or other elements.

Building Information Modeling. The use of technology has had impacts on the roles of participants in the construction process. *Building information modeling (BIM)* is an integrated, electronically managed system that aligns all working drawings, structural drawings, and shop drawings into a consistent system. **See Figure 1-6.** Architects, engineers, contractors, subcontractors, and suppliers all contribute to the electronic information for the computerized model. This sharing of information across various professions creates the need for close integration and collaboration of the participants on a construction project. One of the features of a BIM system provides clash reports that allow for avoidance of on-site conflicts between various building components. For example, a clash report on a BIM project may indicate that a duct system being designed by the mechanical contractor will run into a piping system designed by the fire suppression contractor. Resolution of these clashes prior

to ordering and fabricating materials for the project helps reduce costs and increase on-site productivity.

Vico Software, Inc.

Figure 1-6. Building information modeling is an integrated, electronically managed system that aligns all working drawings, structural drawings, and shop drawings into a consistent system and allows the sharing of information across various professions.

Materials and Methods

Many recent technological developments in construction materials and methods have created the opportunity for structures to be built that could not have been built in the past. Some of the technological developments include newly developed materials and new uses for existing materials, new uses for technology in many aspects of the construction process and the finished building systems, and new construction procedures and techniques based on environmental and sustainability concerns and advances in computer technology, tools, and equipment.

Materials. Engineering developments in the design and properties of new materials have provided an abundance of alternatives for structural engineers and architects. The timely exchange of the most current information between engineers, product designers and manufacturers, building owners, and architects is essential to ensuring that the most appropriate materials become part of the structure. Catalogs and computerized information media available on the Internet and in a variety of electronic formats provide a wealth of information about the latest developments in materials. **See Figure 1-7.**

Planners of modern structures determine the proper uses of many different types and configurations of a broad range of materials, such as concrete blends and products, steel shapes and types, wood and wood products, metals, gypsum products, plastics, glass types, and fiberglass. Where the use of sustainable construction techniques and materials is required by the owner, additional information concerning the environmental characteristics of materials is taken into account.

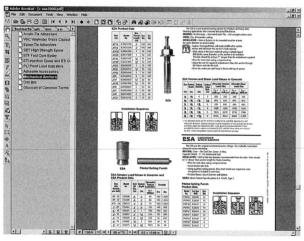

Figure 1-7. Manufacturer catalogs are commonly distributed via the Internet or on CD-ROM.

In addition to new materials, careful planning of mechanical and electrical equipment systems is required for automated building systems. A variety of devices are installed in a building to control lighting, landscape watering, heating and cooling, ventilation, communication, and low-voltage systems such as computer networks, security systems, and fire alarm systems. Many of these enhancements in an automated building system are installed to increase energy efficiency and meet environmental standards. Workers with a high level of technical expertise and printreading knowledge are required to ensure proper installation and operation of automated building systems.

Methods. New construction methods have significantly changed the length of time required to complete a construction project, the skills required of the workers involved in the project, and the procedures for building construction. Environmental issues, computerization, and prefabrication play an ever-increasing role in building construction. The use of the LEED® green building system is becoming more prevalent. The LEED green building system is a certification system created by the U.S. Green Building Council (USGBC) to establish common standards of measurement for green building components, to attempt to impact design practices, and to recognize environmental leadership in building construction through competitions and increasing consumer awareness.

The goal of green building design, often called sustainable design, is to create high-performance buildings. The green building concept evolved from various concerns and experiences including the awareness for increased energy efficiency, an emphasis on reducing waste through recycling efforts, water scarcity in various sections of the country, and the "sick building syndrome."

There are six criteria evaluated for LEED certification. The six criteria are sustainable sites, water efficiency, energy and atmosphere, materials and resources, indoor environmental quality, and innovation and design process.

Sustainable site factors include erosion and sedimentation control, stormwater management, and reduced site disturbance. Water efficiency factors include water use reduction, water-efficient landscaping, and innovative wastewater technologies. Energy and atmosphere factors include sources of renewable energy, optimizing energy performance, and amount of ozone depletion. Material and resource factors include construction waste management, resource reuse, recycled content, and the use of rapidly renewable materials. Indoor environmental quality factors include the use of low-emitting paints; types of carpet, composite wood, adhesives, and solvents; thermal comfort; and minimum indoor air quality performance standards. Innovation in design factors include the use of new and original design ideas that promote green building concepts.

Points are awarded for each factor successfully fulfilling the desired goal. The points are totaled, and the overall building performance is rated based on four levels of LEED certification. The four levels of LEED certification from lowest to highest are Certified, Silver, Gold, and Platinum. A LEED certification checklist is provided by the U.S. Green Building Council to measure attainment of these standards. **See Figure 1-8.**

New methods of scheduling and sharing information between the owner, architect, engineers, contractors, subcontractors, material suppliers, and others involved in the construction project are changing building construction. The use of project-specific web sites allows building team members to quickly share information and collaborate via the Internet.

A project-specific web site may include digital photos of construction progress, requests for information (RFIs) from various building team members, schedules, change order requests, and other information specific to the construction project. This system for managing information flow throughout the construction team increases productivity throughout the project. Video cameras may be installed at various locations around the job site to allow building team members to view the building in real time. At the completion of a construction project, the information from a project-specific web site can be electronically archived to create a record for future building owners and tenants.

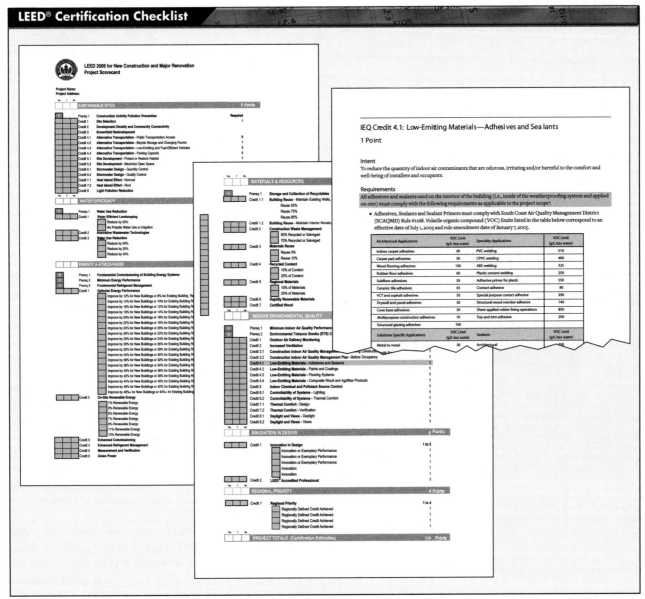

U.S. Green Building Council

Figure 1-8. A LEED® certification checklist defines certification levels that can be achieved by meeting a minimum number of required and optional credits.

Lean construction is another project management system for building projects. *Lean construction* is a project management system that utilizes various information and planning processes adopted from manufacturing to maximize value and minimize waste. Lean construction analyzes the construction process and seeks to increase efficiency and minimize waste at every level of the construction process including planning, scheduling, and material and work flow.

The use of large prefabricated units has shortened the time necessary to complete many construction projects. As BIM systems provide more accurate models of a project that include information from all members of the building team, larger and more complex prefabricated units can be constructed without fear of the need for on-site modification. Developments in lifting and transportation equipment have allowed larger components to be set in place as total units.

New tools and procedures have increased the skill levels needed by tradesworkers on a construction site. For example, the use of various types of laser equipment for operations as diverse as surveying, excavating, and finish ceiling installation requires knowledge and skills that were not necessary in the past. **See Figure 1-9.** Procedures such as the installation of tilt-up concrete panels and large precast components place additional demands on the expertise of all tradesworkers involved in a construction project.

Leica Geosystems

Figure 1-9. A rotary laser level combined with a laser detector mounted on an excavator arm allows precise control of excavation depth.

Geographic Requirements

Planners must take a number of regional variations into account in the design and construction of heavy commercial buildings. Variations in climate, soil conditions, building codes, construction materials, construction methods, and environmental legislation have a significant impact on the design and construction of a building.

Climate. Climatic conditions, such as variations in temperature, rainfall, and snow loads impact the design of structural frames and exterior protective systems. Soil conditions vary from area to area, and in many cases, can change dramatically on a single construction site. Climatic conditions impact foundation planning, shoring requirements, and drainage structures. For sustainable building design, the position of a structure on a building site may take variables such as wind speed, wind direction, and sunlight patterns into consideration to maximize energy efficiency.

Building Codes/Legislation. The International Building Code® (IBC) is a model code pertaining to commercial building systems. The IBC was developed by the International Code Council (ICC) through the joint efforts of other code organizations, including the Building Officials and Code Administrators (BOCA), the Southern Building Code Congress International (SBCCI), and the International Council of Building Officials (ICBO). As a model code, the IBC establishes minimum and maximum requirements for commercial building systems using prescriptive and performance-related provisions. The IBC may be adopted by state, county, or local jurisdictions and modified to meet the needs of their residents. The IBC may vary from building codes in effect in specific jurisdictions. **See Figure 1-10.**

Environmental legislation also creates a need for careful and thorough planning of any structure. Construction requirements impacted by environmental legislation may include noise abatement during construction or proper stormwater runoff protection. The requirements vary from jurisdiction to jurisdiction. Planners and designers must be aware of recent legislation and the materials available to meet environmental requirements.

The Americans with Disabilities Act (ADA) is another legislative requirement that impacts design and construction of public and commercial structures. The ADA specifies accommodations that must be provided in various building areas for accessibility by individuals with physical disabilities. Examples of handicapped-accessible accommodations include signage, alarm systems, ramps, door widths and hardware requirements, and installation heights for various fixtures.

Legislative requirements that apply to the allowable types of project delivery systems may also be in place. Depending on the type of building construction and the location of the project, local jurisdictions may either require or ban certain types of project delivery systems. For example, a local school district may require use of the design-bid-build system and not allow design-build project delivery for construction projects.

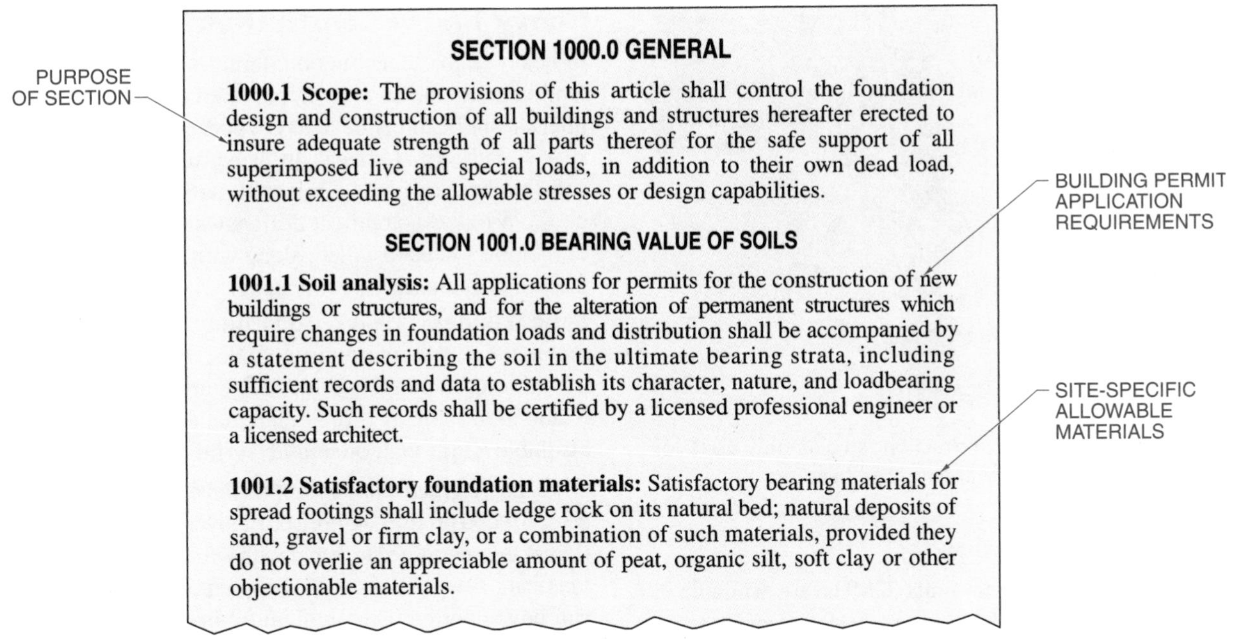

PURPOSE OF SECTION

SECTION 1000.0 GENERAL

1000.1 Scope: The provisions of this article shall control the foundation design and construction of all buildings and structures hereafter erected to insure adequate strength of all parts thereof for the safe support of all superimposed live and special loads, in addition to their own dead load, without exceeding the allowable stresses or design capabilities.

SECTION 1001.0 BEARING VALUE OF SOILS

1001.1 Soil analysis: All applications for permits for the construction of new buildings or structures, and for the alteration of permanent structures which require changes in foundation loads and distribution shall be accompanied by a statement describing the soil in the ultimate bearing strata, including sufficient records and data to establish its character, nature, and loadbearing capacity. Such records shall be certified by a licensed professional engineer or a licensed architect.

1001.2 Satisfactory foundation materials: Satisfactory bearing materials for spread footings shall include ledge rock on its natural bed; natural deposits of sand, gravel or firm clay, or a combination of such materials, provided they do not overlie an appreciable amount of peat, organic silt, soft clay or other objectionable materials.

BUILDING PERMIT APPLICATION REQUIREMENTS

SITE-SPECIFIC ALLOWABLE MATERIALS

Figure 1-10. Construction projects in a specific geographic area must conform to the building codes adopted for that area.

Materials. Building material availability is not consistent in all areas of the world. Building design may be dictated to some extent by the materials that are reasonably available at a particular construction site. Designs for structures in remote locations may require special considerations when necessary building materials must be moved over long distances. The distance of travel required for various building materials may also have an impact on the sustainability rating for a structure.

Methods. Construction methods and terminology vary greatly from location to location. For example, the term "jack" in one part of the country or in one trade may have an entirely different meaning in another location or in another trade. Knowledge of local building practices ensures that all tradesworkers and construction project participants are communicating effectively.

CONSTRUCTION SYSTEMS

After considering all the factors involved in a construction project, the owner, architect, and engineers must select the type of construction system that is best suited for a particular project. Construction systems include wood or metal framing, masonry, heavy timber, structural steel, and reinforced concrete. Each construction system has specific applications and advantages. Most structures combine several of the systems to achieve their overall design and purpose. Sustainable construction methods can be integrated into

any of these systems. Knowledge of printreading for all the various construction systems is necessary to combine the systems in a single structure.

TECH TIP

Steel is the world's most recycled material. More than 82 million tons of steel is recycled or exporrted for recycling by the United States each year.

Framed Construction

Wood or metal framing may be used in all types of buildings but is especially appropriate for small structures, such as dwellings and small commercial buildings. In heavy commercial construction, wood or metal framing is commonly used for interior partitions. Smaller wood or metal components, including studs, plates, runners, braces, trimmer studs, and cripple studs, are joined together to form a rigid framework. **See Figure 1-11.** Openings between framing members allow for installation of insulation, mechanical piping, and electrical systems. After the framework is constructed from wood or metal framing members, and mechanical and electrical systems and insulation are installed, the framework is covered with finish materials, such as gypsum board, masonry veneer, wood products, or wood or metal siding. Wood- and metal-framed structures provide a flexible design and are more economical than many other building systems.

Figure 1-11. Framed construction is commonly used for interior partitions in heavy commercial buildings.

Masonry Construction

Brick and concrete masonry units (CMUs) are available in many sizes and shapes. Brick and CMUs are made from various materials such as clay, concrete, and glass. Stone is also used for structural and decorative applications.

Masonry units are joined with mortar and may be reinforced with various designs of steel ties or other reinforcement. Large structures and structures requiring significant load-bearing capacity can be constructed using masonry units. Masonry units are highly fire resistant and are often used to create a fire break between adjoining areas of a structure. Accommodations must be made for piping and ductwork for mechanical and electrical systems during installation of masonry units.

Masonry is also commonly used as a surface veneer for other types of construction, such as wood- or metal-framed, structural steel, or reinforced concrete construction. Ties anchor the masonry veneer to the framing. **See Figure 1-12.**

Figure 1-12. A masonry surface veneer may be installed over a framed structure with metal ties securing the masonry veneer to the framing.

Heavy Timber Construction

In heavy timber construction, large solid timber or glued laminated members span large, open areas. Exterior and interior applications of heavy wood timbers create open spaces with visual appeal. In heavy timber construction, solid timbers or glued laminated members are indicated on the prints. An architect designates the sizes and types of members to be installed, along with their properties, on the prints and in the specifications.

Solid Timbers. Heavy wood timbers are solid wood members with a minimum nominal thickness of 5″ and a minimum nominal width of 5″, with lengths starting at 6′ and increasing in 1′ or 2′ increments. Solid timbers are available with rough or smooth surfaces.

Glulam. A *glued laminated (glulam) timber* is an engineered wood product comprised of layers of wood members (lams) that are joined together with adhesives to form larger members. Long distances can be spanned, and heavier loads can be supported using glulam timbers than can be supported using single solid wood members. Glulam timbers can be manufactured as straight or curved wood members. **See Figure 1-13.** Floor and roof beams, columns, and trusses are manufactured using glulam timbers.

APA – The Engineered Wood Association

Figure 1-13. Glulam timbers may be specified by architects to span large, open areas.

Design specifications for glulam timbers include specific wood species and sizes and grades of lumber. Wood species and grade are based on span and load-bearing requirements and on appearance. Design specifications also include the number of laminations, adhesive to be used, and the overall thickness, width, and length of the timber. Faces of laminated pieces to be joined are smoothed. The individual lams comprising an individual layer are set in their approximate locations relative to each other. An adhesive is applied, and the lams are clamped together with high-pressure clamping devices. After the adhesive sets, the exposed surfaces of the glulam timber are surfaced according to the design specifications.

Structural Steel Construction

Many of the largest buildings in the world are constructed using structural steel. Lightweight and heavy structural steel members may be specified in the design of a building. Lightweight members may be used to build industrial buildings and storage structures. Large beam and truss assemblies may be used to construct high-rise towers and bridges. In structural steel construction, horizontal girders, beams, and trusses are joined to vertical columns to create the framework for structures incorporating large open areas and for design flexibility. **See Figure 1-14.**

Figure 1-14. Structural steel construction allows for almost an unlimited number of variations in design and construction.

In structural steel construction, special consideration must be given to rigging requirements, working at heights, lifting heavy members into place, and safety for all tradesworkers on the job site. Structural steel buildings use a variety of bracing systems that provide great resistance to exterior stresses such as high winds, earthquakes, and other imposed loads. Many types of finish cladding can be attached to the exterior of structural steel buildings including glass, metal, masonry, stone, and precast concrete.

Structural steel is also commonly used for long-span roof truss systems and bridges. Careful engineering and designing are performed to ensure that proper steel shapes, sizes, and connections are used to meet all loading requirements.

Reinforced Concrete Construction

Cast-in-place concrete and precast concrete are two types of reinforced concrete construction systems. *Cast-in-place concrete* is concrete formed using a system of wood, earth, fiberglass, or metal forming materials that are set in their specified locations and shapes and act as molds for the fresh concrete. Reinforcing steel, or rebar, is positioned in the forms, and concrete is placed into the forms around the rebar. The concrete provides compressive strength, and the rebar provides tensile strength for the cast-in-place member creating a long-lasting structure. **See Figure 1-15.** The specifications for a construction project typically include information about removal of forms.

Portland Cement Association

Figure 1-15. Large cast-in-place members or structures are reinforced with steel to provide additional tensile strength.

TECH TIP

Concrete is a mixture of portland cement, fine aggregates (sand), coarse aggregates (gravel), and water.

Precast concrete is concrete that is formed, placed, and cured to a specific strength at a location other than its final location. A variety of precast members are available including beams, pipes, walls, floor sections, and exterior cladding. When designing a precast concrete structure, transporting and lifting the large, heavy members into place must be carefully considered. **See Figure 1-16.**

Portland Cement Association

Figure 1-16. Precast members are lifted into place with a crane and are carefully positioned by tradesworkers.

Tilt-up concrete construction is a type of precast concrete construction in which concrete for large wall sections is placed on a flat slab adjacent to its final position. A bond breaker is applied to the slab prior to placing the concrete to ensure the wall sections do not bond with the slab. Reinforcing steel and lifting inserts are installed in the wall prior to concrete placement. After the concrete is placed, finished, and cured, the walls are tilted up into place using a crane.

Road and Bridge Construction

Highway, road, and bridge construction requires specialized printreading skills and knowledge of construction processes related to these types of structures. Specifications for road and bridge construction commonly include requirements that describe limits for climatic conditions under which pavement may be placed. Proper temperature and curing conditions must exist to ensure proper performance for traffic surfaces. New roads and bridges often require extensive excavation and surface preparation. **See Figure 1-17.**

Erosion control requirements are included in local building codes or the plans and specifications. Paving operations begin after the grading and subgrade preparation are complete. For concrete and asphalt paving, the roadbed subgrade is carefully prepared to ensure proper compaction and the ability to withstand imposed loads. Edge forms and reinforcing steel are set in place for concrete roadways. String lines or targets are set to specific elevations along the edge of the roadway to guide the heavy equipment placing the paving materials.

Concrete or asphalt paving must be placed according to the elevations and slopes shown on the prints. Specifications describe the specific types of paving materials to be utilized. Proper surface finishing, scoring, and curing of the paving material are necessary to ensure that proper compressive strength and overall design requirements are achieved. Finish grading, shoulder work, guardrails, signage, and fencing are additional items needed to complete the construction.

Bridge building is based on reinforced concrete or structural steel construction. Piers, columns, and abutments are commonly constructed from reinforced concrete. Beams, which support the bridge deck, may be precast concrete, structural steel, or cast-in-place concrete. **See Figure 1-18.**

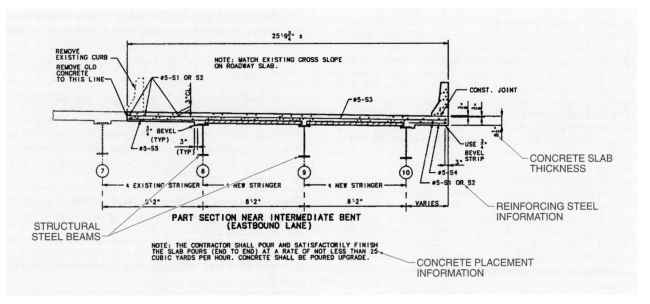

Figure 1-17. Prints for building roads include information about all aspects of the building process, including grading, paving, and finishing.

Portland Cement Association

Figure 1-18. Bridge building typically involves reinforced concrete and structural steel construction.

PRINT DIVISIONS AND PRINTREADING SKILLS

Tradesworkers and other team members on a construction project require a common source of information that is legally and functionally reliable. Prints must be accurately read and interpreted to ensure the final structure conforms to building codes and fulfills the needs of the owner and the design of the architect. Working drawings included on a set of prints are divided into separate categories to help readily locate information. The use of symbols and abbreviations and the ability to interpret portions of prints across the categories are necessary skills in construction.

Print Divisions

The title block of a print sheet contains a sheet number. For heavy commercial structures, the sheet number commonly contains a prefix of an uppercase letter denoting the specific category of the print. Six categories of prints are commonly used for heavy commercial structures: architectural, structural, mechanical, electrical, civil, and landscaping. Architectural prints are noted by an uppercase "A" followed by the page number. In a similar manner, structural prints are denoted by an "S", mechanical prints by an "M", electrical prints by an "E", civil prints by a "C", and landscaping prints by an "L". **See Figure 1-19.** Architects may use additional divisions depending on the nature of the building project.

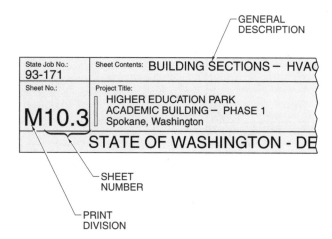

Figure 1-19. Sheets in a set of prints for heavy commercial construction projects are organized by print divisions and are placed in numerical sequence.

Computer-aided design (CAD) is the primary method of producing working drawings for use by all of the construction team members.

Sheet numbers begin with the number "1" within each set. For example, the architectural prints may be numbered from Sheet A1 to Sheet A45. The structural prints begin with Sheet S1. Subcategories within these prints may be denoted as S1.1, S1.2, and S1.3 where several pages apply to a similar print category.

Architectural (A). General construction information, floor plans, elevations, sections, and details are included as architectural prints. Architectural prints commonly comprise the largest number of sheets in a set of prints for a construction project.

Structural (S). Structural prints provide information about sizes, styles, and placement of foundations, beams, columns, joists, and other framing and load-bearing members. Structural prints contain information about structural wood, masonry, reinforced concrete, and steel members of a building.

Mechanical (M). Plumbing, heating, ventilating, and air conditioning information, including ductwork, piping, and equipment placement and sizes, is part of the mechanical prints. Information about fire protection systems may also be a part of the mechanical prints or may be provided in a separate section.

Electrical (E). The electrical prints indicate the capacity and placement of power plant systems, luminaires (lighting fixtures), cable trays, conduit and panel schedules, finish fixtures, wiring, switches, and any other electrical installations. Low-voltage electrical systems may be included on separate sheets depending on the complexity of the system.

Civil (C) and Landscaping (L). The overall site layout, grading, elevations, and topographical information are included on the civil prints. Civil prints also include site drainage, paving designs, and parking layout. Landscaping plans may also be a part of civil prints or they may be provided in a separate division. Where landscaping plans are provided as a separate division, they are designated with the letter "L".

Specifications

Written specifications for a large construction project are organized based on a format, known as the Master-Format™, developed by the Construction Specifications Institute (CSI) and Construction Specifications Canada (CSC). *MasterFormat™* is a uniform system of numbers and titles for organizing information about construction requirements, products, and activities into a standard sequence. The MasterFormat organization allows all building team members to easily identify and access information regarding the construction project. **See Figure 1-20.**

The MasterFormat is revised on a regular basis to reflect changes in the construction industry. The latest edition of the MasterFormat, entitled *MasterFormat 2004,* reflects the rapidly growing volume and complexity of information generated for nonresidential building projects.

TECH TIP

The number of divisions in MasterFormat™ increased from 16 in *MasterFormat 95* to 34 active and 16 reserved divisions in *MasterFormat 2004*. Divisions 03 through 14 covering building construction work remain basically unchanged, while the new divisions allow more flexibility for specifying civil, process, and other engineering work.

Symbols

Various symbols are used on a set of prints to represent building materials and components and to represent various dimensions and locations. Architects working on large construction projects commonly use unique or modified symbols to represent various building materials, components, and specialty items. **See Figure 1-21.** A legend, including symbols that are unique to the set of prints, is typically included on a large set of plans. The symbols used by an architect on a set of prints supersede any other common interpretations for the symbols.

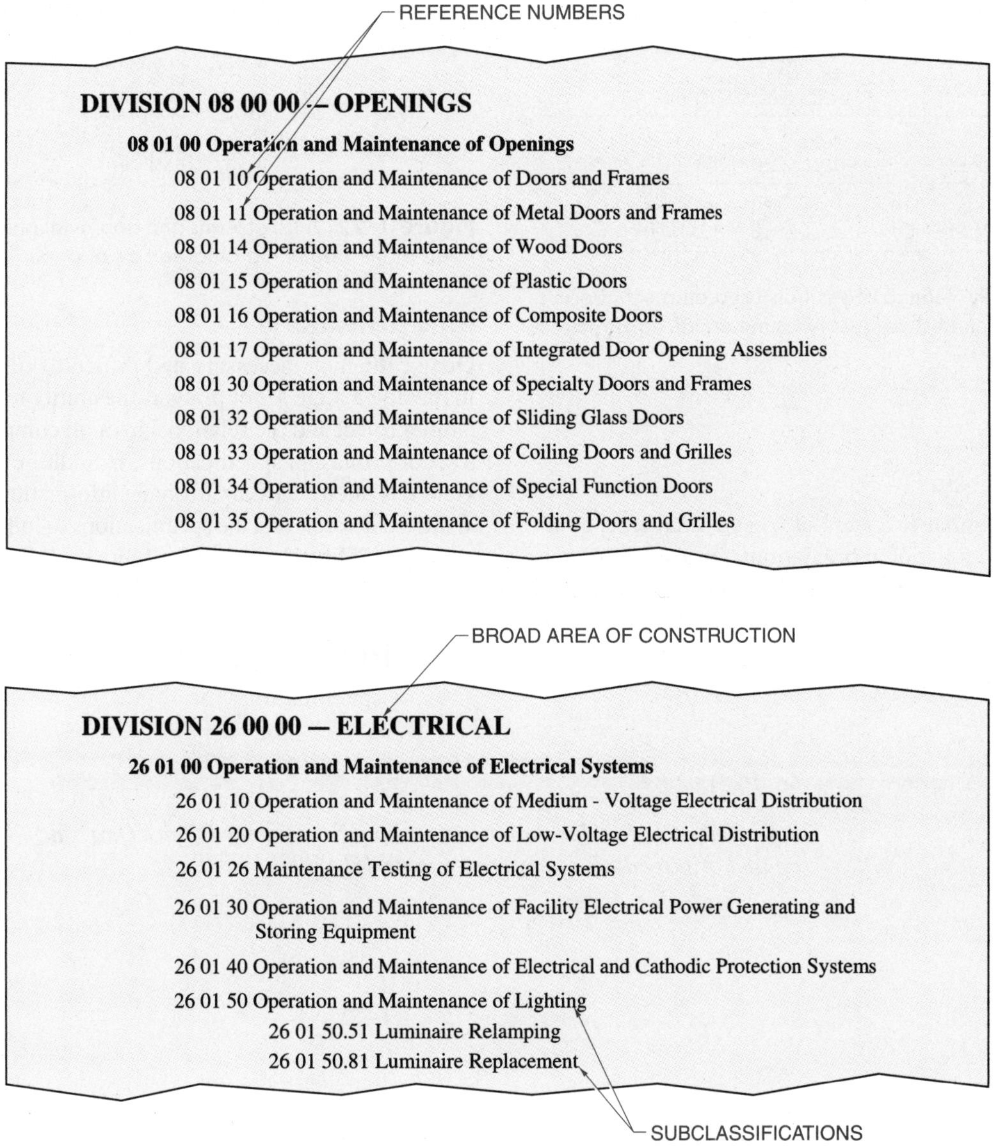

REFERENCE NUMBERS

DIVISION 08 00 00 — OPENINGS

08 01 00 Operation and Maintenance of Openings

08 01 10 Operation and Maintenance of Doors and Frames

08 01 11 Operation and Maintenance of Metal Doors and Frames

08 01 14 Operation and Maintenance of Wood Doors

08 01 15 Operation and Maintenance of Plastic Doors

08 01 16 Operation and Maintenance of Composite Doors

08 01 17 Operation and Maintenance of Integrated Door Opening Assemblies

08 01 30 Operation and Maintenance of Specialty Doors and Frames

08 01 32 Operation and Maintenance of Sliding Glass Doors

08 01 33 Operation and Maintenance of Coiling Doors and Grilles

08 01 34 Operation and Maintenance of Special Function Doors

08 01 35 Operation and Maintenance of Folding Doors and Grilles

BROAD AREA OF CONSTRUCTION

DIVISION 26 00 00 — ELECTRICAL

26 01 00 Operation and Maintenance of Electrical Systems

26 01 10 Operation and Maintenance of Medium - Voltage Electrical Distribution

26 01 20 Operation and Maintenance of Low-Voltage Electrical Distribution

26 01 26 Maintenance Testing of Electrical Systems

26 01 30 Operation and Maintenance of Facility Electrical Power Generating and Storing Equipment

26 01 40 Operation and Maintenance of Electrical and Cathodic Protection Systems

26 01 50 Operation and Maintenance of Lighting

26 01 50.51 Luminaire Relamping

26 01 50.81 Luminaire Replacement

SUBCLASSIFICATIONS

Figure 1-20. The MasterFormat™ is a uniform system of numbers and titles for organizing information about construction requirements, products, and activities into a standard sequence.

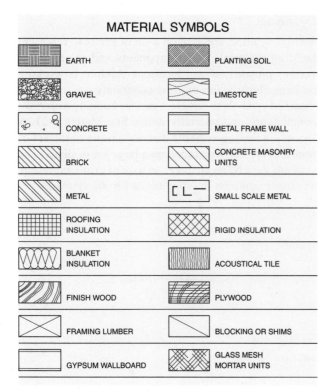

MATERIAL SYMBOLS

Symbol	Name	Symbol	Name
	EARTH		PLANTING SOIL
	GRAVEL		LIMESTONE
	CONCRETE		METAL FRAME WALL
	BRICK		CONCRETE MASONRY UNITS
	METAL		SMALL SCALE METAL
	ROOFING INSULATION		RIGID INSULATION
	BLANKET INSULATION		ACOUSTICAL TILE
	FINISH WOOD		PLYWOOD
	FRAMING LUMBER		BLOCKING OR SHIMS
	GYPSUM WALLBOARD		GLASS MESH MORTAR UNITS

Figure 1-21. A legend of symbols used on a set of prints is typically included on heavy commercial construction prints.

Abbreviations

In a manner similar to a symbol legend, architects commonly include a list of abbreviations for a set of prints. The list is not a complete list of abbreviations but a partial list of common terms and specialty terms. **See Figure 1-22.**

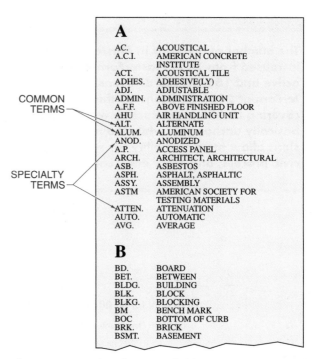

A

AC.	ACOUSTICAL
A.C.I.	AMERICAN CONCRETE INSTITUTE
ACT.	ACOUSTICAL TILE
ADHES.	ADHESIVE(LY)
ADJ.	ADJUSTABLE
ADMIN.	ADMINISTRATION
A.F.F.	ABOVE FINISHED FLOOR
AHU	AIR HANDLING UNIT
ALT.	ALTERNATE
ALUM.	ALUMINUM
ANOD.	ANODIZED
A.P.	ACCESS PANEL
ARCH.	ARCHITECT, ARCHITECTURAL
ASB.	ASBESTOS
ASPH.	ASPHALT, ASPHALTIC
ASSY.	ASSEMBLY
ASTM	AMERICAN SOCIETY FOR TESTING MATERIALS
ATTEN.	ATTENUATION
AUTO.	AUTOMATIC
AVG.	AVERAGE

B

BD.	BOARD
BET.	BETWEEN
BLDG.	BUILDING
BLK.	BLOCK
BLKG.	BLOCKING
BM	BENCH MARK
BOC	BOTTOM OF CURB
BRK.	BRICK
BSMT.	BASEMENT

COMMON TERMS

SPECIALTY TERMS

Figure 1-22. A list of terms and abbreviations is commonly listed at the beginning of large sets of prints.

Interpretations

One of the most necessary and difficult skills to develop in reading a large set of prints is the ability to visualize an entire project and the relationship of all components from a set of prints and specifications. In addition to an overall view, it is often necessary to obtain information from several different sheets and the specifications to fully understand a single building element. The ability to obtain information from a variety of sources and combine this information into a common understanding is the biggest challenge in reading prints for a large construction project.

Refer to the CD-ROM in the back of the book for Chapter 1 Quick Quiz® and related printreading and reference material.

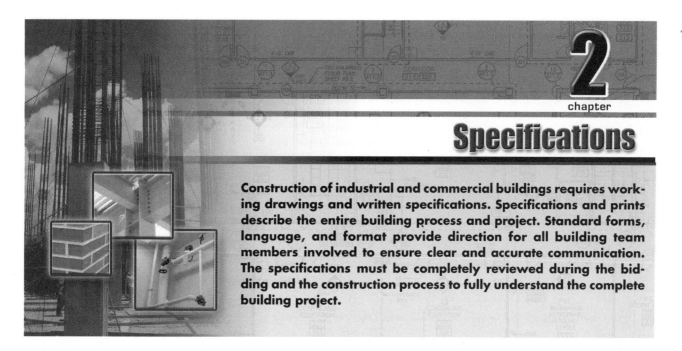

chapter 2

Specifications

Construction of industrial and commercial buildings requires working drawings and written specifications. Specifications and prints describe the entire building process and project. Standard forms, language, and format provide direction for all building team members involved to ensure clear and accurate communication. The specifications must be completely reviewed during the bidding and the construction process to fully understand the complete building project.

SPECIFICATIONS

Specifications are written supplements to working drawings that provide additional construction information. Specifications contain information related to building materials, construction procedures, quality control, and the legal issues of building construction. Bidding information, contract requirements, and information on all phases of construction are provided in the specifications in an organized, standardized manner. Standard forms, language, and format provide direction for all involved to ensure clear and accurate communication. Specifications must be completely and carefully reviewed during both the bidding and construction process to fully understand the complete building project.

MASTERFORMAT™

The MasterFormat™ is a uniform system of numbers and titles for organizing information about construction requirements, products, and activities into a standard sequence. The MasterFormat is typically used as a base for organizing written specifications. Since its introduction, the MasterFormat has been widely accepted as standard practice in the United States and Canada.

The MasterFormat is published jointly by the Construction Specifications Institute (CSI) and Construction Specifications Canada (CSC). The CSI and CSC are organizations of individuals and trade groups such as architects, engineers, constructors, writers, and suppliers of construction products in the United States and Canada, respectively.

The MasterFormat was first published as part of the *CSI Format for Construction Specifications*, which was later used as the basis for the *Uniform System for Construction Specifications, Data Filing, and Cost Accounting—Title One Buildings*. The *Uniform System* was developed and endorsed by the American Institute of Architects (AIA), the Associated General Contractors of America (AGC), the Council of Mechanical Specialty Contracting Industries Inc., the American Society of Landscape Architects, the National Society of Professional Engineers, and the Construction Specifications Institute. A similar effort in Canada published the *Building Construction Index*. The *Uniform System* and *Building Construction Index* were merged into a single format in the early 1970s and published as the *Uniform Construction Index*. In the late 1970s, the CSI and the CSC jointly published the first edition of the MasterFormat.

The MasterFormat is revised on an ongoing basis. There are currently two versions of the CSI MasterFormat in use, *Master Format 95* and *Master Format 2004*. Because *MasterFormat 95* had achieved widespread acceptance as a construction standard for many years, some construction project specifications continue to be written, organized, and produced according to the 16-division structure that was in use prior to 2004. In 2004, the CSI approved a significant revision of the MasterFormat for the first time in over 40 years. The *MasterFormat 2004* edition includes 50 divisions instead of 16, allowing for greater detail and expansion of construction standards. The use of *MasterFormat 2004* continues to increase as the old standard is phased out. Individuals reading the specifications must be knowledgeable about both versions of the CSI MasterFormat to accurately read and interpret the specifications and locate all the items necessary. **See Figure 2-1.**

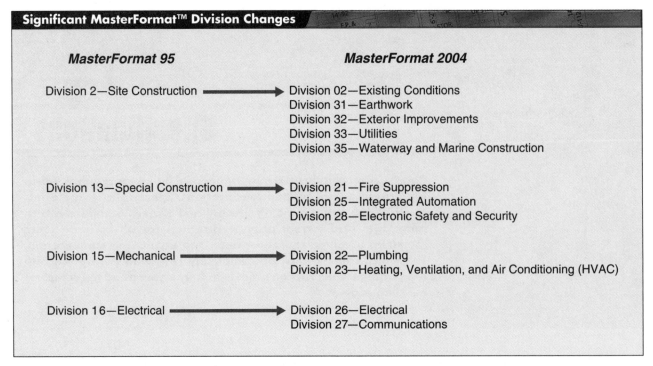

Figure 2-1. Four *MasterFormat 95* divisions have undergone significant changes for the conversion to *MasterFormat 2004*. Many level two and level three titles, especially in the area of electrical, have changed as well.

Groupings

A large portion of the work on a construction project, including the processes of bidding, bonding, and signing of contracts, occurs before any on-site work occurs. The MasterFormat is the standard for titling and organizing project manuals containing bidding and contracting requirements and specifications. Groupings prefacing the MasterFormat divisions in project manuals primarily pertain to the legal aspects of a building project. The groupings provide the information necessary to ensure that all members of the building team understand the legal steps that must be completed before construction begins. These groupings include introductory information, bidding requirements, contracting requirements, facilities and spaces, and systems and assemblies.

Bidding Information. An advertisement for bids lists both general information and bidding information. The legal name of the project, owner, bid range, accepting date and time, and bid opening date and time are listed in the advertisement. **See Figure 2-2.** In many instances, a pre-bid meeting is conducted to allow all interested parties to meet with the architect, program manager, or project manager to discuss the building project. Attendees may include general contractors, subcontractors, building trades union officials, government officials, and community representatives. Plans and specifications may be available for review or purchase at a number of locations. Any registrations, bonds, and guarantees required of bidders are explained at the pre-bid meeting.

Conditions of Construction. Contractor and owner responsibilities and duties are described in the conditions of construction. Contractors who bid on the project accept certain responsibilities in diverse areas, such as complete site exploration and disadvantaged business enterprise participation. Other state or local governmental requirements are listed. Owners and architects commonly include a list of the items for which they will not be held responsible. These items must be part of the contractor's portion of the construction bid.

MasterFormat 95 Groupings

The numbers and titles in *MasterFormat 95* are grouped under the following general headings:
- Introductory Information (00001 to 00099)
- Bidding Requirements (00100 to 00499)
- Contracting Requirements (00500 to 00999)
- Facilities and Spaces (no numbering)
- Systems and Assemblies (no numbering)
- Construction Products and Activities (Division 1 to 16)

SECTION 00020

ADVERTISEMENT FOR BIDS

A.   **GENERAL INFORMATION**

   1.   Project Name: Phase I - Classroom Building — LEGAL NAME OF PROJECT
                                   Spokane, Washington

   2.   Owner: Joint Center for Higher Education (JCHE) — OWNER
                                     Spokane, Washington

   3.   State Project Number: 93-171

   4.   Estimated Bid Range: $11,400,000 - $12,000,000 — BID RANGE

   5.   Pre-Bid Meeting: To be held at 1 p.m. on Tuesday, June 28, in the JCHE
        PRE-BID MEETING   Board Conference Room, Suite 245, at the Riverpoint One Office
        INFORMATION   Building, North 501 Riverpoint Blvd., Spokane, WA, for the
                                     purpose of answering questions from bidders and interested parties
                                     relating to the project.

   NOTE: **Attendance at the Pre-Bid Meeting is <u>MANDATORY</u> for all General Contractors intending to bid this project. A roster of those attending the Pre-Bid Meeting will be issued to all planholders by addendum following the meeting. Any bid received from any General Contractor whose name does not appear on this roster will be considered nonresponsive and will be rejected.**

   6.   Architect Contact: INTEGRUS Architecture, P.S.
                                       Gordon E. Ruehl
                                       Gary D. Joralemon

   7.   JCHE Contact: R.K. "Butch" Slaughter
                                         Physical Plant Manager, JCHE

B.   **BIDDING** — BID ACCEPTING DATE AND TIME

   1.   Sealed bids for construction of the Phase I - Classroom Building, Spokane, Washington, will be received by the Joint Center for Higher Education, North 501 Riverpoint Blvd., Suite 245, Spokane, WA 99202-1649, on Tuesday, July 12. Part I - Price Proposals must be received on or before 6 p.m. Part II - Price Proposals must be received on or before 7 p.m. — BID OPENING TIME AND DATE

   2.   All bids received will remain sealed until 7 p.m. on July 12, when they will be opened and read aloud in the JCHE Board Conference Room, Suite 245, North 501 Riverpoint Blvd., Spokane, WA 99202-1649.

   3.   Proposals received after times and dates set for opening will not be considered.

Figure 2-2. Architects, contractors, and subcontractors use specifications to define bidding procedures.

MasterFormat 95 Organization

The *MasterFormat 95* organizes related construction products and activities into 16 level one titles, called divisions. These divisions define the broad areas of construction.

See Figure 2-3. Each of the 16 divisions is designed to provide complete written information about individual construction requirements for building and materials needs. The numbers and titles of the divisions are as follows:

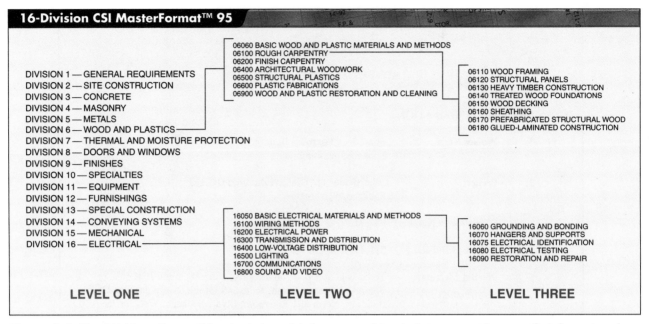

Figure 2-3. The CSI *MasterFormat 95* is divided into 16 divisions, with each division providing specific information about various materials.

- Division 1—General Requirements
- Division 2—Site Construction
- Division 3—Concrete
- Division 4—Masonry
- Division 5—Metals
- Division 6—Wood and Plastics
- Division 7—Thermal and Moisture Protection
- Division 8—Doors and Windows
- Division 9—Finishes
- Division 10—Specialties
- Division 11—Equipment
- Division 12—Furnishings
- Division 13—Special Construction
- Division 14—Conveying Systems
- Division 15—Mechanical
- Division 16—Electrical

Each of the divisions is divided into subclassifications of level two titles. Level two titles identify clusters of products having an identifying characteristic in common. For example, Division 16—Electrical has eight level two titles. Electrical Power, Lighting, and Communications are three examples. Each level two title has a reference number for ease of identification. For example, the reference number for Electrical Power is 16200. Level two titles usually end with two zeros.

Each level two title is subdivided into level three titles and numbers. Level three titles are presented as the last three digits of the five-digit designation. Level three is the last level to be indicated using both numbers and titles. Level three numbers typically end with a single zero.

The MasterFormat suggests level four titles, but does not indicate numbers. Users can create additional level numbers by interpolating between assigned level three numbers.

Certain divisions, for example Division 14—Conveying Systems, maybe not apply to some construction projects. When this occurs, the architect can write the specifications using a condensed form and arrange the divisions to fit the particular building being planned.

MasterFormat 2004 Groupings

Similar to *MasterFormat 95,* the numbers and titles of *MasterFormat 2004* are grouped into the following general headings:

- Procurement and Contracting Requirement Group (Division 00)
- Specifications Group (Divisions 01 to 49)
- General Requirements Subgroup (Division 01)
- Facility Construction Subgroup (Divisions 02 to 19)
- Facility Services Subgroup (Divisions 20 to 29)
- Site and Infrastructure Subgroup (Divisions 30 to 39)
- Process Equipment Subgroup (Divisions 40 to 49)

MasterFormat 2004 Organization

The current edition of the *MasterFormat 2004* contains 50 divisions that define broad areas of construction. **See Figure 2-4.** Assigned divisions are numbered 00 to 14; 21 to 23; 25 to 28; 31 to 35; 40 to 45; and 48. Divisions such as 15 to 20 were not assigned to specific current construction areas to allow for future expansion.

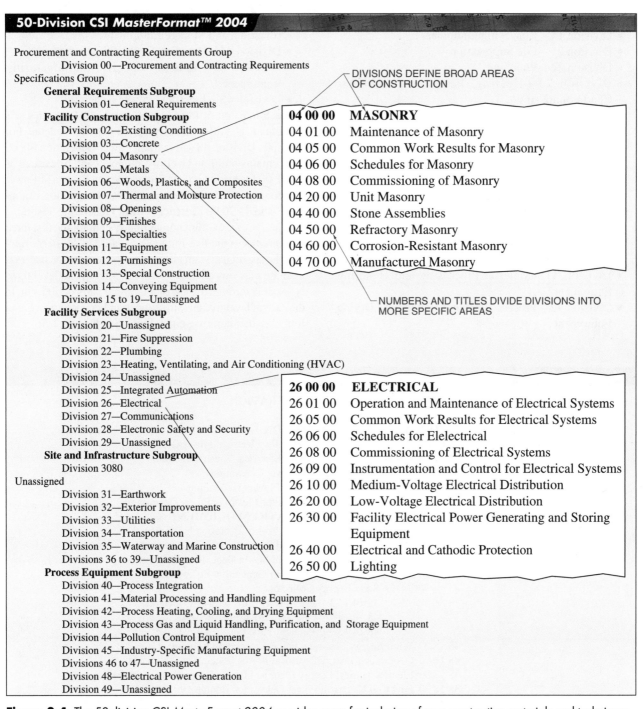

50-Division CSI *MasterFormat*™ **2004**

Procurement and Contracting Requirements Group
    Division 00—Procurement and Contracting Requirements
Specifications Group
    General Requirements Subgroup
        Division 01—General Requirements
    Facility Construction Subgroup
        Division 02—Existing Conditions
        Division 03—Concrete
        Division 04—Masonry
        Division 05—Metals
        Division 06—Woods, Plastics, and Composites
        Division 07—Thermal and Moisture Protection
        Division 08—Openings
        Division 09—Finishes
        Division 10—Specialties
        Division 11—Equipment
        Division 12—Furnishings
        Division 13—Special Construction
        Division 14—Conveying Equipment
        Divisions 15 to 19—Unassigned
    Facility Services Subgroup
        Division 20—Unassigned
        Division 21—Fire Suppression
        Division 22—Plumbing
        Division 23—Heating, Ventilating, and Air Conditioning (HVAC)
        Division 24—Unassigned
        Division 25—Integrated Automation
        Division 26—Electrical
        Division 27—Communications
        Division 28—Electronic Safety and Security
        Division 29—Unassigned
    Site and Infrastructure Subgroup
        Division 3080
Unassigned
        Division 31—Earthwork
        Division 32—Exterior Improvements
        Division 33—Utilities
        Division 34—Transportation
        Division 35—Waterway and Marine Construction
        Divisions 36 to 39—Unassigned
    Process Equipment Subgroup
        Division 40—Process Integration
        Division 41—Material Processing and Handling Equipment
        Division 42—Process Heating, Cooling, and Drying Equipment
        Division 43—Process Gas and Liquid Handling, Purification, and Storage Equipment
        Division 44—Pollution Control Equipment
        Division 45—Industry-Specific Manufacturing Equipment
        Divisions 46 to 47—Unassigned
        Division 48—Electrical Power Generation
        Division 49—Unassigned

DIVISIONS DEFINE BROAD AREAS OF CONSTRUCTION

04 00 00	**MASONRY**
04 01 00	Maintenance of Masonry
04 05 00	Common Work Results for Masonry
04 06 00	Schedules for Masonry
04 08 00	Commissioning of Masonry
04 20 00	Unit Masonry
04 40 00	Stone Assemblies
04 50 00	Refractory Masonry
04 60 00	Corrosion-Resistant Masonry
04 70 00	Manufactured Masonry

NUMBERS AND TITLES DIVIDE DIVISIONS INTO MORE SPECIFIC AREAS

26 00 00	**ELECTRICAL**
26 01 00	Operation and Maintenance of Electrical Systems
26 05 00	Common Work Results for Electrical Systems
26 06 00	Schedules for Elelectrical
26 08 00	Commissioning of Electrical Systems
26 09 00	Instrumentation and Control for Electrical Systems
26 10 00	Medium-Voltage Electrical Distribution
26 20 00	Low-Voltage Electrical Distribution
26 30 00	Facility Electrical Power Generating and Storing Equipment
26 40 00	Electrical and Cathodic Protection
26 50 00	Lighting

Figure 2-4. The 50-division CSI *MasterFormat 2004* provides room for inclusion of new construction materials and techniques.

The numbers and titles of the CSI *MasterFormat 2004* are as follows:
- Division 00—Procurement and Contracting Requirements
- Division 01—General Requirements
- Division 02—Existing Conditions
- Division 03—Concrete
- Division 04—Masonry
- Division 05—Metals
- Division 06—Wood, Plastics, and Composites
- Division 07—Thermal and Moisture Protection
- Division 08—Openings
- Division 09—Finishes
- Division 10—Specialties
- Division 11—Equipment
- Division 12—Furnishings

- Division 13—Special Construction
- Division 14—Conveying Equipment
- Division 21—Fire Suppression
- Division 22—Plumbing
- Division 23—Heating, Ventilating, and Air Conditioning (HVAC)
- Division 25—Integrated Automation
- Division 26—Electrical
- Division 27—Communications
- Division 28—Electronic Safety and Security
- Division 31—Earthwork
- Division 32—Exterior Improvements
- Division 33—Utilities
- Division 34—Transportation
- Division 35—Waterway and Marine Construction
- Division 40—Process Integration
- Division 41—Material Processing and Handling Equipment
- Division 42—Process Heating, Cooling, and Drying Equipment
- Division 43—Process Gas and Liquid Handling, Purification, and Storage Equipment
- Division 44—Pollution Control Equipment
- Division 45—Industry-Specific Manufacturing Equipment
- Division 48—Electrical Power Generation

Each division of the *MasterFormat 2004* includes sets of numbers and titles that correspond to levels of detail. For example, Division 48—Electrical Power Generation (48 00 00) contains multiple level two numbers and titles such as 48 01 00 Operation and Maintenance for Electrical Power Generation, 48 10 00 Electrical Power Generation Equipment, and 48 70 00 Electrical Power Generation Testing.

The use of a six-digit numbering system for the first three levels replaces the five-digit system used in *MasterFormat 95*. The six-digit system allows for more expansion per level than the previous edition. Classification beyond level three is possible through the use of additional pairs of digits or, in the case of level five classification, numbers or letters. Level four classifications use eight digits. **See Figure 2-5.**

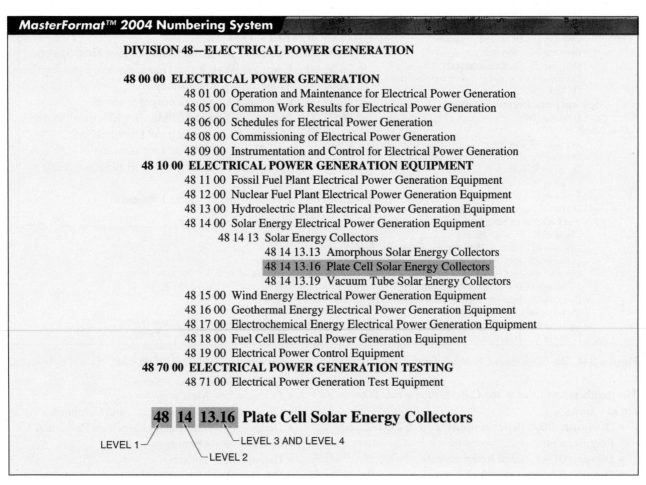

Figure 2-5. Level four classification of *MasterFormat 2004* uses eight digits arranged as three pairs of digits followed by a decimal point and then another pair of digits.

For example, Plate Cell Solar Energy Collectors is at the fourth level of classification. CSI number and title 48 00 00 Electrical Power Generation is the first level of classification. CSI number and title 48 10 00 Electrical Power Generation Equipment is the second level of classification. CSI number and title 48 14 00 Solar Energy Electrical Power Generation Equipment is also the second level of classification, but is a subset of 48 10 00. CSI number and title 48 14 13 Solar Energy Collectors is the third level of classification, and CSI number and title 48 14 13.16 Plate Cell Solar Energy Collectors is the fourth level of classification. Generally, level four classifications and above are used for specifications and project manuals while level five classifications are usually done by individual users for their internal use.

Since Sweet's Network provides product and material information following the *MasterFormat 2004* organization, references to products are made with relative ease. Specifications written according to the MasterFormat organization are similar in the arrangement of content. However, specifications for small buildings with light construction may not always follow the MasterFormat organization.

Division 1—General Requirements

Division 1 of a set of specifications includes the forms that must be submitted by contractors for payment and for documentation of various project requirements pertaining to materials, change orders, and substitutions. Health and safety, quality control, and contract closeout are also described in Division 1.

Procedures and Schedules. The first section of Division 1 provides a description of the overall construction project using an index of all the prints and several paragraphs on the project in general. **See Figure 2-6.** General descriptions of alternates are provided along with the bidding procedures for construction projects that have several alternate additional construction items.

TECH TIP

Specifications may only be several pages long for a small project, or they may be over several hundred pages for large, complex projects like power plants or manufacturing facilities.

Procedures for filing for approval and payment of change orders and unit pricing are listed. Certain procedures in this section must be followed in order for the contractor to receive payment. Standard invoice forms are commonly used for payment requests and are included in Division 1. **See Figure 2-7.**

The completion time requirements for a construction project are indicated in Division 1. Depending on the project, there may be financial penalties called liquidated damages assessed if the project is not completed within a certain period of time. The format for submission of the construction schedule to the owner and others involved in the building project may also be included in the specifications.

GENERAL OVERVIEW

1.4 DESCRIPTION OF THE WORK

A. The SIRTI facility is located on property in the Riverpoint area of downtown Spokane, bordered on the north by the Spokane River, south by Riverpoint Boulevard and on the east by Trent Avenue. To the west of the site there is additional undeveloped property.

B. In general, the structure is concrete slab on grade, structural concrete frame, dome pan (waffle slab) floor system, open web steel joists with fluted steel decking, rigid insulation and a single ply roof membrane. The exterior walls are constructed of structural steel studs, sheathing and brick veneer. Mechanical systems include gas-fired boilers; water-cooled screw compressor chillers; centrifugal fan, open evaporative cooling tower; air handling units; VAV terminal boxes; direct digital electronic control system, and automatic fire sprinkler system. Electrical systems include power distribution, interior and exterior lighting, raceways for power, data and communications, and fire alarm system.

C. Site work includes excavation, soil improvement, on-site drainage, paving, curbs, walks, site lighting, landscaping and irrigation system, as well as coordinated connections to existing sanitary sewer, water, power, gas, and telephone, provided by others.

Figure 2-6. A general statement concerning the size and scope of the building project is included in Division 1 of a set of specifications.

FORM SF 8254 REV 09		STATE OF WASHINGTON **APPLICATION AND CERTIFICATE FOR PAYMENT ON CONTRACT**

CERTIFICATE FOR (Partial/Final) PAYMENT. For period from _____ to _____

Contract For:

Location:

Contractor: _____   Contract No. _____

Original Contract Amount $

Change Order No.

Adjusted Contract Amount $ _____

ITEM NO.	DETAIL	ESTIMATED COST	AMOUNT		PREVIOUSLY CLAIMED	THIS ESTIMATE
			TOTAL EARNED	%		

(*) Items Sales Tax Exempt TOTALS						
Sales Tax on Applicable Items						
TOTAL						
Less Retainage . . . . %	X X X X					
NET	X X X X					
Less Previous Payments	X X X X				X X X X	X X X X
Adjustments (specify)		X X X X			X X X X	
Amount Due This Estimate	X X X X				X X X X	

This is to certify that, the contractor having complied with the terms of the above mentioned contract, there is due and payable from the State of Washington, the amount set after "Amount due this Estimate."

_____
(Contracting firm)

_____
(Supervising Engineers or Architects)

By _____

By _____

-GEN-1202-   3

Figure 2-7. Specifications include many standard forms to be used throughout the construction process.

Procedures for completion of the project are listed. These procedures include the final cleaning and preparation of the building; submission of all documents, such as contract drawings, specifications, addenda, change orders, shop drawings, and warranties; and operation and maintenance data.

Quality. Regular meetings between the architect, owner, contractor, construction manager, program manager, superintendent, and major subcontractors help to ensure that the project is properly completed and on schedule. Meeting procedures are detailed in Division 1 of a set of specifications, including preconstruction conferences and progress meetings.

A variety of industry association and governmental standards exist for construction. Items such as field engineering, hazardous material handling, equipment installations, and health and safety should be performed according to standards listed in the specifications.

The contractor is responsible for employing qualified workers, providing samples and mock-ups, and using approved testing agencies. These methods help the contractor ensure the quality of work meets or exceeds the quality requirements set forth in the construction documents. Failure to meet the minimum quality standards could lead to work having to be redone or cause a project to fall behind schedule.

Divisions 2, 31, 32, 33, 34, and 35 — Site and Transportation Construction

Divisions 2, 31, and 32 of the MasterFormat include information concerning items below ground, such as foundations, tunnels, pipes, and piers, as well as items on the ground, such as landscaping, fencing, and paving. The contractor's responsibilities for subsurface exploration, excavation, compaction, and disposal of excavated materials are also included in these divisions.

Site Preparation. The specifications detail the contractor's responsibilities for testing and handling of the construction site soils, fill, and backfill materials. Bearing capacities for the subsurface must be achieved and measured according to industry standards, such as those published by ASTM International (formerly the American Society for Testing and Materials). Removal and use of surface soils and necessary soil improvements are indicated. Where required, specific information may be provided concerning surface geomembrane installation and other soil properties. See Figure 2-8. Information regarding utilities, drainage piping, access holes, and inlets is included in Division 33.

Paving. Paving materials and standards for their use and installation are described in Division 32. The specifications

include the weather conditions under which paving may or may not be installed. Slopes and smoothness requirements are also indicated.

Landscaping. Soil mixtures, trees and shrubs, grasses, finish grading, fertilizers, sods, ground covers, and mulches are detailed in Division 32. Installation, protection, and maintenance procedures for plant materials are provided. Piping and connection information is included where irrigation systems are installed.

Transportation. Division 34 contains information specifically related to transportation systems. These systems include railways, roadways, bridges, and airfields. This type of construction is very specialized and commonly performed by contractors and subcontractors who are familiar with the paving, marking, scheduling, and procurement processes related to transportation work. Waterway and marine construction in Division 35 is also highly specialized and in many instances related to transportation construction.

TECH TIP

ASTM International has over 10,000 standards in 129 different technical areas. The ASTM was formed in 1898 and its first standard, "Structural Steel for Bridges" was written in 1901 by ASTM's first technical committee on steel.

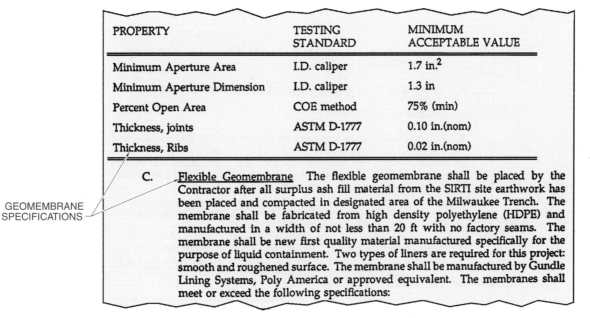

PROPERTY	TESTING STANDARD	MINIMUM ACCEPTABLE VALUE
Minimum Aperture Area	I.D. caliper	1.7 in.^{2}
Minimum Aperture Dimension	I.D. caliper	1.3 in
Percent Open Area	COE method	75% (min)
Thickness, joints	ASTM D-1777	0.10 in.(nom)
Thickness, Ribs	ASTM D-1777	0.02 in.(nom)

GEOMEMBRANE SPECIFICATIONS

C. Flexible Geomembrane The flexible geomembrane shall be placed by the Contractor after all surplus ash fill material from the SIRTI site earthwork has been placed and compacted in designated area of the Milwaukee Trench. The membrane shall be fabricated from high density polyethylene (HDPE) and manufactured in a width of not less than 20 ft with no factory seams. The membrane shall be new first quality material manufactured specifically for the purpose of liquid containment. Two types of liners are required for this project: smooth and roughened surface. The membrane shall be manufactured by Gundle Lining Systems, Poly America or approved equivalent. The membranes shall meet or exceed the following specifications:

Figure 2-8. Soil and slope finish information includes details on geomembranes, which may be installed to inhibit erosion.

Division 3—Concrete

Division 3 of the MasterFormat provides information concerning concrete including procedures for placing, curing, and finishing concrete, formwork construction, removal, materials, reinforcing methods, and other related information. Precast concrete members are also described in Division 3.

Materials. The primary material described in Division 3 is concrete. Concrete ingredient information includes the type of cement to be used, aggregate, admixtures, and water quality. Reinforcing steel is described according to ASTM International references concerning strength and shape. Accessories to be placed in concrete, such as chairs and bolsters for supporting reinforcing steel, dowels, and anchors, may be described according to a specific manufacturer's name and product number. **See Figure 2-9.** Form materials, form-release agents, grout, joint fillers, waterstops, and curing compounds are also specified in the specifications.

Methods. Concrete must be properly placed and cured to ensure that it meets final design requirements. For cast-in-place concrete, the specifications describe the methods of erecting formwork, placing embedded accessories such as dowels and waterstops, placing reinforcement, placing concrete, curing concrete, removing formwork, finishing, and patching. Testing methods for concrete strength and slump are noted. For precast concrete members, the methods of installation are provided.

Division 4—Masonry

Division 4 of the MasterFormat details components of masonry construction including masonry units, mortar, reinforcement, and accessories. **See Figure 2-10.**

Figure 2-10. Division 4 of the MasterFormat addresses components of masonry construction.

Materials. Different types and blends of mortars are used in various applications, such as load-bearing masonry walls, non-load-bearing masonry walls, and tuckpointing. Mortar components are detailed in the specifications, including cement, aggregate, water quality, bonding agents, coloring, and admixtures. **See Figure 2-11.**

2. PART 2   PRODUCTS

REINFORCING STEEL DESCRIBED ACCORDING TO ASTM INTERNATIONAL REFERENCES

2.1   REINFORCEMENT

A. Reinforcing Steel: As noted on the Structural Drawings. ASTM A615, 60 ksi yield grade; deformed billet steel bars.

B. Welded Steel Wire Fabric: ASTM A185 Plain Type in flat sheets.

2.2   ACCESSORY MATERIALS

A. Tie Wire: Minimum 16 gage annealed type, or patented system as approved.

B. Chairs, Bolsters, Bar Supports, Spacers: Sized and shaped for strength and support of reinforcement during concrete placement conditions including load bearing pad on bottom to prevent vapor barrier puncture at slab on grade.

MANUFACTURER NAME AND PRODUCT NUMBER

C. Special Chairs, Bolsters, Bar Supports, Spacers Adjacent to Weather Exposed Concrete Surfaces: Plastic coated steel type; size and shape as required.

D. Dowel Flanged Couplers (DFC): Williams Form Engineering Corp. CD2 couplings with CD2 indicators, Dayton-Superior D-50 DBR, Richmond Screw Anchor Co. Inc. DB-SAE splicer and DB-S indicator, or approved. Provide in size to meet or exceed rebar capacity. System may be used as substitutions for dowel bars.

Figure 2-9. Specifications commonly mention specific manufacturer names and product numbers to define the types and qualities of materials and hardware.

PORTLAND CEMENT/LIME MORTARS				
Type	Description	Proportions by Volume		
		Portland Cement	Hydrated Lime or Lime Putty	Sand
M	Mortar of high compressive strength (at lest 2500 psi) after curing 28 days and with greater durability than some other types; used for masonry below ground and in contact with the earth, such as foundations, retaining walls, and paving; Type M mortar withstands severe frost action and high lateral loads.	1	¼	3
S	Mortar with a fairly high compressive strength (at least 1800 psi) after curing 28 days; used in reinforced masonry and for standard masonry where maximum flexural strength is required; also used when mortar is the sole bonding agent between facing and backing units.	1	½	4¼
N	Mortar with a medium compressive strength (at least 750 psi) after curing 28 days; used for exposed masonry aboveground and where high compressive strength or lateral masonry strengths are not required.	1	1	6
O	Mortar with a low compressive strength (at least 350 psi) after curing 28 days; used for general interior walls; may be used for load-bearing walls of solid masonry if axial compressive stress does not exceed 100 psi and wall is not exposed to weathering or freezing.	1	2	9

Figure 2-11. Mortar components include cement, aggregate, water, bonding agents, coloring, and admixtures.

TECH TIP

The Brick Industry Association (BIA) is the national trade association representing distributors and manufacturers of clay brick and suppliers of related products and services.

Sizes and colors of face brick and concrete masonry units may be described according to a specific manufacturer. Where stone is supplied, a specific supplier may be identified to ensure stone quality and uniformity. Other masonry construction information includes details on metal ties and anchors, flashing, and control joints.

Methods. In a manner similar to concrete, certain temperature requirements exist for masonry construction. Precautions must be taken to protect new masonry construction from excessive cold or hot, dry conditions. Environmental requirements for masonry construction are defined in the specifications. **See Figure 2-12.**

Other aspects of masonry construction described in the specifications include testing of mortar and grout mixes; type of bond used for brick placement; anchor placement; flashing installation; mortar joint style; tolerances for positioning and variations from plumb and level; cleaning and sealing of the finished surface; preparation of surfaces to be patched; vibration of grout materials; and examination of the final masonry installation.

1.6 ENVIRONMENTAL REQUIREMENTS

A. Maintain materials and surrounding air temperatures to minimum 50 degrees F prior to, during, and 48 hours after completion of masonry work. Comply with UBC Section 2404 requirements where more restrictive.

1.7 MIX TESTS

A. Test mortar and grout in accordance with Section 01400.

B. Testing of Mortar Mix: In accordance with ASTM C780.

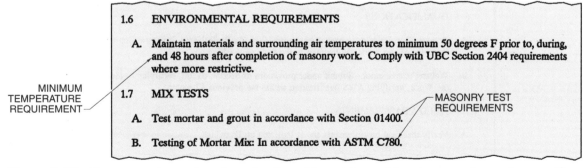

MINIMUM TEMPERATURE REQUIREMENT

MASONRY TEST REQUIREMENTS

Figure 2-12. Masonry work may be adversely affected by extreme cold and extreme heat unless specific precautions are taken.

TECH TIP

Steel is the world's most recycled material. The United States recycles or exports for recycling more than 82 million tons of steel each year. The use of recycled structural steel may contribute to LEED® certification credits.

Division 5 — Metals

Division 5 of the MasterFormat provides information about metals used on a construction project including structural steel members, such as columns, beams, and joists; steel decking for floors, walls, and roofs; light-gauge metal framing members; metal stairs; and ornamental metals, such as handrails, ladders, and expansion joints. Metal shingles, roof tiles, roof panels, wall panels, siding, and flashings are described in Division 7. Metal duct, pipe, and conduit specifications are described in Divisions 21, 22, 23, 26, 27, and 40.

Materials. Steel material specifications commonly refer to ASTM International standards for structural steel shapes, coatings, and connectors such as bolts, nuts, and washers. See Figure 2-13. The shape, diameter, type of metal, and pipe schedule specifications are provided for railings and other ornamental iron.

Methods. The *Structural Welding Code*, published by the American Welding Society (AWS), is the reference that provides information for acceptable methods, certification requirements, and tolerances for welded connections. See Figure 2-14. Metal and steel construction methods described in Division 5 include fabrication procedures; installation tolerances for squareness, plumb, and alignment; fastener locations; and finishing processes such as grinding, shot or sand blasting, and paint priming.

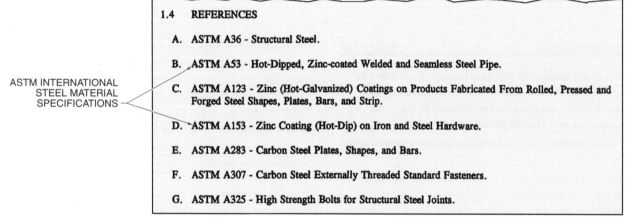

ASTM INTERNATIONAL STEEL MATERIAL SPECIFICATIONS —

1.4 REFERENCES

A. ASTM A36 - Structural Steel.

B. ASTM A53 - Hot-Dipped, Zinc-coated Welded and Seamless Steel Pipe.

C. ASTM A123 - Zinc (Hot-Galvanized) Coatings on Products Fabricated From Rolled, Pressed and Forged Steel Shapes, Plates, Bars, and Strip.

D. ASTM A153 - Zinc Coating (Hot-Dip) on Iron and Steel Hardware.

E. ASTM A283 - Carbon Steel Plates, Shapes, and Bars.

F. ASTM A307 - Carbon Steel Externally Threaded Standard Fasteners.

G. ASTM A325 - High Strength Bolts for Structural Steel Joints.

Figure 2-13. Standard structural steel coatings and shapes are used by mills and fabricators to ensure consistency and quality.

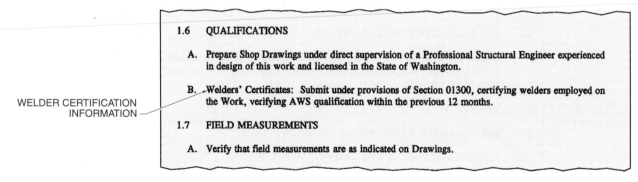

WELDER CERTIFICATION INFORMATION —

1.6 QUALIFICATIONS

A. Prepare Shop Drawings under direct supervision of a Professional Structural Engineer experienced in design of this work and licensed in the State of Washington.

B. Welders' Certificates: Submit under provisions of Section 01300, certifying welders employed on the Work, verifying AWS qualification within the previous 12 months.

1.7 FIELD MEASUREMENTS

A. Verify that field measurements are as indicated on Drawings.

Figure 2-14. The *Structural Welding Code,* published by the American Welding Society (AWS), is the reference that provides information for acceptable methods, certification requirements, and tolerances for welded connections.

Division 6—Wood, Plastics, and Composites

Division 6 of the MasterFormat includes information about rough wood framing, finish woodworking, and plastic materials such as plastic laminate. In large commercial projects, there is often a limited amount of wood framing. Division 6 of the MasterFormat also provides additional information concerning heavy timber framing when applicable.

Materials. As with other building materials, standards from various industry groups, such as the American Forest & Paper Association (AF&PA), the Western Wood Products Association (WWPA), and APA—The Engineered Wood Association (formerly the American Plywood Association), are referenced for lumber products. Finish wood materials are related to standards provided by the Architectural Woodwork Institute (AWI). **See Figure 2-15.** Wood-related materials described include lumber and lumber treatments, such as fireproofing; softwood and hardwood plywood; finish woods, such as oak or maple; and fasteners, such as nails, bolts, and screws. Some finish casework-related information provided includes plastic laminate grades and various hardware. The majority of wood casework information is described in Division 12.

Methods. Rough framing methods and requirements are typically defined in the building code and are not part of the specifications. Methods included in Division 6 pertain to items such as applications for treated lumber, fabrication and finishing of wood casework and finish materials, and installation tolerances for finish woodwork.

Division 7—Thermal and Moisture Protection

Division 7 of the MasterFormat includes information about a wide range of construction products including asphalt roofing, rubberized roofing, mastics, waterproof coatings, vapor barriers, sheet metal flashings, insulation materials, fireproofing materials, and joint sealants. Products specified in Division 7 stop moisture movement or provide thermal insulation to the structure.

Materials. Due to the specialized nature of many thermal and moisture protection products, information in Division 7 relies heavily on manufacturer names and product numbers. **See Figure 2-16.** The specification information is performance based where common materials are used. For example, specifications for expanded polystyrene insulation board may include the required board density, thermal resistance, and compressive strength. Application schedules for various materials may be included to assist in properly installing the materials identified in the specifications.

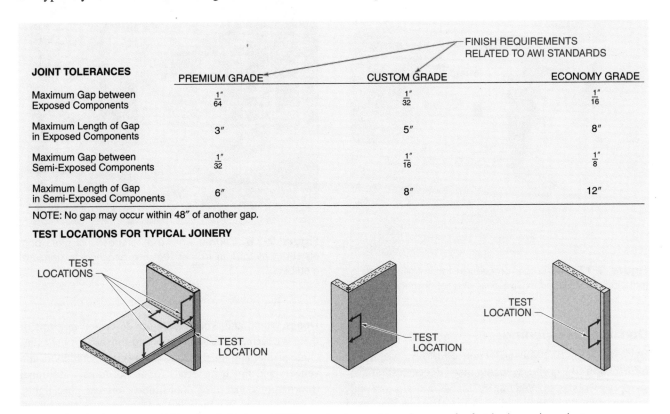

Figure 2-15. Architectural Woodwork Institute (AWI) standards provide references for finished wood products.

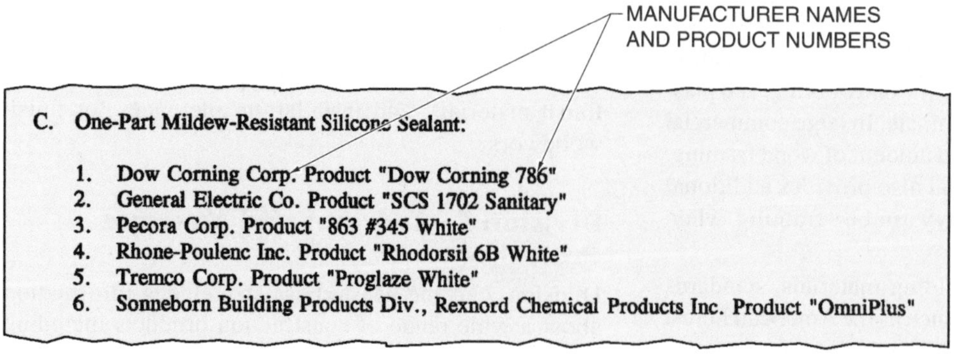

MANUFACTURER NAMES
AND PRODUCT NUMBERS

C. One-Part Mildew-Resistant Silicone Sealant:

    1. Dow Corning Corp. Product "Dow Corning 786"
    2. General Electric Co. Product "SCS 1702 Sanitary"
    3. Pecora Corp. Product "863 #345 White"
    4. Rhone-Poulenc Inc. Product "Rhodorsil 6B White"
    5. Tremco Corp. Product "Proglaze White"
    6. Sonneborn Building Products Div., Rexnord Chemical Products Inc. Product "OmniPlus"

Figure 2-16. Many manufacturers provide thermal and moisture protection products that may be designated in building specifications.

Methods. Proper thermal and waterproofing performances rely on proper installation. Division 7 of a set of specifications provides information concerning proper surface preparation, seaming and overlap requirements, mastic and fastener applications, proper number of coatings of a particular product, and final quality inspection. **See Figure 2-17.**

is a non-load-bearing, prefabricated or job-built, glass or metal panel supported by metal frame members or set with various clip systems that are attached to structural members. **See Figure 2-18.** Division 8 also includes door and window schedules and information about associated hardware.

Figure 2-17. The proper procedures for installing roofing materials are included in Division 7 of a set of specifications.

Figure 2-18. Curtain walls are composed of glass panels set in metal trim frames that are attached to structural members.

Division 8—Openings

Division 8 of the MasterFormat contains extensive information concerning swinging metal doors and frames, swinging wood doors and frames, access doors, doors and grilles, glass doors, and sliding doors. Windows are also described, including glass curtain walls. A *curtain wall*

Doors. Metal and wood doors are described in terms of their fire rating, core materials, and finishes. Finishes include primer and finish paint coatings for metal doors and veneers and finish materials for wood doors. Additional information concerning door frames, louvers, glass lights, and astragals is also included in Division 8. A door schedule providing information about each door shown on the working drawings is often provided in the specifications. **See Figure 2-19.** Tolerances for door warp, bow, and variation from plumb and level are also included in the door specifications.

Windows. Types and finishes of window frames, glazing, and weather stripping comprise a large portion of the window specifications. Window types may be specified by manufacturer product codes. Window performance requirements such as the ability to withstand wind pressures, deflection, air leakage, thermal performance, and water leakage may also be included in the specifications. Drawings in the specifications may provide additional details for large or complicated window installations. **See Figure 2-20.** The specifications may also include a schedule for placement of various glazing materials, such as glass types and glazing compounds.

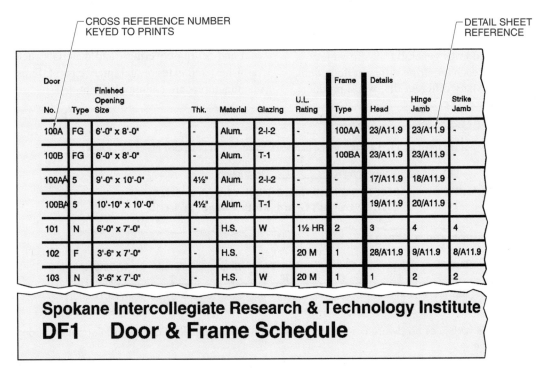

CROSS REFERENCE NUMBER KEYED TO PRINTS

DETAIL SHEET REFERENCE

Door No.	Type	Finished Opening Size	Thk.	Material	Glazing	U.L. Rating	Frame Type	Details Head	Hinge Jamb	Strike Jamb
100A	FG	6'-0" x 8'-0"	-	Alum.	2-I-2	-	100AA	23/A11.9	23/A11.9	-
100B	FG	6'-0" x 8'-0"	-	Alum.	T-1	-	100BA	23/A11.9	23/A11.9	-
100AA	5	9'-0" x 10'-0"	4½"	Alum.	2-I-2	-	-	17/A11.9	18/A11.9	-
100BA	5	10'-10" x 10'-0"	4½"	Alum.	T-1	-	-	19/A11.9	20/A11.9	-
101	N	6'-0" x 7'-0"	-	H.S.	W	1½ HR	2	3	4	4
102	F	3'-6" x 7'-0"	-	H.S.	-	20 M	1	28/A11.9	9/A11.9	8/A11.9
103	N	3'-6" x 7'-0"	-	H.S.	W	20 M	1	1	2	2

Spokane Intercollegiate Research & Technology Institute
DF1   Door & Frame Schedule

Figure 2-19. Door schedules provide detailed information about doors in a format that is keyed to door locations on the working drawings.

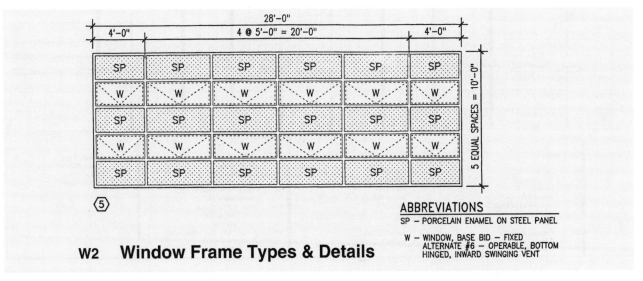

28'-0"
4'-0"   4 @ 5'-0" = 20'-0"   4'-0"

5 EQUAL SPACES = 10'-0"

⟨5⟩

ABBREVIATIONS
SP – PORCELAIN ENAMEL ON STEEL PANEL

W – WINDOW, BASE BID – FIXED
ALTERNATE #6 – OPERABLE, BOTTOM HINGED, INWARD SWINGING VENT

W2   Window Frame Types & Details

Figure 2-20. Shop drawings may be included in specifications to clarify complicated installations.

Hardware. Hardware components are typically described in the specifications with information about types and locations for installation. Manufacturer names, designs, sizes, finishes, and functions may be included as part of the hardware specifications in Division 8. The hardware schedule for each door may include hinges, locks, door closers, door pulls, pushplates, kickplates, stops, bolts, thresholds, weather stripping, and door seals. **See Figure 2-21.** Spacing of hinges and locations for locksets are indicated.

Division 9—Finishes

Large commercial buildings generally have a variety of rooms or usage areas, each requiring different floor, wall, and ceiling finishes. Many different tile, wood products, gypsum products, cementitious materials, paints, and special treatments are required to meet the varying demands of each room or area. Division 9 of the MasterFormat details specific information concerning finish materials to be applied in each room or area. **See Figure 2-22.**

Materials. Finish materials for each area of a large commercial building may be denoted on a room finish schedule. Finish materials included in Division 9 are metal lath and plaster, gypsum products, metal stud framing and accessories, ceramic floor and wall tile, resilient flooring, carpeting, wood flooring, suspended ceiling systems, special wall and ceiling coverings, and materials such as stains, varnishes, and exterior and interior paint. Finish material manufacturers are commonly identified.

2.09 THRESHOLDS

Types are listed as T1, T2, etc., in the hardware schedule and described under "T" below:

"T"	Factory Number	Description
T6	814CP-32D	Nat'l Guard, width x length as details
T7	838CP-S1A	Nat'l Guard, width x length as details

TYPES

2.12 HARDWARE SCHEDULE

DOOR HARDWARE SPECIFICATIONS

THRESHOLD TYPES

Door	Butts B	Locks L	Closers C	Kick Pl K	Stops S	Thresh T	W-Strip W
100AA	B8	L24				T6,7	W10,11
100BA	B8	L24				T6,7	W10,11

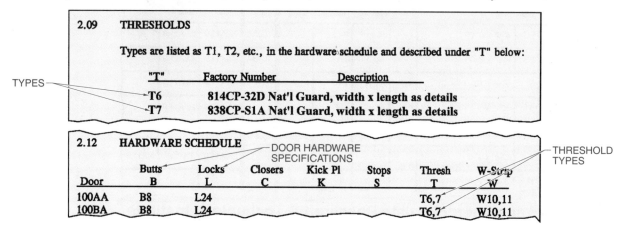

Figure 2-21. A wide variety of hardware components are described in specifications, with information about types and locations for installation.

ROOM FINISH SCHEDULE

Room		Floor			Base			Walls North			South			East			West			Ceiling				Notes
No.	Name	Mat.	Fin.	Col.	Mat.	Fin.	Col.	Mat.	Fin.	Col.	Mat.	Fin.	Col.	Mat.	Fin.	Col.	Mat.	Fin.	Col.	Mat.	Fin.	Col.	Ht.	
000	LOBBY	14	F3		21	---		33	F8	P1	33	F8	P1	32	F8	P1	33	F8	P1	41	---		11'-½"	
000A	STAIR 1	10	F1		20 23	F17 F17		33	F8	P2	33	F8	P2	33	F8	P2	33	F8	P2	41	---		11'-½"	
001	CORRIDOR	14	F3		21	---		32	F8	P1	33	F8	P1	32	F8	P1	33	F8	P1	41	---		11'-½"	
009	MEN'S TOILET	13	F2		22	F2		32	F11 F6	P4	32	F11 F6	P4	32	F11 F6	P4	32	F11 F6	P4	42	F8	P4	8'-0"	
009A	VESTIBULE	13	F2		22	F2		32	F8	P4	32	F8		32	F8		32	F8		42	F8	P4	8'-0"	
010	WOMEN'S TOILET	13	F2		22	F2		32	F11 P6	P4	32	F11 F6	P4	32	F11 F6	P4	32	F11 F6	P4	42	F8	P4	8'-0"	

Materials

Floor Materials

10 Concrete (CONC)
11 Slate Pavers (SP)
12 Not Used
13 Ceramic Mosaic Tile (CMT)
14 Vinyl Composition Tile (VCT)

Base Materials

20 Hardwood (Maple)
21 Vinyl Cove
22 Ceramic Mosaic Tile

Wall Materials

31 Concrete
32 Gypsum Wallboard (Single Layer) on Metal Studs
33 Gypsum Wallboard (Single Layer) on Metal Furring

Ceiling Materials

41 Exposed Concrete
42 Gypsum Wallboard

Finish

F1 Concrete Hardener/Sealer
F2 Clean & Seal
F3 Clean, Wax, & Buff
F4 Tackable Wall Covering
F5 Vinyl Fabric Wall Covering
F6 Ceramic Mosaic Tile
F7 Paint - Masonry System
F8 Paint - GWB System
F9 Traffic Membrane
F10 Rubber Treads, Risers, Sheet Rubber
Nonslip Tile @ Landing
F11 High Build Glazed Coating

Color

P1 Columbia #5391W
P2 Columbia #5393M
P3 Columbia #5394D
P4 Columbia #5390W

Figure 2-22. A large commercial project will have many different floor, wall, and ceiling finishes described on the room finish schedule.

Methods. To ensure that the proper floor, wall, and ceiling finishes are obtained, the architect may specify the qualifications of the installers, proper material handling prior to installation, environmental requirements, and sequence of installation. **See Figure 2-23.** Cleaning of the finished areas and protection of the final product are also indicated in the specifications.

TECH TIP

Access Flooring was reclassified and renumbered from a level two title of Division 10 (10270) in MasterFormat 95 to a level two title of Division 9 (09 69 00) in *MasterFormat 2004.*

Division 10—Specialties

Division 10 of the MasterFormat includes specialty items that may or may not be part of a large commercial building. Division 10 includes information on specialty items such as chalkboards, signage, display cases, informational kiosks, toilet compartments, demountable partitions, corner guards, fire extinguishers and cabinets, lockers, mailboxes, storage shelving, and flagpoles. **See Figure 2-24.** In a

manner similar to other divisions, manufacturer names and product numbers are included in Division 10 of a set of specifications.

Figure 2-24. Lockers are one of the many specialized types of construction included in Division 10 of the MasterFormat.

    B. Erect single layer non-rated gypsum board in most economical direction, with ends and edges occurring over firm bearing.

    C. Erect single layer fire rated gypsum board vertically, with edges and ends occurring over firm bearing.

    D. Use screws when fastening gypsum board to metal furring or framing.

    E. Double Layer Applications: Use gypsum backing board for first layer, placed over framing or furring members. Use fire rated gypsum backing board for 2-hr. fire rated partitions. Place second layer perpendicular to first layer. Offset joints of second layer from joints of first layer.

    F. Double Layer Applications: Secure second layer to first in manner required by code for indicated fire rating. Apply adhesive in accordance with manufacturer's instructions.

    G. Treat cut edges and holes in moisture resistant gypsum board and glass mesh mortar units with sealant.

    H. Place control joints consistent with lines of building spaces.

    I. Place corner beads at external corners. Use longest practical length. Place casing beads where gypsum board abuts dissimilar materials.

3.9   JOINT TREATMENT

    A. Tape, fill, and sand exposed joints, edges, and corners to produce smooth surface ready to receive finishes.

    B. Feather coats onto adjoining surfaces so that camber is maximum 1/32 inch.

Figure 2-23. Architects provide detailed information concerning procedures for finishes to ensure a quality product.

Division 11—Equipment

Many commercial buildings require equipment to be built-in during construction. Projection screens, dishwashers, ovens, water treatment equipment, and a variety of industrial equipment may be installed in a structure during construction. The equipment is part of the initial construction bid. The general contractor and subcontractors are responsible for obtaining and installing the specialized equipment according to the information provided in Division 11 prior to completion of the project.

Division 12—Furnishings

Division 12 of the MasterFormat provides information concerning metal and wood cabinetry and countertops, including material types, finishes, hardware, fabrication, and installation. **See Figure 2-25.** Other information commonly provided in Division 12 includes installed seating, such as for a theater or classroom, as well as window blinds and curtains.

Division 13—Special Construction

Division 13 of the MasterFormat includes highly specialized types of construction for various commercial projects. Common equipment and systems included in Division 13 are seismic controls and controlled environment rooms with specialized controls for such items as lighting, fire protection, and heating and cooling systems. Due to the highly specialized nature of items included in Division 13, manufacturer information is critical in the specifications for this division.

Division 14—Conveying Systems

Machinery for moving people, materials, and equipment is included in most large construction projects. Elevators are essential in large commercial buildings. Retail stores often rely on elevators and escalators to move customers and materials. In manufacturing facilities and power plants, conveyors, and monorail systems move materials and products from one area to another. Division 14 of the MasterFormat includes information on all types of conveying systems.

TECH TIP

Division 13 of MasterFormat 2004 has undergone a considerable amount of reorganization from the previous edition. Information regarding special facility components, such as hot tubs, precedes information dealing with special-purpose rooms and specialty structures.

Elevators. The most common conveying system included in Division 14 is elevators. Elevator specifications include the rated net capacity, rated speed, travel distance, number of stops, car size and interior finish, power requirements, and automatic control systems. **See Figure 2-26.** A variety of industry standards and governmental regulations apply to elevator construction, including standards and regulations of the American National Standards Institute (ANSI), the Americans with Disabilities Act (ADA), the National Electrical Code® (NEC®), and Underwriters Laboratories Inc. (UL).

Material Handling Systems. A variety of material handling systems may be included in commercial buildings. Overhead lifts and monorail systems are typically electrically controlled and are described in the specifications according to their lifting capacity, speed of lift and movement, and height of lift required. Conveying systems move items such as cans, coal, and pieces of mail. Conveyor type, speed, and capacity are included as part of Division 14.

C. Cabinet Bodies:

1. Fabricate, assemble and finish each cabinet as a complete and self-supporting unit. Unless otherwise shown, counter and tall storage units to be 24 inch minimum overall depth.
2. Ship counter tops separate for field installation.
3. Provide backs on all cabinets and boxed into rabbets at all edges.
4. Fabricate bottom and top panels of lock-joint construction, glued and fastened to end panel-simple butted joint not permitted. Top and bottom of cabinets to be 3/4" thickness. Fasten partitions and fixed shelves into tops, bottoms or ends, as applicable; partitions and fixed shelves 3/4" thick except shelves 4'0" or wider to be 1" thick.
5. Unless otherwise shown, provide toe-space on floor-mounted units, 4-1/4" high and 3" deep.
6. Adjustable Metal Feet: Except for open shelving, equip bases of fixed floor-mounted cabinets 24" width and over, with adjustable metal feet at each corner, with adjustment being made with tool after removal of insert cap in cabinet bottom.
7. Adjustable Shelving: Support each shelf with four shelf clips permitting adjustment on 1-inch centers.

Figure 2-25. Manufactured cabinetry and casework are built and installed according to project specifications.

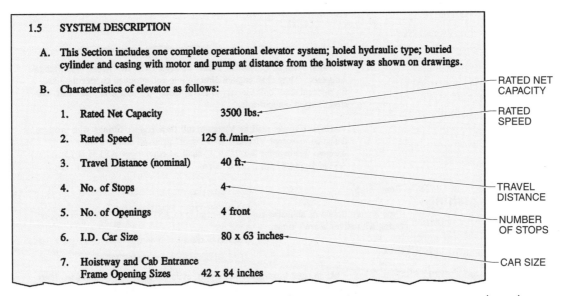

Figure 2-26. Division 14 of a set of specifications provides information about conveying systems such as elevators.

Divisions 21, 22, 23, and 40—Mechanical

Mechanical systems for large commercial buildings are extremely complex. Architects, mechanical engineers, electrical engineers, and the general contractor and subcontractors must coordinate their work to ensure proper installation and operation of mechanical systems. General conditions in these divisions include product and equipment warranties, maintenance instructions, commissioning, and balance and test run reports. Basic materials and methods information is provided on pipe types, hangers, and supports; duct and pipe connectors; insulation; gauges; flow control and measurement devices; electric motors and starters; and fuel tanks. The control systems installed on mechanical systems are also described in Division 23 and 25.

Fire Protection. Agencies that regulate fire protection equipment and develop standards and codes include the National Fire Protection Association (NFPA), the International Building Code (IBC), the Occupational Safety and Health Administration (OSHA), and local fire authorities. Fire protection equipment and systems are designed and installed according to the standards and codes. Specifications describe pipes, sprinkler heads, check valves, and fire department connections for wet pipe systems. Dry pipe systems include much of this same information in addition to air compressor and air pressure requirements.

Plumbing. Division 22 applies specifically to plumbing systems. Water supply and treatment information details connections to available water supplies and fire protection service pipes, valves, meters, and hydrants. Wastewater disposal and treatment information is provided for waste pipe and fittings and stormwater drains. Pipe and valves are specified by pipe type, size, material, and ability to withstand a certain amount of pressure. **See Figure 2-27.** Pipe and fittings for various process piping installations such as natural gas, compressed air, and vacuum systems are included in Division 22 and 40 of the MasterFormat. Tanks used for storing or heating water are also described in Division 22. Plumbing fixture information for sinks, water closets, drinking fountains, and faucets is provided in the specifications.

HVAC. Heating, ventilating, and air conditioning (HVAC) systems generate heat from sources such as boilers, solar systems, or natural gas heaters, provide cooling with systems of refrigerants, chillers, and compressors, and distribute air using various air handlers, motors, fans, and duct systems. **See Figure 2-28.** In hot and cold water systems, specifications are provided for pipes, circulating pumps, and heating and cooling transfer equipment. Information is also included in Division 23 about other heating and cooling systems, such as solar thermal heating systems, unit heaters, and air conditioners for special conditions. Air-handling equipment used to filter, remove, and exhaust fumes and smoke is also included in this division. Division 23 also provides information related to fuel piping for oil, gasoline, natural gas, and liquefied petroleum gas.

TECH TIP

The American Society of Heating, Refrigerating and Air-Conditioning Engineers (ASHRAE) is an international organization that creates uniform methods of testing, design requirements, and standard practices for HVAC systems.

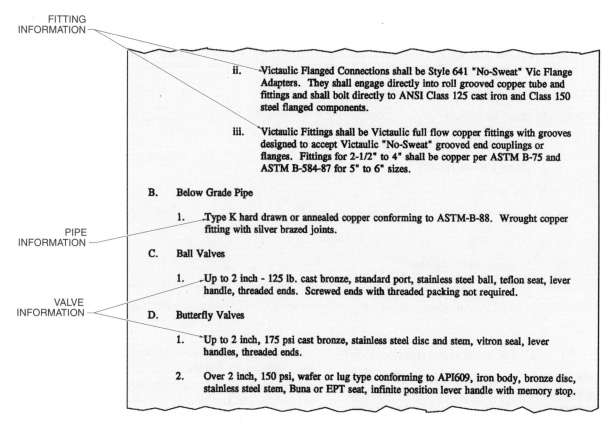

FITTING INFORMATION

ii.   Victaulic Flanged Connections shall be Style 641 "No-Sweat" Vic Flange Adapters. They shall engage directly into roll grooved copper tube and fittings and shall bolt directly to ANSI Class 125 cast iron and Class 150 steel flanged components.

iii.   Victaulic Fittings shall be Victaulic full flow copper fittings with grooves designed to accept Victaulic "No-Sweat" grooved end couplings or flanges. Fittings for 2-1/2" to 4" shall be copper per ASTM B-75 and ASTM B-584-87 for 5" to 6" sizes.

B.   Below Grade Pipe

1.   Type K hard drawn or annealed copper conforming to ASTM-B-88. Wrought copper fitting with silver brazed joints.

PIPE INFORMATION

C.   Ball Valves

1.   Up to 2 inch - 125 lb. cast bronze, standard port, stainless steel ball, teflon seat, lever handle, threaded ends. Screwed ends with threaded packing not required.

VALVE INFORMATION

D.   Butterfly Valves

1.   Up to 2 inch, 175 psi cast bronze, stainless steel disc and stem, vitron seal, lever handles, threaded ends.

2.   Over 2 inch, 150 psi, wafer or lug type conforming to API609, iron body, bronze disc, stainless steel stem, Buna or EPT seat, infinite position lever handle with memory stop.

Figure 2-27. Plumbing specifications provide information about all piping components including valves and supports.

Figure 2-28. Ductwork carries heated or cooled air throughout the structure in forced-air installations.

Divisions 25, 26, 27, and 28—Electrical

Divisions 25 through 28 provide information regarding automation systems, wiring, equipment, and finish of electrical systems, including descriptions of electrical sitework, raceways and conduits, panelboards, lighting, communication, data and telephone systems, and electrical safety and security systems. The NEC® is updated every three years and is the model code on which many of the specifications are based. For example, 700.9(D)(1) of the 2008 NEC® is related to assembly occupancies for greater than 1000 individuals or for buildings over 75′ tall with any of the following occupancies: assembly, educational, residential, detention and correctional, business, or mercantile. Feeder-circuit wiring installed in these locations must be protected by approved automatic fire suppression systems, have a listed protective system with a 1 hr fire rating, be protected by a listed thermal barrier system, be protected by a fire-rated assembly to achieve a minimum 1 hr fire rating, be embedded in not less than 2″ of concrete, or be a cable listed to maintain circuit integrity for a minimum of 1 hr when installed according to listing requirements. **See Figure 2-29.**

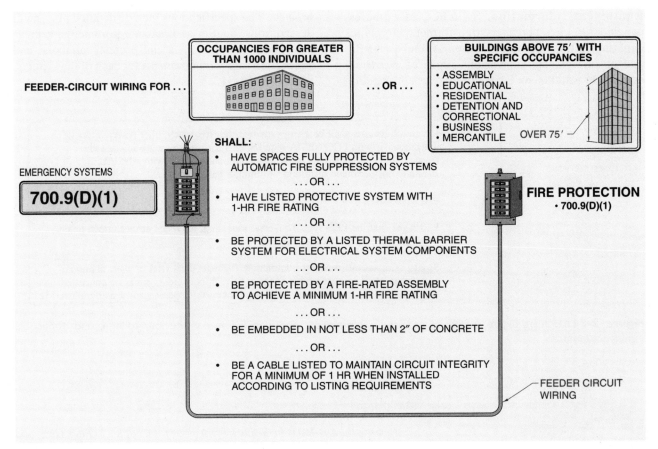

Figure 2-29. Many specifications and local building codes rely on information from the National Electrical Code® (NEC®).

Wiring Information. Electrical wiring considerations in the specifications include the type and placement of conductor and cable raceways, cable trays, conduit, junction boxes, and fittings. Proper procedures for trenching and sitework for placement of conductors and cables are provided. Types and sizes of electrical conductors and cable are described according to assembly, wire gauge, type of insulation and outer jacket, and diameter. **See Figure 2-30.** Procedures for splicing, connecting, and terminating conductors and cables are also described.

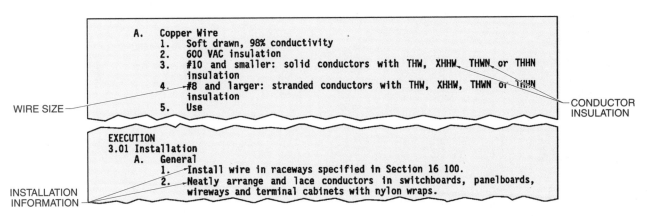

Figure 2-30. Division 16 of *MasterFormat 95* and Division 26 of *MasterFormat 2004* include information concerning conductors and cables.

Equipment Information. Switches, receptacles, panelboards, circuit breakers, transformers, luminaires (lighting fixtures), communication systems, and fire and alarm systems are various types of electrical equipment installed on large commercial construction projects. The specifications include detailed placement descriptions, quality assurance requirements, acceptable manufacturer names and product numbers, and connection safety requirements for each of these items. **See Figure 2-31.**

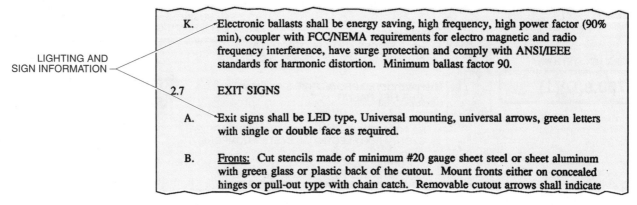

LIGHTING AND
SIGN INFORMATION

K.   Electronic ballasts shall be energy saving, high frequency, high power factor (90% min), coupler with FCC/NEMA requirements for electro magnetic and radio frequency interference, have surge protection and comply with ANSI/IEEE standards for harmonic distortion. Minimum ballast factor 90.

2.7   EXIT SIGNS

A.   Exit signs shall be LED type, Universal mounting, universal arrows, green letters with single or double face as required.

B.   Fronts: Cut stencils made of minimum #20 gauge sheet steel or sheet aluminum with green glass or plastic back of the cutout. Mount fronts either on concealed hinges or pull-out type with chain catch. Removable cutout arrows shall indicate

Figure 2-31. Lighting fixture specifications include information regarding fixture construction, voltage, and applicable industry standards.

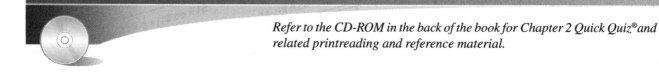

Refer to the CD-ROM in the back of the book for Chapter 2 Quick Quiz® and related printreading and reference material.

Sitework

Site plans illustrate the existing building site conditions and the layout of the planned final construction. Site plans include topographical, paving, landscape, and detail drawings for drainage and underground piping. Site plans may also indicate locations for traffic control and security features.

The planned locations of new construction are determined and shown on site plans after existing conditions are determined. Site plans contain extensive information about sitework that must be completed before building construction begins.

SITEWORK

Site plans illustrate the existing building site conditions and the layout of the planned final construction. Having detailed measurements of a building site and the planned building is necessary to ensure the proper placement of the building on the site. Horizontal elevations and vertical distances are measured by civil engineers and surveyors to accurately describe the property on which the building is located. Previously existing surface and subsurface conditions are determined by various methods and indicated on the site plans. Site plans for a building site identify property lines, rock and soil strata (layers of rock or soil), buried items from prior construction such as foundations, footings, and utilities, existing landscape features, and the possible existence of hazardous materials.

Site plans are referred to by a variety of names including site surveys, site maps, site drawings, and civil drawings. Site plans include topographical, paving, landscape, and detail drawings for drainage and underground piping. Site plans are commonly identified with a "C" prefix, representing "civil." For example, Sheet C1.1 is the first page of a set of site plans.

After the existing conditions are identified, the planned locations of new buildings are determined and shown on the site plans. In addition to building locations, new construction information includes utility location and information regarding surface grading, paving, curbs, walkways, and landscaping.

For heavy commercial buildings, architects and engineers provide legends on site plans to describe commonly used lines, symbols, and abbreviations. A site plan scale is used that is large enough for large property areas to be shown on a single sheet, for example, 1″ = 50′ or larger. A symbol pointing toward the north is also included on site plans. A portion of a city, county, or state map may be included on the site plans to assist in locating the building site.

CIVIL ENGINEERING

Civil engineers are trained and licensed to measure and describe property. Based on existing elevation and property line information, measurements are taken on the building site to verify existing information, establish new points where existing information is inaccurate or incomplete, and provide complete location information for all new construction operations. **See Figure 3-1.**

Legal Information

Land surveyors are licensed by their location of practice, which may be by state, county, or city. A legal system of property description begins with plat plans that are recorded with state, county, or city governmental agencies and provide legal boundaries for a property. A *plat plan*, or plat, is a drawing of a parcel of land providing its legal description and showing existing and planned construction. New surveys rely on the previously recorded plat plans as starting points for topographical and property line information. The United States Army Corps of Engineers (USACE) has established a series of monuments that provide clearly defined legal elevations for property description. **See Figure 3-2.**

Leica Geosystems

Figure 3-1. Surveyors and civil engineers measure the dimensions and elevations of construction sites.

```
                    NOTES

1)  Property lines shown on this drawing were generated
by using the record bearings and distances from the
RIVERPOINT TWO SHORT PLAT # CITY 88-12 (Inst.#
8811140373 ).NO pin were found or tied in during this
survey:

2)  Basis of Bearing being N 48 59'48" W along the
centerline of Riverpoint Boulevard from the monument at
its intersect with Trent Avenue to the monument at the
start of a curve.

3)  Basis of Vertical Datum being Elevation 1915.14 on
the top of a U.S. Army Corps Brass cap monument in the
southwest corner of the Division Street bridge.

4)  Manhole inverts shown on this drawing are from the
Riverpoint Boulevard phase two plans and were not
verified due to the bolt down lids.

5)  Established Temporary Bench Marks
    TBM #1 = South bolt on top of a fire hydrant at the
northwest corner of Parcel C. Elev.= 1903.84
    TBM #2 = South bolt on top of a fire hydrant at the
southwest corner of Parcel C. Elev.= 1915.35

BENCHMARKS┘              DATUM ┘

     ⊙    Control Monument

     ○    Property Corner

     ⬡    Fire Hydrant

     Ⓔ    Electric Box

     Ⓜ    Water Meter Box

     Ⓣ    Telephone Riser

     ✳    Light Pole

     ▣    Catch Basin

     ⊛    Sewer Manhole

     ⊕    Water Gate Valve

     ×    Spot elevation at this point
```

Figure 3-2. Notes on topographical drawings provide information concerning datum points and benchmarks.

Property Description. Boundaries for construction projects are measured from street centerlines or from township lines established by the United States Coast and Geodetic Survey. Intersecting points of street centerlines provide surveying stations in metropolitan areas. Township boundaries measured from principal meridians are used where street intersection points are unavailable.

Electronic surveying equipment is used to ensure accuracy when measuring long and irregular shapes. **See Figure 3-3.** Theodolites and total station instruments are commonly used for electronic distance measurement (EDM) to quickly and accurately measure distances. A *theodolite* is a precision surveying instrument that establishes and verifies vertical and horizontal angles through electronic distance measurement. A *total station instrument* is a precision surveying instrument used to perform leveling, plumbing, and horizontal and vertical measurement operations. Theodolites and total station instruments combine digital data processing with surveying technology. Information can be uploaded from a computer to theodolites and total station instruments, or downloaded from them to a computer and then a computer-aided design (CAD) system for the modification and/or production of site plans.

Figure 3-3. Electronic surveying equipment helps ensure accuracy when measuring distances and elevations on large construction sites.

Satellite technology and the use of Global Positioning System (GPS) information provides additional information for surveying crews to accurately locate property information and share this information with other parts of the construction team. GPS information can be electronically downloaded and linked directly into a building information modeling (BIM) program to create models that allow for virtual location of a building and excavations on a piece of property. **See Figure 3-4.**

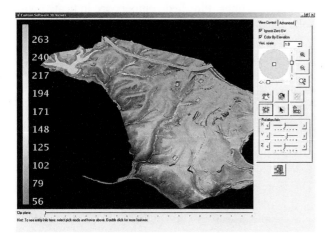

Carlson Software/The Engineering Groupe, Inc.

Figure 3-4. GPS information can be electronically downloaded and linked directly to a building information modeling (BIM) program to create models that allow for virtual location of a building and excavations on a piece of property.

TECH TIP

The Global Positioning System (GPS) is a U.S. space-based radionavigation system that consists of a group of 24 operating satellites that transmit one-way signals that give the current GPS satellite position and time.

Property lines or reference lines for commercial property commonly consist of straight and curved segments. Long, straight or curved property and reference lines are easily described when divided into smaller segments. For this purpose, a series of stations may be established at regular intervals along the property or reference lines. Straight property or reference lines are indicated using compass directions and distances from recognized points. For example, a reference line may be described as N35° 36′ 24″E 190.52′. This description indicates that the reference line between two particular surveying stations is 35 degrees, 36 minutes, and 24 seconds east of a true north/south line and that the line continues in this direction for a distance of 190 feet and 52 hundredths of a foot. **See Figure 3-5.** Compass direction and distance are restated at the next point and between the next two surveying stations.

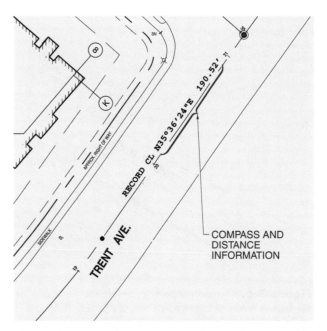

Figure 3-5. Straight property or reference lines are indicated on site plans by a compass reading and a distance dimension.

Earth moving equipment with laser detection devices quickly and accurately bring a building site to the required finished grades.

Curved property or reference lines are described in terms of change in compass direction, length of the radius for calculating the amount of curvature, and length of the curved segment. For example, a notation of $\Delta = 9° 32′ 5″$, R = 1800.02′, L = 300.00′ is shown on a site plan. **See Figure 3-6.** The delta sign (Δ) indicates the change of compass direction, the R indicates the length of the radius of the curve, and the L indicates the length of the arc section. In the example, the compass direction changes by 9 degrees, 32 minutes, and 5 seconds. The length of the radius for calculating the amount of the curvature is 1800 feet and 2 hundredths of a foot, and the length of the curved segment is 300 feet between two surveying stations or other reference points.

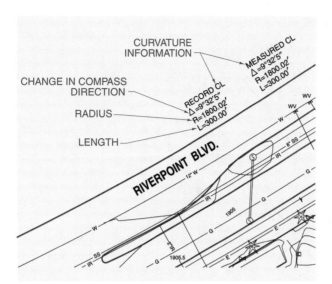

Figure 3-6. Curved property or reference lines are described in terms of change in compass direction, length of radius, and length of curved segment.

Site Layout

Legal boundaries and monuments serve as starting points for the remainder of the site layout. The existing grade of the property is measured. Building and reference lines for new construction are established. Easements and right-of-ways are located as noted on the site plans.

A clear layout of an entire site may not fit on one drawing sheet if the building and/or property are large. A match line may be shown on the working drawings to act as a reference line between individual sheets. A *match line* is an aligning mark on a print that is used when a drawing is too large to be contained on one sheet. **See Figure 3-7.**

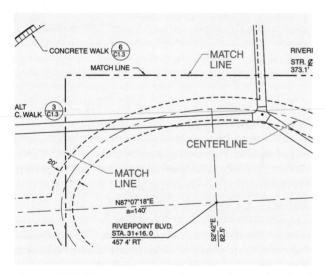

Figure 3-7. Two or more site plans may be easily related to each other with the use of match lines.

Topographic Description. Property elevations are indicated by a series of contour lines. Contour lines may be at 1′ or 2′ intervals in elevation, depending on the severity of the slope of the property. Each contour line has a number along its length indicating the elevation along that line. **See Figure 3-8.** Existing contours are indicated with dashed lines. Finished contours are indicated with solid lines.

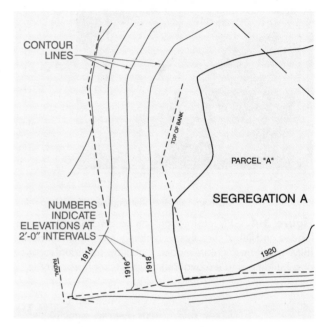

Figure 3-8. Site topographical information is shown with contour lines indicating the elevations.

Building Lines. Site plans provide the dimensions for locating a new building on a property. During surveying operations, reference stations are placed on the property. Building layout proceeds from these stations. The stations may be in the form of a building reference corner or line and are established by surveyors for use throughout the entire construction project.

Utilities. Information regarding utilities shown on site plans includes locations for electrical connections, electrical light standards, natural gas piping, water supply piping, fire hydrant(s), stormwater and wastewater drainage piping, and telephone and data cables. **See Figure 3-9.** Locations of existing underground utilities, such as electrical wiring and piping, are based on prior surveys or information provided by utility companies. Underground utilities are commonly indicated with a dashed line or with a solid line that is broken for placement of a letter indicating the type of utility. For example, a solid line with a "W" indicates the location of a water pipe. Identifying the locations of utilities before excavation will help prevent severing the existing utility lines.

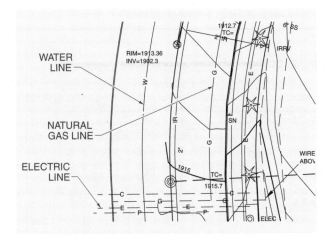

Figure 3-9. Existing underground utility locations are indicated with dashed or solid lines and letters noting the type of utility service.

The width and location of utility easements are indicated on site plans. An *easement* is a strip of land, commonly along the perimeter of property lines, that is used for placement and maintenance of utilities.

SITE PREPARATION

Site plans contain extensive information about the sitework that must be completed before the construction process begins. Soil engineers determine the bearing capacities and quality of the strata below grade level. Operating engineers grade, remove, add, and compact the soil to prepare it for new construction. Underground drainage systems are laid out and installed.

The general contractor and construction manager use the site plans to lay out the entire construction area. The location for positioning construction equipment and storing materials is based on information shown on the site plan. Temporary construction trailers are located where they will not interfere with construction. Staging areas for delivery of materials must be determined. Areas on the building site where concrete is to be placed must be accessible. Accessibility must be provided for cranes when large members are to be lifted into place. Plans must be made for trees and shrubs that will be either removed or protected.

Soil Engineering

Soil engineers analyze samples taken from the building site. **See Figure 3-10.** The composition of each sample is determined. A report from the soil engineer to the owner and architect describes the various materials below the surface at specific points on the construction site. Determining the composition of subsurface materials provides valuable information for structural engineers when designing foundation systems. Structural engineers calculate the live and dead loads that the structure places on the bearing strata at the building site. This loading information is compared to the soil samples to determine the steps necessary to ensure that the footings and foundation provide adequate support. Subsurface information is also used by various contractors bidding on drilling, soil stabilization, or excavation work.

Figure 3-10. Core samples are taken and analyzed to determine the bearing capacities of the soil on the building site.

TECH TIP

Soil is classified into four major groups determined by soil particle size: clay (up to and including 0.0002″), silt (greater than 0.0002″ to 0.003″), sand (greater than 0.003″ to 0.08″), and gravel (greater than 0.08″ to 3″). Soil particles greater than 3″ are classified as boulders.

Test Boring. A soil engineer determines the layout for test boring at the building site. A core drill is used to drill holes into the earth at these predetermined points to extract core samples. The core sample evaluations include measuring the depth of the various layers of soil, rock, and other subsurface materials and analyzing the layers for composition. **See Figure 3-11.** Test borings are often made to determine the depth at which solid bearing strata can be reached. The locations of the test boring holes may be shown on a site plan or included on an additional drawing provided by the soil engineering firm.

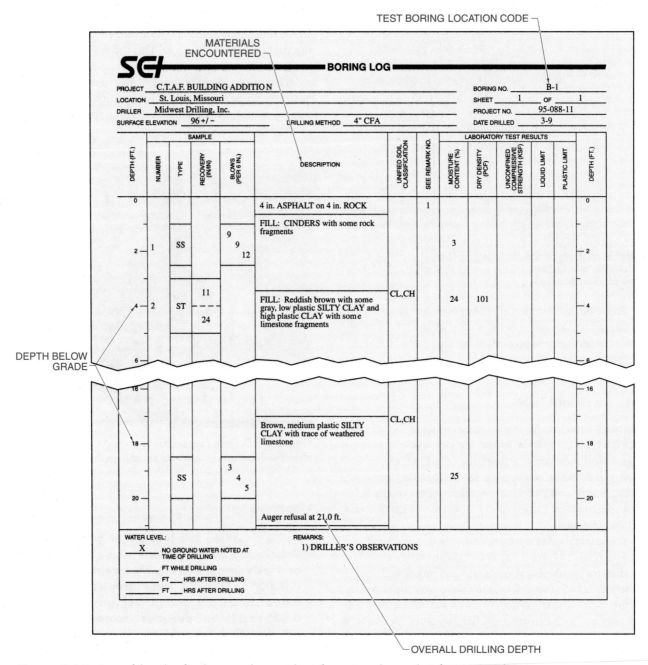

Figure 3-11. A careful study of soil test results provides information about subsurface materials.

Subsurface Materials. Many types of materials are found underground, including rock, decayed rock, loose rock, boulders, gravel, sand, clay, silt, and soil. Solid rock is typically considered to provide stable bearing for a foundation. Decayed rock may be compact and hard or fully decayed and soft. Loose rock is rock that at one time became detached from the rock layer in which it was originally formed. Boulders are rocks that have been transported by geological action from the site of their formation to their current location. Gravel is composed of pieces of rock smaller than boulders and larger than sand. Sand is classified as fine, medium, or coarse based on grain size. Clay is a mixture of silica, alumina, and water that expands and contracts greatly based on water volume. Soil is measured for its depth at the site and its amount of compaction.

Other nongeologic materials may be discovered, such as abandoned foundations, wells, caves, and tunnels. Each presents different foundation design requirements for the engineer and architect.

Hazardous Materials. Soil analysis may indicate the existence of hazardous materials on a construction site. Hazardous materials may include solid waste, toxic chemicals, and radioactive waste. The Environmental Protection Agency (EPA) has identified specific hazardous materials and classified them according to corrosivity, ignitability, reactivity, and toxicity.

Special remediation processes must be taken to contain or safely remove hazardous materials when they are encountered. Some hazardous materials are isolated using containment structures. The materials are collected into an isolated area that is protected from the elements and removed from human contact. For smaller amounts of hazardous materials, the soil and contaminants may be loaded into drums or other containers and sealed for removal to larger or safer containment areas. Remediation processes require approval and licensing by the EPA and any applicable local or state departments of natural resources. Tradesworkers involved in the removal and containment process must meet all federal and local licensing and certification requirements and must follow guidelines relating to the use of personal protective equipment for hazardous materials.

Information indicated on site plans that relates to hazardous materials includes the locations of contaminated soil areas and locations for installation of geotextiles and groundwater monitoring wells. **See Figure 3-12.** *Geotextiles* are sheets or rolls of material that stabilize and retain soil or earth in position on slopes or in other unstable conditions. Geotextiles divert groundwater away from contaminated areas to keep hazardous materials from leeching into drinking water supplies. Groundwater monitoring wells enable regular checks of the subsurface water and ensure lack of contamination.

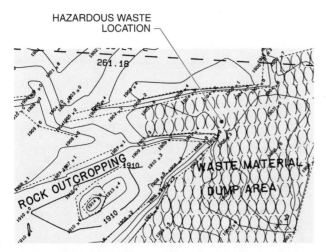

Figure 3-12. Site plans indicate the location and plans for containing or removing hazardous materials on a building site.

Earthwork

Earthwork is digging and excavating operations. Various plan views and sections provide information regarding earthwork. Contour lines indicate the grading required for placement of new construction including buildings, paved areas, landscaping, and drainage.

Layout. Existing and planned elevations are shown on site plans using contour lines. **See Figure 3-13.** Dashed lines represent existing elevations while solid lines represent planned elevations. Surveyors lay out and mark elevations at several points on the building site following the elevations and slopes indicated on the site plans. Elevation stakes are marked with elevations and measurements for use as guides for operating engineers. Operating engineers refer to the markings on the layout stakes to determine the amount of cut and/or fill necessary. It may also be necessary to refer to architectural and foundation plans to determine the depth of excavations for foundations, footings, and piers. Additional layout information for subsurface piping may be found on the mechanical prints.

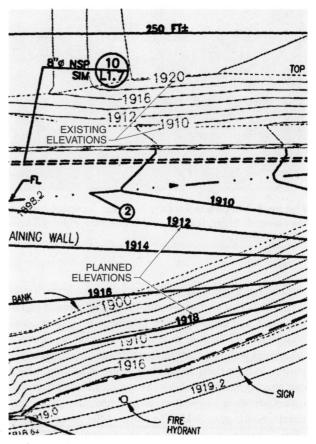

Figure 3-13. Existing and planned elevations are shown on site plans using contour lines.

Grading. A variety of excavation equipment, including drilling machines, backhoes, bulldozers, and scrapers, may be required to excavate and move earth and rock around the construction site. The excavation equipment used depends on the size of the site and the types of earth or rock to be moved. In some instances, the use of GPS technology on excavation equipment helps determine various grade elevations. **See Figure 3-14.** Development of clean air guidelines and emission requirements also has an impact on equipment usage at a construction site. Excavating contractors need to check project specifications, environmental requirements, and local building code requirements when determining the appropriate excavation equipment for a project.

Figure 3-15. Proper compaction of layers of subgrade materials is achieved using a sheepsfoot compactor.

Figure 3-14. GPS technology is used to accurately control the elevation of grading by using information obtained from orbiting satellites.

In some situations, blasting may be required to loosen rock. **See Figure 3-16.** To achieve this, a series of holes is drilled into the rock layers at regular intervals and explosive charges are carefully placed in the holes. The spacing, number of holes, and amount of explosives in each hole is determined by an analysis of the type and strength of the rock. After blasting, the loose rock is removed using the appropriate excavating equipment.

Depending on the soil type and surface conditions, it may be necessary to control the amount of dust created and to minimize soil erosion during grading. The proper finish elevations and amount of compaction must be obtained where soil is the primary surface and subsurface material. Failure to fully compact soil may result in settling, which can change finish elevations. Sheepsfoot compactors and rollers are used to compact surface and subsurface materials. **See Figure 3-15.** The slope of the finish grade is often indicated as a percentage. For example, a 2% grade indicates a change in elevation equivalent to 2′ over a distance of 100′.

Figure 3-16. Controlled blasting loosens rock by using explosive charges placed below ground level.

Stormwater Erosion Control. During excavating operations, large quantities of loose soil are exposed and can be washed by rain into surrounding creeks and water supplies. Federal and state requirements define measures that must be taken to prevent stormwater and silt runoff. Based on the requirements, a stormwater pollution prevention plan must be developed and implemented for construction sites. A *stormwater pollution prevention plan (SWPPP)* is a written plan detailing the pollution control measures that will be taken on a construction site to prevent stormwater and silt runoff.

A sediment containment system is outlined in an SWPPP and includes details for barriers, drain/inlet protection, and other nonstructural items. A *silt fence* is a barrier used to contain soil sediment that consists of geotextiles or straw bales secured in place with wood or metal stakes. **See Figure 3-17.** Silt fences and other barriers slow the flow of stormwater as it runs off the grading site and cause soil sediment to settle and become trapped.

Figure 3-17. Stormwater pollution prevention plans (SWPPP) typically require silt fences around the perimeters of construction sites.

Silt fences must be checked frequently during excavation operations to ensure no breaks have developed that would allow stormwater to run off the site unimpeded. Drain/inlet protection consists of straw bales or rock barriers that divert soil sediment away from drains and inlets while allowing stormwater to enter the drainage system. In some situations, landscaping fabric may be wrapped around a drain or inlet opening to allow only stormwater to enter the drainage system. Nonstructural components of a stormwater pollution prevention plan may include temporary vegetation, mulch, drainage channels, or rock or stone structures.

Drainage Systems

Methods and materials for the removal of surface water and wastewater are indicated on site plans. These methods and materials include surface drainage plans, catch basins, trenches, piping, and geomembranes.

Storm Drainage. The flow direction of surface water into drainage systems is shown on site plans. Proper finish grade elevations are designed to channel water away from buildings and into surface drains. Catch basins and connecting pipes are some of the first items installed on a building site. A *catch basin* is a reservoir or tank in a surface water drainage system that is used to obstruct the flow of objects that will not readily pass through a sewer while allowing surface water to freely flow. Locations and elevations of catch basins are indicated on site plans. **See Figure 3-18.** Catch basin information includes the elevation at the rim, elevation at the bottom of inlet pipes, and the slope of inlet and outlet pipes. The *invert elevation* is the inside elevation at the bottom of the inside flow line of a pipe. Slopes are commonly indicated as the percent of grade. Excavation for the installation of catch basins and piping is based on the provided elevations.

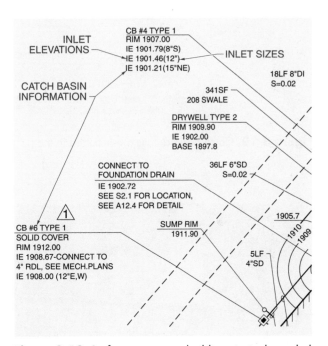

Figure 3-18. Surface water on a building site is channeled into catch basins at regular intervals.

Catch basins and their connecting pipes are commonly made of precast concrete. Details for catch basins are included as part of the prints and provide dimensions and material information. Support materials for catch basins include cast-in-place concrete and gravel. Proper compaction of subsurface materials is required to ensure that catch basins and pipes maintain their designed elevations. The design and material for surface grates is also part of the details. **See Figure 3-19.**

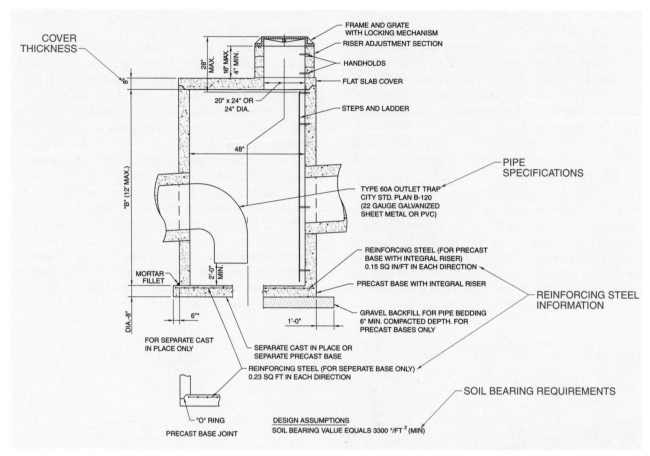

Figure 3-19. Catch basin details are included in the part of the site plan pertaining to site drainage.

A pipe laser is used to ensure that precast concrete drainage pipe is set at the slope indicated on the site plans.

If roof drains are included in the stormwater drainage plan, elevations for the connections of the roof drainage piping to underground piping are indicated. The diameters and type of pipe may also be shown.

Retention or detention ponds may be used to control stormwater runoff to minimize the amount of load on the storm drainage system during heavy or continuous rains. A *retention pond* is stormwater treatment system designed to hold a specific amount of water indefinitely. A *detention pond* is a low-lying area that is designed to temporarily hold stormwater while slowly draining to another location. Retention or detention ponds can also be used to capture and recycle stormwater for other uses at the building site, such as irrigation for vegetation and plantings. **See Figure 3-20.**

Waste Drainage. A partial site plan may be provided as part of the mechanical prints indicating the location of the connection of the drainage systems to existing stormwater and wastewater utility pipes. Additional information about waste piping for a structure is included on the mechanical prints.

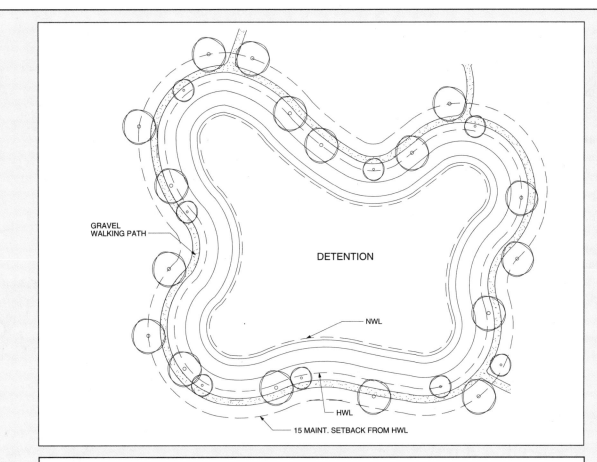

GRAVEL
WALKING PATH

DETENTION

NWL

HWL

15 MAINT. SETBACK FROM HWL

LEED 2009 for New Construction and Major Renovation
Project Checklist

8	4	2		**Sustainable Sites**	Possible Points:	26
Y	N	?				
Y			Prereq 1	Construction Activity Pollution Prevention		
	1		Credit 1	Site Selection		1
	1		Credit 2	Development Density and Community Connectivity		5
	1		Credit 3	Brownfield Redevelopment		1
	1		Credit 4.1	Alternative Transportation — Public Transportation Access		6
1			Credit 4.2	Alternative Transportation — Bicycle Storage and Changing Rooms		1
1			Credit 4.3	Alternative Transportation — Low-Emitting and Fuel-Efficient Vehicles		3
1			Credit 4.4	Alternative Transportation — Parking Capacity		2
1			Credit 5.1	Site Development — Protect or Restore Habitat		1
1			Credit 5.2	Site Development — Maximize Open Space		1
		1	Credit 6.1	Stormwater Design — Quantity Control		1
1			Credit 6.2	Stormwater Design — Quality Control		1
		1	Credit 7.1	Heat Island Effect — Non-roof		1
1			Credit 7.2	Heat Island Effect — Roof		1
1			Credit 8	Light Pollution Reduction		1

U.S. Green Building Council

Figure 3-20. A vegetated swale and a detention pond promote the infiltration, capturing, and treatment of stormwater runoff, which lowers environmental impact and contributes to LEED® certification.

FINISH

Toward the completion of the construction project, site plans are used to obtain information about surface finishes. After the majority of the building is constructed, the surrounding areas must be finished to provide the appropriate access and landscaping. Streets, parking lots, walkways, and curbs are detailed on the site plans. Landscaping locations and materials may also be part of site plans.

Driveways

Access to a building is commonly provided by vehicular-access driveways. The widths and locations for the placement of vehicular-access driveways are indicated on the site plans. **See Figure 3-21.**

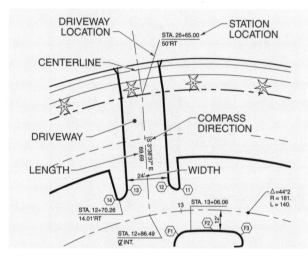

Figure 3-21. Driveway locations on site plans include the compass direction of the centerline, station location, length, and width.

Driveway Paving. Two common paving materials for driveways include asphalt and concrete. *Impervious paving* is paving that is watertight and does not allow for water to penetrate the surface of the paved area. *Pervious paving* is a surface treatment that allows stormwater to run through the paved area into the ground below, thereby increasing the absorption of stormwater at the building site and minimizing stormwater runoff. Pervious paving may consist of either asphalt or various concrete designs and mixes and can be used for driveways, walkways, or parking lots. The materials and designs of the paved areas depend on their planned usage. Driveways designed to withstand heavy loads require additional reinforcement and are typically thicker. Details in the site plans indicate the surface and subsurface paving materials required. Suitable pavement performance requires proper compaction of the subgrade, installation and compaction of structural fill materials, and application of surface materials. **See Figure 3-22.**

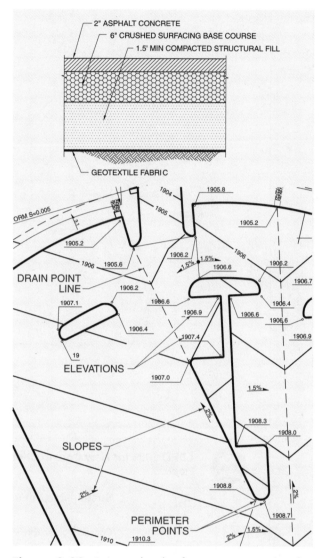

Figure 3-22. Section details of pavement provide information regarding the thicknesses and materials for each layer. Parking lot slopes and elevations are indicated with a series of perimeter points, percentage slopes, and drain point lines.

The proper pavement slopes, elevations, and dimensions are required to ensure proper drainage and accessibility. Site plans indicate elevations to the top edges of paved areas and the percentage of slope. For irregularly shaped paved areas, a schedule of paving designs and curvatures may be provided on the prints or in the specifications. A parking layout plan indicates the size and number of parking spaces provided in each paved area. Painting and signage in the paved parking areas are also shown.

Driveway Curbs. Various curb designs may be used depending on the need to match existing curbs or protect against damage in heavy-usage conditions. Site plans indicate where existing curbs must be removed to provide for new driveways. Details on the site plans show the design and materials for curbs. **See Figure 3-23.**

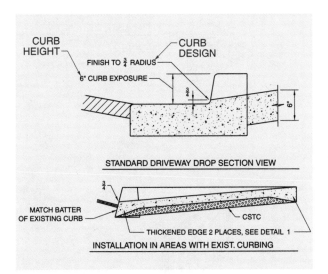

Figure 3-23. The amount of exposure and radius designs for curbs are provided on details.

Walkways

Paved walkways are made of many different materials and in many different designs. The proper layout and finishing of walkways is indicated on the site plans. Specifications for walkways include dimensions, finish, width, length, and direction. **See Figure 3-24.**

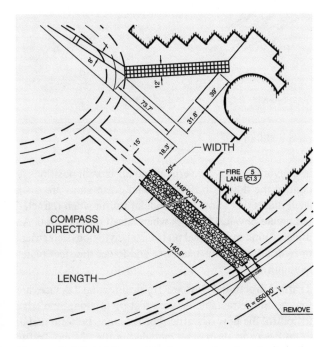

Figure 3-24. Walkway locations on site plans include the compass direction of the centerline, length, and width.

Walkway Paving. Walkways must be finished in a manner so that walking is easy and slipping is prevented. Cast-in-place concrete, precast concrete, and asphalt are commonly used for walkway materials. Walkways may be formed with impervious or pervious materials in a manner similar to driveways. As with paved areas, details on site plans provide subsurface and surface finish information.

Walkway Curbs. Various curb and ramp designs that minimize tripping and provide handicapped access are used where walkways and paved areas meet. **See Figure 3-25.** Lowered curb areas provide small, low ramps for handicapped access to meet the guidelines of the Americans with Disabilities Act (ADA). Special curbing around walkways near planters or grassy areas may be necessary.

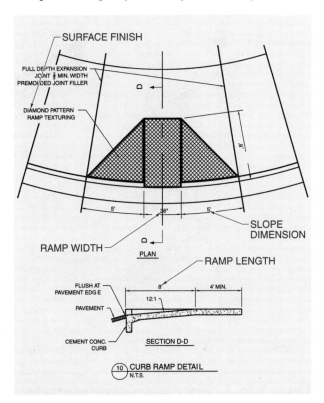

Figure 3-25. Sloping ramps provide a transition and safe access from paved driveways to walkways.

Security

Many commercial buildings require special security installations to control access to buildings by vehicular traffic. These security installations are designed to provide only authorized access to the building or building site. Security installation locations are shown on site plans and may include bollards, vehicle barriers, or traffic directional devices. **See Figure 3-26.** In addition, guard structures may be required for highly secured locations. A variety of precast concrete barriers and metal fences may also be installed as temporary or permanent vehicle barriers.

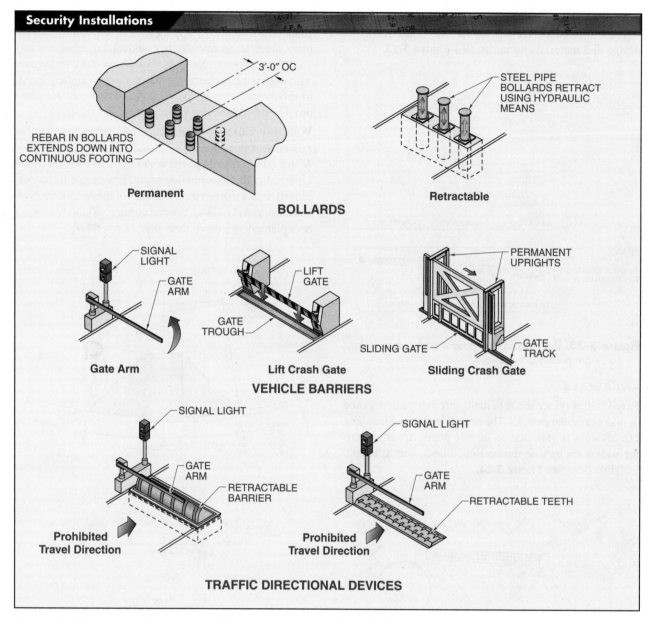

Security Installations

3'-0" OC

REBAR IN BOLLARDS EXTENDS DOWN INTO CONTINUOUS FOOTING

Permanent

STEEL PIPE BOLLARDS RETRACT USING HYDRAULIC MEANS

Retractable

BOLLARDS

SIGNAL LIGHT

GATE ARM

Gate Arm

LIFT GATE

GATE TROUGH

Lift Crash Gate

PERMANENT UPRIGHTS

SLIDING GATE

GATE TRACK

Sliding Crash Gate

VEHICLE BARRIERS

SIGNAL LIGHT

GATE ARM

RETRACTABLE BARRIER

Prohibited Travel Direction

SIGNAL LIGHT

GATE ARM

RETRACTABLE TEETH

Prohibited Travel Direction

TRAFFIC DIRECTIONAL DEVICES

Figure 3-26. A variety of security installations may be located on a building site.

Bollards. A *bollard* is a metal post, reinforced stone post, or concrete-filled metal post placed in a manner that inhibits vehicular traffic. Bollards may be fixed or movable. Fixed bollards commonly consist of steel pipes placed in a concrete pier or foundation and are filled with concrete. Movable bollards are equipped with locking devices that allow for retraction or removal when necessary. Bollard spacing across walkways or other vehicular traffic areas is designed to prohibit vehicular access but allow for pedestrian access.

Vehicle Barriers. Gate arms and lift or crash gates are common vehicle barriers for driveways. Gate arms are wood, metal, or fiberglass members that are supported at one end

and pivot between the vertical and horizontal positions to control the flow of traffic. Lift and crash gates are thick steel plates that allow for vehicular traffic when retracted but create physical barriers when raised into position. An underground mechanical pit is installed to house electrical or hydraulic systems that raise and lower the steel plates as required.

Traffic Directional Devices. Traffic directional devices are low-profile retractable devices that pivot along one edge to control the flow of traffic. The face may be solid metal, or spikes or teeth may be included in the design. Traffic directional devices are typically used in conjunction with a gate arm and signal light.

Landscaping

Many local building codes require that a certain amount of landscaping and vegetated open space be provided on new construction projects. LEED® certification credits may be awarded for maximizing vegetated open space within the project boundaries. **See Figure 3-27.** The use of appropriate landscaping may also increase the environmental quality of the overall building site. The paving portion of the site plan provides the dimensions and locations for planters and open areas. Landscaping plans provide information about the types of plants, planting methods, and final surface treatments.

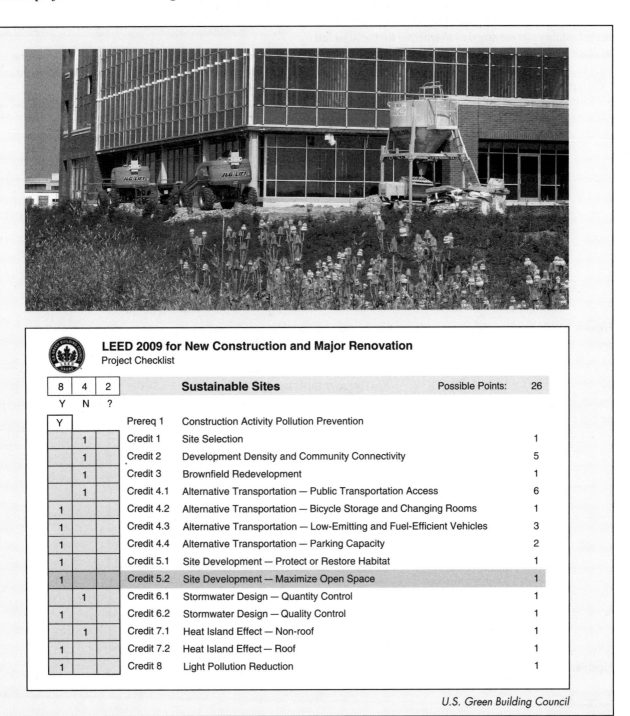

LEED 2009 for New Construction and Major Renovation
Project Checklist

8	4	2		**Sustainable Sites**	Possible Points:	26
Y	N	?				
Y			Prereq 1	Construction Activity Pollution Prevention		
	1		Credit 1	Site Selection		1
	1		Credit 2	Development Density and Community Connectivity		5
	1		Credit 3	Brownfield Redevelopment		1
	1		Credit 4.1	Alternative Transportation — Public Transportation Access		6
1			Credit 4.2	Alternative Transportation — Bicycle Storage and Changing Rooms		1
1			Credit 4.3	Alternative Transportation — Low-Emitting and Fuel-Efficient Vehicles		3
1			Credit 4.4	Alternative Transportation — Parking Capacity		2
1			Credit 5.1	Site Development — Protect or Restore Habitat		1
1			Credit 5.2	Site Development — Maximize Open Space		1
	1		Credit 6.1	Stormwater Design — Quantity Control		1
1			Credit 6.2	Stormwater Design — Quality Control		1
	1		Credit 7.1	Heat Island Effect — Non-roof		1
1			Credit 7.2	Heat Island Effect — Roof		1
1			Credit 8	Light Pollution Reduction		1

U.S. Green Building Council

Figure 3-27. To obtain LEED® certification points, the ratio of vegetated open space to developed space is maximized to provide open space and biodiversity.

If an irrigation system is to be installed, landscaping plans will indicate the water source, locations and types of sprinkler heads, and piping. **See Figure 3-28.** Details show pipe connections, valve boxes, and sprinkler head details. Locations for exterior signage may also be shown on details or on the topographical portion of a set of site plans.

Plants. Various types of trees, shrubs, and ground cover are indicated on landscaping plans. Previously existing plants at the building site are noted as to whether they will be left in place or removed. A plant schedule for new plantings may be included. The plant schedule provides the size of each type of plant and the scientific name, common name, location, and number of each plant. **See Figure 3-29.**

Details may be included showing the planting methods and materials to be used, including the planting depth, soil modifications, and mulch type.

Open Areas. Unpaved and unlandscaped areas are sodded or seeded for grass or left in their natural state. Treatment of these areas is indicated on the landscape plans.

TECH TIP

Gray water irrigation systems can reduce potable water use and may contribute to LEED® certification points.

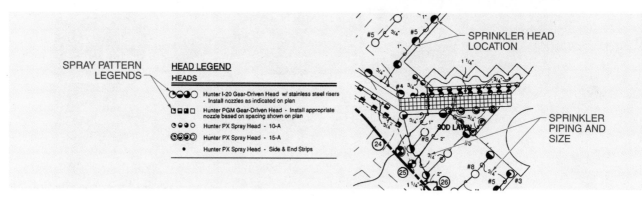

Figure 3-28. Landscaping plans indicate the locations of sprinkler heads and piping.

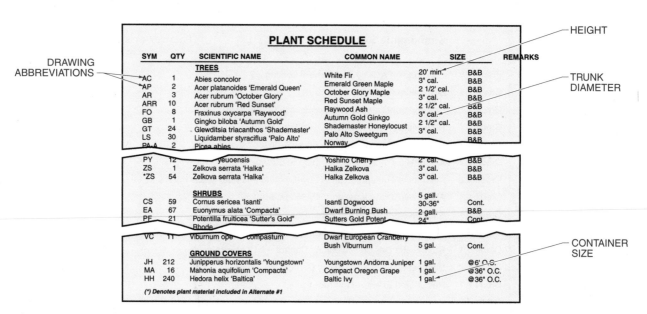

Figure 3-29. A plant schedule may be included as part of the landscaping plans.

Refer to the CD-ROM in the back of the book for Chapter 3 Quick Quiz® and related printreading and reference material.

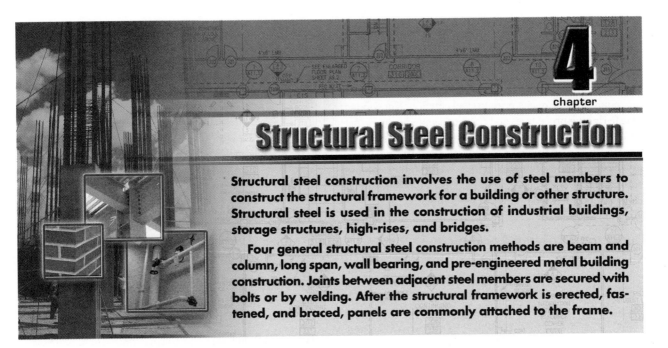

chapter 4

Structural Steel Construction

Structural steel construction involves the use of steel members to construct the structural framework for a building or other structure. Structural steel is used in the construction of industrial buildings, storage structures, high-rises, and bridges.

Four general structural steel construction methods are beam and column, long span, wall bearing, and pre-engineered metal building construction. Joints between adjacent steel members are secured with bolts or by welding. After the structural framework is erected, fastened, and braced, panels are commonly attached to the frame.

STRUCTURAL STEEL CONSTRUCTION

Structural steel construction involves the use of light to heavy gauge steel members to construct the structural framework for a building or other structure. Light gauge steel members may be used in the construction of industrial buildings and storage structures. Heavy beams and truss assemblies are used to construct high-rises and bridges. Steel may be used as the sole material for the framework or it may be combined with other construction materials such as masonry or reinforced concrete.

Several construction operations are common when building with structural steel. A structural engineer determines the steel to be used and sizes and shapes of steel members required based on the loads that will be imposed on the structure. BIM techniques are commonly used to move the steel designs from the design phase into the fabrication and production shop. Structural steel members are fabricated according to electronic files or shop drawings at a fabrication facility. The structural steel members are transported to the job site where the members are unloaded, or shaken out. *Shaking out* is the process of unloading steel members in a planned manner to minimize moving of members during erection.

Structural steel members are lifted into place with various lifting equipment. **See Figure 4-1.** Structural steel members are erected, braced, and secured together to create the structural framework. This framework is covered with the materials for floors, walls, and roofing.

STRUCTURAL STEEL CONSTRUCTION METHODS

Varying engineering requirements and job site conditions create the need for several structural steel construction methods. Four general structural steel construction methods are beam and column, long span, wall bearing, and pre-engineered metal building construction. Erection plans provide information for proper placement of steel members for each structural steel construction method.

Figure 4-1. Structural steel members are erected, braced, and secured together to create a structural framework.

With each method, the proper handling, lifting, and placement of steel members are essential. Developments in lifting technology allow extremely heavy loads and large members to be lifted safely. Safe use of rigging and lifting equipment may require that local or state licensing or testing requirements are met prior to steel erection. Knowledge of safe rigging procedures and crane capacities ensures that structural steel members are erected efficiently and safely. Tradesworkers working at heights of over 10′ must take proper precautions, tying-off with personal fall protection equipment in accordance with Occupational Safety and Health Administration (OSHA) regulations.

Beam and Column Construction

The most common structural steel construction method is beam and column construction. *Beam and column construction* is a structural steel construction method consisting of bays of framed structural steel that are repeated to create large structures. **See Figure 4-2.** A *bay* is the space between the centers of adjacent columns along exterior walls. Steel columns are erected over foundation members consisting of a series of footings, foundations, piers, or pilings. The columns are set onto the foundation members and secured with anchor bolts and plates.

Horizontal steel beams and girders are fastened to the columns. Steel angles attached to the columns form a seat for the beams, or the angles may be attached to beams that are then bolted to columns. Tie rods, channels, and other braces stabilize the structure. Joists and purlins are placed between the columns and beams to complete the structural framework.

Republic Steel Corporation

Figure 4-2. In beam and column construction, beams and girders support floor and roof loads and distribute the loads to the vertical columns.

Long Span Construction

Long span construction is a structural steel construction method in which large girders and trusses constructed of large horizontal steel members are fastened together (built up) to span large areas. In long span construction, a series of built-up girders and trusses is used for spanning large areas without the need for excessive intermediate columns or other supports. **See Figure 4-3.** Long span construction is commonly used for structures such as bridges and large arenas. In bridge construction, large girders and beams span between bridge abutments, piers, and other supports. Concrete or steel decking is supported by large structural steel members. Details and shop drawings indicate the arrangement of the various angles, channels, and other members used to construct girders and trusses.

Appropriate personal fall-arrest equipment must be used by tradesworkers engaged in steel erection activities with an unprotected side or edge more than 15′ above a lower level.

TECH TIP

Per OSHA 29 CFR 1926.760—*Fall Protection*, connectors (workers involved in securing together structural steel members) must have completed specialized connector training. In addition, at heights over 15′ and up to 30′ above a lower level, connectors must be provided with a personal fall-arrest system, positioning device system, or fall-restraint system and wear the equipment necessary to be able to be tied off. Connectors must be tied off at heights greater than 30′.

Figure 4-3. In long span construction, long distances are spanned with built-up structural steel girders and trusses.

Figure 4-4. In wall bearing construction, horizontal steel beams and joists are supported by other construction materials such as masonry.

Erection Plans

Erection plans include working drawings that provide information about anchor bolt layout, plan views of the structural steel members at each floor level, sections, elevations, and details that include steel member and component information and connections. **See Figure 4-6.** The number of drawings required and their complexity depend on the size of the structure and the amount of steel to be placed.

Wall Bearing Construction

Wall bearing construction is a structural steel construction method in which horizontal steel beams and joists are supported by other construction materials such as masonry and reinforced concrete. Structural steel members span floors and roofs between masonry or reinforced concrete walls. The masonry or concrete walls support the vertical loads and the structural steel beams and joists support the horizontal loads. **See Figure 4-4.** Bearing base plates installed on the masonry or reinforced concrete walls provide proper load distribution where the steel members rest on the walls.

Pre-Engineered Metal Building Construction

Pre-engineered metal building construction is a structural steel construction method consisting of prefabricated structural steel members including beams, columns, girts, and trusses. The width of pre-engineered metal buildings ranges from 10′ to 360′. Basic types of pre-engineered metal buildings include rigid frame, beam and column, and truss frame. **See Figure 4-5.** Interior columns are erected as indicated on the erection plans. Lengths of pre-engineered metal buildings are based on the number of bays. Bays vary in length from 18′ to 30′. Pre-engineered building manufacturers use specialized systems of fasteners, braces, and rafters based on their particular product design.

TECH TIP

Cranes used in steel erection must be inspected prior to each shift to check the overall set-up, operation, and functionality and to check for component defects or problem areas.

On multistory structures, perimeter safety cables must be installed at final interior and exterior perimeters of the floors as soon as the metal floor decking has been installed.

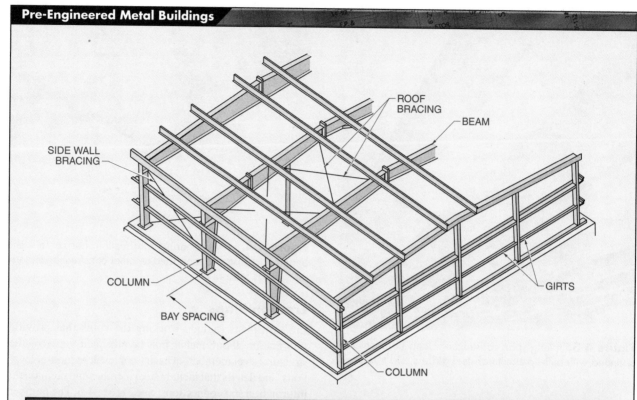

Pre-Engineered Metal Buildings

FRAME SHAPES AND WIDTHS		
Description	**Shape**	**Common Widths**
Rigid frame, high profile		25′ to 120′
Rigid frame, low profile		40′ to 120′
Rigid frame, one-way slope		20′ to 80′
Beam and column, one interior column		80′ to 120′
Beam and column, two interior columns		120′ to 180′
Beam and column, three interior columns		160′ to 240′
Truss frame with straight columns		30′ to 120′

Figure 4-5. Pre-engineered metal buildings consist of prefabricated structural steel members including beams, columns, girts, and trusses.

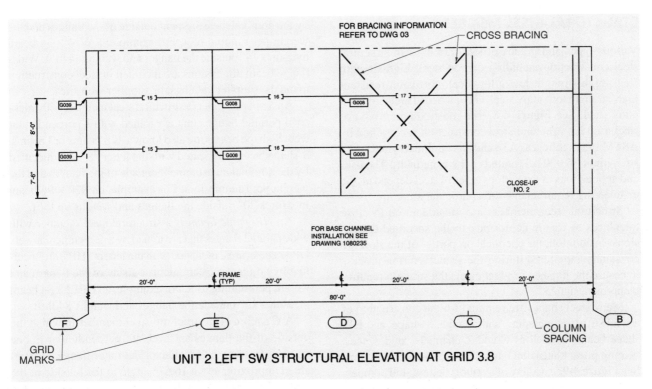

Figure 4-6. Erection plans provide information regarding structural steel construction.

Purposes. Structural steel erection plans are required to ensure that the proper steel members are placed in the specified locations. Erection plans provide directional information so structural steel members can be placed in the proper direction and relation to each other and ensure proper connections are made with other steel members and building components. Connection information includes using bolts, welding, or other required fasteners.

Format. Erection plans include the location of each structural steel member, assembly information, dimensions, the number of steel components comprising a member, and any additional information needed for steel erection. Steel members are placed and oriented based on a letter and number grid shown on plan views. **See Figure 4-7.** The distances between grid lines are shown around the building perimeter on the erection plans. Columns, beams, girders, joists, and braces are identified according to the letter and number grid on the plan views. The letter and number grid system on the plan views is related to all other drawings, including elevations, details, and fabrication drawings. Uppercase letters denote the main structural members, such as "B" for beam or "C" for column. Lowercase letters denote components, such as "b" for bracket.

A plan view is provided for each level of the structure. For buildings, the plan view for the lowest level provides anchor bolt information. The next level of the plan view

indicates column and beam locations. Subsequent plan views indicate floor and roof decking placement. Elevations provide information about columns, beams, and exterior cladding. Various braces are also shown. Details provide connection and fastener information.

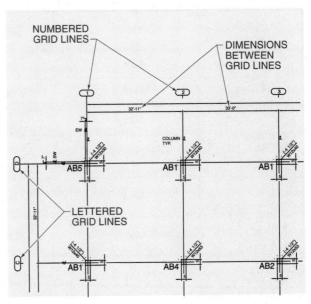

Figure 4-7. A dimensioned grid of letters and numbers provides reference points on erection plans.

STRUCTURAL STEEL MEMBERS

Various types and grades of steel are used in structural steel construction including carbon steel, high-strength steel, high-strength low-alloy steel, corrosion-resistant high-strength low-alloy steel, and quenched and tempered alloy steel. **See Figure 4-8.** The most common type of steel used for structural steel construction is classified by ASTM International as A36 and has minimum yield stress strength of 36,000 psi (pounds per square inch). The uses and applications of the various types of steel depend on the engineering requirements for a particular structure.

Structural steel members are manufactured to close tolerances to ensure conformity to the specified dimensions. In addition, the chemical properties of the steel are constantly monitored during the manufacturing process to ensure the finished product meets the strength requirements as designed.

Many steel shapes are required for the construction of a steel structure including wide-flange, S-shape, and HP-shape beams as well as C and Z channels, angles, tees, bearing piles, plates, flat bars, tie rods, and pipe columns. **See Figure 4-9.** A variety of symbols, letters, and numbers are used to indicate structural steel shapes on prints.

A *wide-flange beam* is a structural steel member with parallel flanges that are joined with a perpendicular web. The intersection of the web and flanges is filleted. Wide-flange beams, indicated with the letter "W", are specified by the nominal measurement outside of the flanges and the weight per running foot. For example, a W14 × 34 beam measures 14″ outside the flanges and weighs 34 lb/ft. Wide-flange beam dimensions are included in steel construction manuals, supplier catalogs, or supplier web sites.

An *S-shape beam* is a structural steel member with thickened parallel flanges that are joined with a perpendicular web. S-shape beams are also known as S beams or I beams. S-shape beams, indicated with the letter "S", are specified by the nominal measurement outside of the flanges and the weight per running foot. For example, an S20 × 96 beam measures 20″ outside the flanges and weighs 96 lb/ft.

An *HP-shape beam* is a structural steel member with wide parallel flanges that are joined by a perpendicular web. HP-shape beams, indicated with the letters "HP", are specified by nominal measurements outside of the flanges and weight per running foot. For example, an HP12 × 84 beam measures 12″ outside the flanges and weighs 84 lb/ft.

A *C channel* is a structural steel member in which the outside of the flanges and web are perpendicular to one another, forming a "C" shape. The inner flange surfaces are at approximately a 16.67° angle to the outside of the flanges. C channels, indicated with the letter "C", are specified by the actual dimension outside of the flanges and the weight per running foot. For example a C12 × 30 channel measures 12″ outside the flanges and weighs 30 lb/ft.

STRUCTURAL STEEL				
Steel Type	**ASTM Designation**	**Minimum Yield Stress***	**Form**	**Remarks**
Carbon steel	A36	36	Plates, shapes, bars, sheets and strips, rivets, bolts, and nuts	For buildings and general structures; available in high toughness grades
	A529	42	Plates, shapes, bars	For buildings and similar construction
High-strength	A440	42 to 50	Plates, shapes, bar	Lightweight with superior corrosion resistance
High-strength low-alloy	A441	40 to 50	Plates, shapes, bars	Primarily for lightweight welded buildings and bridges
	A572	42 to 65	Several types; some available as shapes, plates, or bars	Lightweight, high toughness for buildings, bridges, and similar structures
Corrosion-resistant high-strength low-alloy	A242	42 to 50	Plates, shapes, bars	Lightweight with added durability; weathering grades available
	A588	42 to 50	Plates, shapes, bars	Lightweight, durable in high thicknesses; weathering grades available
Quenched and tempered alloy	A514	90 to 100	Several types; some available as shapes, others as plates	Strength varies with thickness and type

* in KSI (1000 lb)

Figure 4-8. Various types and grades of steel are used in structural steel construction.

COMMON STEEL SHAPES

Description	Pictorial	Symbol	Designation
Wide-flange beam	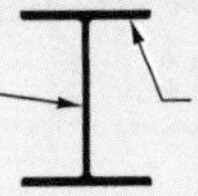 WEB → FLANGE	W	W14 × 34
S-shape beam		S	S15 × 42.9
HP-shape beam	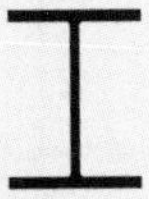	HP	HP10 × 57
C channel	 DEPTH	[	[9 × 13.4
Z channel		Z	Z6 × 3$\frac{1}{2}$ × 15.7
Angle	 LEGS	∠	∠ 6 × 6 × $\frac{1}{2}$
Tee	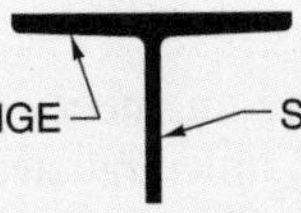 FLANGE — STEM	T	T4 × 9.2
Bearing pile	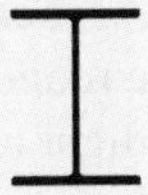	BP	BP14 × 73
Plate	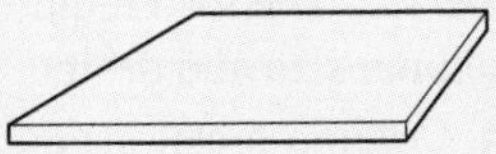	PL	PL 18 × $\frac{1}{2}$ × 2′-6″
Flat bar		BAR	BAR 2$\frac{1}{2}$ × $\frac{1}{4}$
Tie rod	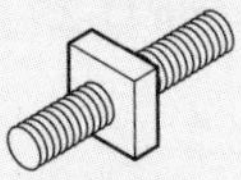	TR	TR $\frac{3}{4}$ Ø
Pipe column		O	O 8 SCH 60

Figure 4-9. A variety of steel shapes are commonly used in structural steel construction. Standard abbreviations and designations are included on erection plans to indicate structural steel members.

A *Z channel* is a structural steel member in which the flanges extend parallel from the web but in opposite directions. Z channels, indicated with the letter "Z", are specified by the channel depth, flange widths, and web thickness. For example, a Z3 × 2¹¹⁄₁₆ × 2¹¹⁄₁₆ × ¼ channel is 3″ deep, has 2¹¹⁄₁₆″ wide flanges, and is ¼″ thick.

A *steel angle* is a structural steel member with an L-shaped cross section with equal- or unequal-width legs. Steel angles, indicated with the ∠ symbol, are specified by the width of the legs and the thickness of the legs. For example, a ∠6 × 6 × ⅞ steel angle measures 6″ along the back of each leg and is ⅞″ thick.

A variety of steel designs is available to meet load requirements for columns, beams, girders, joists, braces, plates, or other building members. Factors in steel designs include the steel composition, shape, thickness, weight, and length. To prevent failure in case of fire, structural steel members may be coated with several different spray cement mixtures and may be encased in concrete, masonry, or gypsum.

Columns

A *column* is the principle vertical load-bearing member in a steel structure. Columns are supported by and secured to foundations or footings. Columns are typically the first members erected for beam and column construction. Shop drawings indicate the overall column height, spacing of holes for attaching beams and braces, column locations, steel angles to support beams, shear tabs for connections to beams and girders, and base plate information such as plate thickness and size. **See Figure 4-10.**

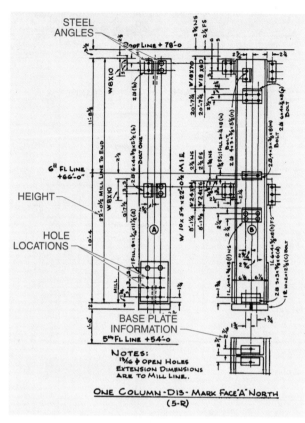

Figure 4-10. Shop drawings provide detailed information required for the fabrication of structural steel members.

Architects and engineers use standard tables to determine sizes of structural steel members to be used for columns. M-, S-, or wide-flange shapes are commonly used for steel columns. **See Figure 4-11.** Structural steel columns may also be round steel pipe or square steel tubing where relatively light loads will be imposed. The size and design of the columns may be provided on a schedule and related details. The load requirements for the column determine column size and design.

Installation. Column locations are indicated on erection plans using a grid of letters and numbers. References to the letters and numbers identify each column. For example, a column located at the intersection of grid lines D and 2 is referred to as column D-2.

Columns are the principal load-bearing members in a steel structure.

TECH TIP

Routes for suspended loads must be planned to ensure no tradesworker is required to work directly below the load except for workers making the initial connection of steel members or workers required to hook or unhook loads.

M-SHAPES

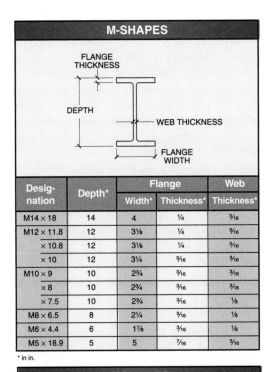

Desig-nation	Depth*	Flange Width*	Flange Thickness*	Web Thickness*
M14×18	14	4	¼	3/16
M12×11.8	12	3⅛	¼	3/16
×10.8	12	3⅛	¼	3/16
×10	12	3¼	3/16	3/16
M10×9	10	2¾	3/16	3/16
×8	10	2¾	3/16	3/16
×7.5	10	2¾	3/16	⅛
M8×6.5	8	2¼	3/16	⅛
M6×4.4	6	1⅞	3/16	⅛
M5×18.9	5	5	7/16	5/16

* in in.

S-SHAPES

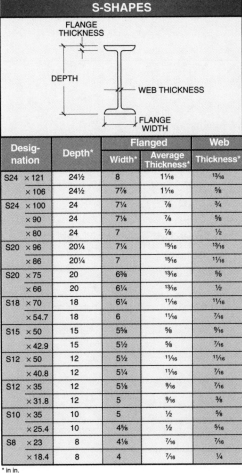

Desig-nation	Depth*	Flanged Width*	Flanged Average Thickness*	Web Thickness*
S24 ×121	24½	8	1 1/16	13/16
×106	24½	7⅞	1 1/16	⅝
S24 ×100	24	7¼	⅞	¾
×90	24	7⅛	⅞	⅝
×80	24	7	⅞	½
S20 ×96	20¼	7¼	15/16	13/16
×86	20¼	7	15/16	11/16
S20 ×75	20	6⅜	13/16	⅝
×66	20	6¼	13/16	½
S18 ×70	18	6¼	11/16	11/16
×54.7	18	6	11/16	7/16
S15 ×50	15	5⅝	⅝	9/16
×42.9	15	5½	⅝	7/16
S12 ×50	12	5½	11/16	11/16
×40.8	12	5¼	11/16	7/16
S12 ×35	12	5⅛	9/16	7/16
×31.8	12	5	9/16	⅜
S10 ×35	10	5	½	⅝
×25.4	10	4⅝	½	5/16
S8 ×23	8	4⅛	7/16	7/16
×18.4	8	4	7/16	¼

* in in.

WIDE-FLANGE SHAPES

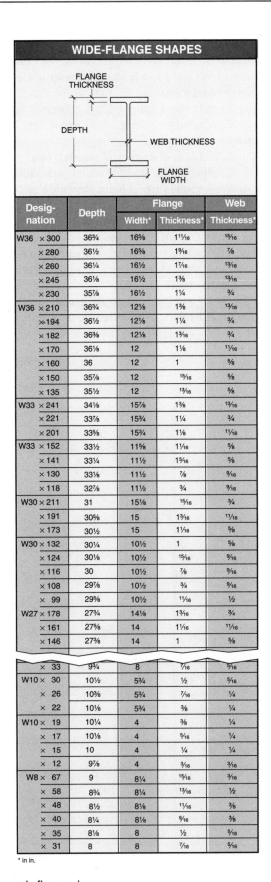

Desig-nation	Depth	Flange Width*	Flange Thickness*	Web Thickness*
W36 ×300	36¾	16⅝	1 11/16	15/16
×280	36½	16⅝	1 9/16	⅞
×260	36¼	16½	1 7/16	13/16
×245	36⅛	16½	1⅜	13/16
×230	35⅞	16½	1¼	¾
W36 ×210	36¾	12⅛	1⅜	13/16
×194	36½	12⅛	1¼	¾
×182	36⅜	12⅛	1 3/16	¾
×170	36⅛	12	1⅛	11/16
×160	36	12	1	⅝
×150	35⅞	12	15/16	⅝
×135	35½	12	13/16	⅝
W33 ×241	34⅛	15⅞	1⅜	13/16
×221	33⅞	15¾	1¼	¾
×201	33⅝	15¾	1⅛	11/16
W33 ×152	33½	11⅝	1 1/16	⅝
×141	33¼	11½	15/16	⅝
×130	33⅛	11½	⅞	9/16
×118	32⅞	11½	¾	9/16
W30 ×211	31	15⅛	15/16	¾
×191	30⅝	15	1 3/16	11/16
×173	30½	15	1 1/16	⅝
W30 ×132	30¼	10½	1	⅝
×124	30⅛	10½	15/16	9/16
×116	30	10½	⅞	9/16
×108	29⅞	10½	¾	9/16
×99	29⅝	10½	11/16	½
W27 ×178	27¾	14⅛	1 3/16	¾
×161	27⅝	14	1 1/16	11/16
×146	27⅜	14	1	⅝
×33	9¾	8	7/16	5/16
W10× 30	10½	5¾	½	5/16
×26	10⅜	5¾	7/16	¼
×22	10⅛	5¾	⅜	¼
W10× 19	10¼	4	⅜	¼
×17	10⅛	4	5/16	¼
×15	10	4	¼	¼
×12	9⅞	4	3/16	3/16
W8× 67	9	8¼	15/16	3/16
×58	8¾	8¼	13/16	½
×48	8½	8⅛	11/16	⅜
×40	8¼	8⅛	9/16	⅜
×35	8⅛	8	½	5/16
×31	8	8	7/16	5/16

* in in.

Figure 4-11. Columns are commonly constructed using M-, S-, or wide-flange shapes.

The design, size, and weight of each column may also be noted at the intersection of the grid lines. For example, a notation of W12 × 53 indicates a column made of a wide-flange beam measuring 12″ outside the flanges and weighing 53 lb/ft. **See Figure 4-12.** A column schedule may also be provided with design, web, and weight information about each column.

The nominal inside diameter and schedule (wall thickness) of the pipe is indicated for round pipe columns. For example, a round pipe column shown on prints as ○ 8 SCH 60 indicates an 8″ diameter pipe column with schedule 60 wall thickness, in this case ½″. For square tubing columns, the outside dimensions of the tubing and steel wall thickness are provided. For example, a print notation of □ 3 × 3 × ¼ indicates a 3″ square steel tube with ¼″ wall thickness.

Anchor bolts and base plates are prepared prior to setting columns in place. When concrete for footings and foundations is placed, anchor bolts for steel columns are installed in the concrete based on specific print dimensions for spacing and height. Templates may be installed over the anchor bolts and grouted in place to ensure exact locations and elevations for the base plates. Base plates are welded onto small columns at a fabricating shop. Large column base plates are set separately; columns are then erected and bolted or welded to the base plates.

When erecting columns, the columns must be oriented in the proper direction to ensure beams and girders connect correctly. Columns are commonly marked with a directional indicator or other notation to ensure proper placement. For example, an "N" marked on one side of each column at the fabricating shop indicates that the marked side should face north. For multistory buildings, the floor on which the column is to be set may also be noted on the column.

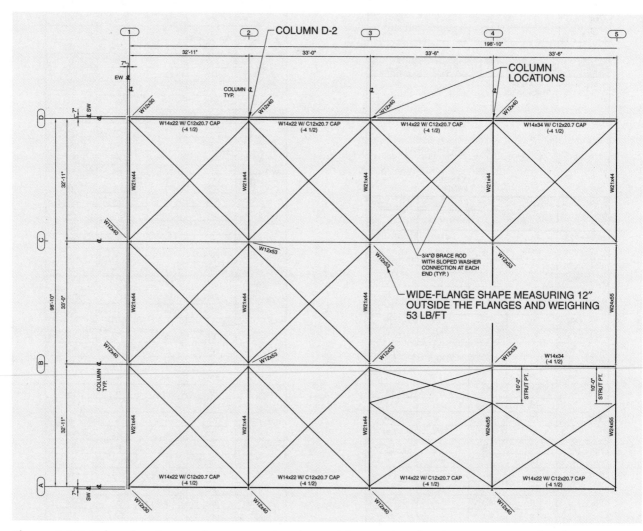

Figure 4-12. Column locations are shown on erection plans with letter and number designations. Specific information about web depth and weight (in lb/ft) is provided for each column.

Columns are rigged, lifted into place with a crane or other lifting device, and set onto anchor bolts. A hole may be provided in the web at the top of the column for attaching lifting hardware. Tall columns may require two or more structural steel members placed end to end, set in place individually, and secured together with splice plates. Splice plates are secured to the beams by using bolts and/or by welding.

Columns are temporarily braced in position with guy wires and turnbuckles until beams and girders are secured in position. Elevations on erection plans provide information concerning elevations at the tops of columns or connection points. The time and sequence for completing connections using bolts or by welding is determined by the size and design of the structure.

Beams and Girders

Beams and girders are horizontal structural steel members that support imposed loads and are commonly spaced more than 4′ on center (OC). Girders are typically the heaviest horizontal members in a structure and support the loads of beams and joists. Structural steel beams include wide-flange beams, American Standard beams, girders, and lightweight beams.

A "WF" or "W" is used to indicate wide-flange beams on prints along with the nominal depth and weight per linear foot. **See Figure 4-13.** The actual depth of a wide-flange beam is greater than the nominal depth. The actual size depends on the flange and web thicknesses as well as the beam manufacturer specifications. American Standard beams, commonly referred to as I beams, are designated on prints with the letters "S" or "I". The nominal and actual sizes of American Standard beams are equal. Various lightweight beams are shown on erection plans with the letter "B" or "JB" for junior beams.

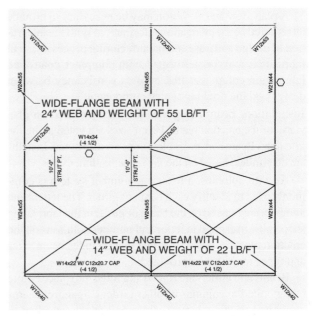

Figure 4-13. Wide-flange beams are identified on erection plans with the letters "W" or "WF".

Wide-flange and I beams are specified in a similar standard format that includes the beam type, nominal dimension outside the flanges, and weight (in lb/ft), and may also include the overall beam length. For example, the notation $W12 \times 29 \times 18'\text{-}3''$ represents a wide-flange beam with a nominal web depth of 12″, a weight of 29 lb/ft, and a length of 18′-3″.

A shop drawing for beam fabrication shows the beam type, weight, length, cutouts to allow for intersection with other structural steel members, dimensions for all beam holes, and any required connecting angles. **See Figure 4-14.**

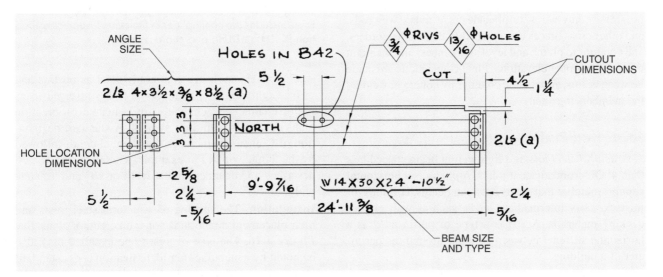

Figure 4-14. Structural steel beams are fabricated based on shop drawings.

A plan view and elevation may be necessary to provide all required beam fabrication information. With the integration of design software and manufacturing processes, steel fabrication may be performed with computer controlled fabrication equipment that ensures consistency between design and the final fabricated component.

Installation. Beam and girder sizes are provided on plan views of the erection plans. Beam sizes are noted along the grid lines. Beam and girder elevations are also noted. Lengths are determined based on the grid spacing dimensions.

As with columns, a letter and number system is commonly used to identify each beam or girder. The letters and numbers are marked on the beam or girder at the fabricating shop and correspond to letter and number notations on the erection plans. A beam schedule that includes beam sizes and types may also be provided.

Beams and girders are rigged and lifted into place before being bolted to columns or other girders. Spandrel beams are bolted to the columns using steel angles attached to the columns and beams. A *spandrel beam* is a beam in the perimeter of a building that spans from column to column. Beams and girders must be oriented in the proper direction to ensure each end of the beam is fastened to the proper column. Tradesworkers setting beams into place are commonly working at heights that require safety nets and/or fall-arrest equipment such as lifelines, harnesses, lanyards, or rope grabs to protect a falling worker. Tie-off requirements are determined by OSHA and must be observed during erection of structural steel.

Columns and beams are initially fastened together using bolts and nuts. Seat lugs or shear plates made of steel angles attached to the columns may be used to support the beam or girder ends during erection. For multimember beams, splice plates may be used to maintain alignment of adjacent beam ends and to secure them together. As construction progresses, guy wires, turnbuckles, sag rods, sway rods, and other cross braces are attached to the beams, girders, and columns to plumb and level all members and maintain them at the proper elevations. Final connections are made between beams, girders, and columns by bolting or welding the members together.

Joists and Purlins

A *structural steel joist* is a lightweight beam spaced less than 4′ OC from adjacent joists. A *purlin* is a horizontal support member that spans between beams, columns, or joists to carry intermediate loads, such as wall or roof decking materials. A *girt* is a type of purlin used as a horizontal stiffener between columns around the perimeter of a building.

Structural steel joists may be formed of a single structural member or built up from smaller steel members.

An *open web steel joist* is a structural steel member constructed with steel angles and bars that are used as chords with steel angles or bars extending between the chords at an angle. **See Figure 4-15.** The standard designation for open web steel joists includes the nominal depth (in in.), span classification (K, LH, or DLH), and chord diameter. For example, an open web steel joist with a designation of 26K7 has a nominal depth of 26″, is a standard K-series joist, and has a chord of #7 steel bar (⅞″ diameter).

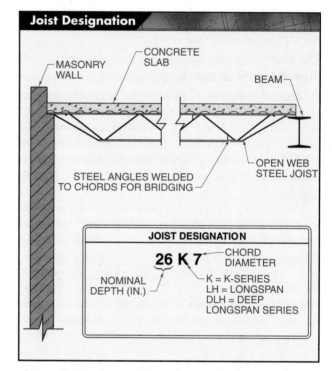

Figure 4-15. Open web steel joists span between beams and girders. The standard designation for open web steel joists includes the nominal depth (in inches), span classification (K, LH, or DLH), and chord diameter.

C or Z channels are commonly used as purlins and girts. Lightweight C channels are also referred to as junior channels. C channels are indicated on erection plans or details with the [symbol. A standard designation of C channels includes the channel depth followed by the flange width. For example, a notation of [10 × 25 describes a C channel with a depth of 10″ and a flange width of 2⅞″.

Installation. The spacing of structural steel joists and their placement direction is noted on erection plans. **See Figure 4-16.** The type of joist to be installed may also be noted by a manufacturer identification code, standard classification format, or fabrication shop code number. An elevation to the top of the joists may be indicated on the plan views or elevations. Openings for stairwells or other access between levels of a structure are shown on a plan view using dashed lines in an "X" pattern. Solid lines in an "X" pattern between joists indicate cross braces.

Structural steel joists are typically welded to the beams and girders with the manner of joist attachment specified on the details. Purlin and girt spacing and direction are indicated for proper installation. Channels or girts are most commonly secured to their supporting members using bolts with the spacing and bolt sizes indicated on details.

Trusses

Steel trusses may be used to span long distances while providing maximum strength with minimal weight. Engineers determine all loads and stresses that the truss must be able to withstand. Various sizes and shapes of steel members, including tees, angles, plates, and bars, are fastened together to provide structural, compressive, and tensile support. Each size and shape is indicated on the truss fabrication drawings. **See Figure 4-17.**

A variety of truss designs are available. The design specified for a building is based on the loads to be supported, truss span, required roof and ceiling pitches, and overall allowable height of the truss. Common steel truss designs include the bowstring, flat, Howe, Pratt, scissors, and Warren.

Open web steel joists may be secured to steel beams or other supporting members such as masonry walls.

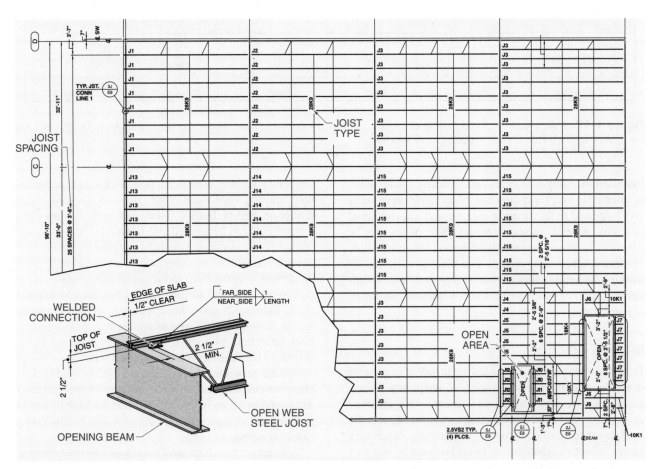

Figure 4-16. Erection plans indicate structural steel joist spacing and installation information. Open web steel joists are secured to beams or other supporting members using bolts or by welding.

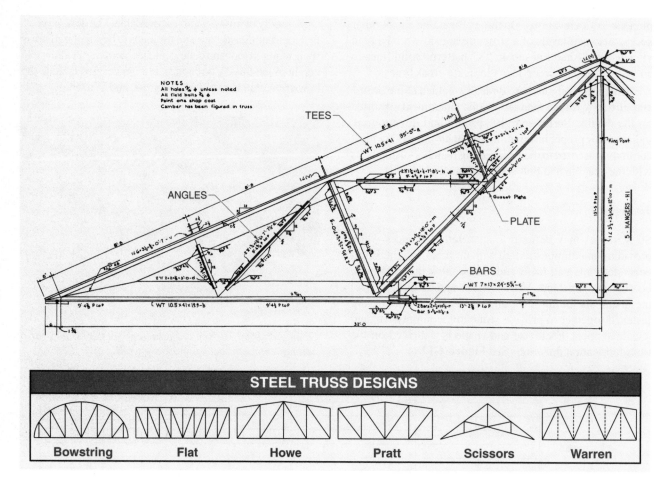

Figure 4-17. A wide variety of structural steel shapes are joined together to form a truss. Common steel truss designs include the bowstring, flat, Howe, Pratt, scissors, and Warren.

Installation. Similar to beams and girders, trusses must be oriented in the proper direction. Steel trusses are usually very large and require care in rigging, erecting, and final fastening. A series of seat lugs and splice plates may be necessary to ensure proper truss fastening and installation. Tradesworkers setting trusses into place are commonly working at heights that require safety nets and/or fall-arrest equipment such as lifelines, harnesses, lanyards, or rope grabs.

Bracing

Temporary and final braces must be installed to support structural steel members. A variety of temporary braces, including wire rope guy wires and rods with attached turnbuckles, are used during construction to support the steel members until construction is complete. Temporary braces are not indicated on the prints but must be installed to secure steel members in position during construction.

Falsework may be erected for large steel structures that include cantilevered areas or long spans. *Falsework* is temporary shoring used to support work under construction. Shoring towers and other falsework are assembled to support girders, beams, and trusses until all supporting components can be erected and fastened to support each other. Falsework construction often requires a separate set of engineering drawings, similar to erection plans, to show the design, construction, and required elevations.

Cross braces and angular braces are used to brace a steel structure. A cross brace consists of two diagonal braces, forming an "X" and extending between structural steel members. An angular brace is one diagonal brace extending between structural steel members. Cross braces extending between columns, beams, girders, and trusses prevent sway, sag, and possible collapse of the structural steel supporting members. A series of angular braces is fastened to the steel structure to resist forces that could cause connections to fail.

A variety of structural steel shapes may be used for final bracing of a steel structure. Steel angles are commonly used for bracing. Angles, indicated by the ∠ symbol, are specified by the width of the longer leg, width of the shorter leg, steel thickness, and length of the steel angle. For example, a notation of $\angle 5 \times 4 \times \frac{1}{2} \times 6'\text{-}5''$ denotes an angle that has one leg that is 5″ wide, one leg that is 4″ wide, and is ½″ thick and 6′-5″ long. Steel angles are commonly used as bridging between open web steel joists and between beams and purlins. **See Figure 4-18.**

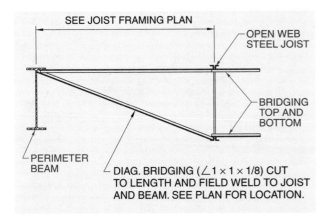

Figure 4-18. Bridging and braces maintain alignment between structural steel columns, beams, and joists and increase resistance to loads.

C channel may also be used for cross or angular braces. Cross braces are commonly attached to beams and columns with a gusset. A *gusset* is a piece of plate or sheet steel that is welded or bolted to all members at the connection point. *Plate steel* is flat steel that is more than ³⁄₁₆″ thick. Sheared

plate steel is trimmed on all edges during manufacturing. Universal plate steel is trimmed only on the ends during manufacturing. *Sheet steel* is flat steel that is ³⁄₁₆″ thick or less. Elevations and sections provide information regarding cross and angular braces.

Tie rods are also used to brace structural steel structures. A *tie rod* is a cylindrical steel member with threads on each end. The ends of tie rods are inserted through holes in opposing members. A beveled washer and nut are attached to each end of the tie rod to secure the structural steel members in position. **See Figure 4-19.** Tie rods are indicated on erection plans with the letters "TR". The rod diameter is specified in inches and fractions of an inch.

Other Steel Shapes

Other structural steel shapes used in construction include tees and bars, which are used for bracing and joist and truss construction. Structural tees are commonly fabricated by cutting standard S-shape beams, wide-flange beams, or smaller beams through the center of their web, forming two tees. Structural tees are indicated on erection plans or details with the letter "T". When structural tees are fabricated from wide-flange beams, they are indicated on erection plans or details with the letters "WT". When fabricated from S-shape beams, structural tees are indicated with the letters "ST". The standard designation for structural tees includes the flange width, nominal depth, and weight (in lb/ft) or the flange width, nominal depth, and steel thickness. For example, a notation of $T4 \times 3 \times \frac{3}{8}$ indicates a tee with a 4″ wide flange, 3″ nominal depth, and ³⁄₈″ thickness.

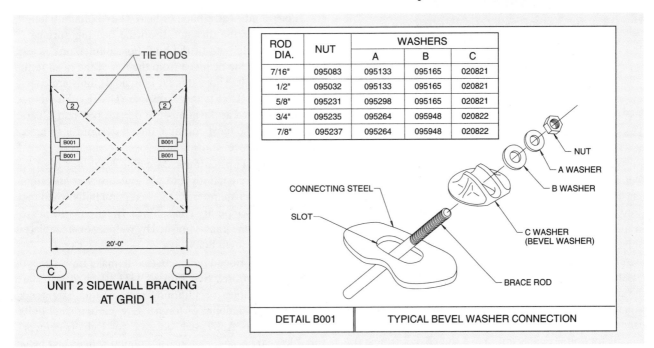

ROD DIA.	NUT	WASHERS		
		A	B	C
7/16″	095083	095133	095165	020821
1/2″	095032	095133	095165	020821
5/8″	095231	095298	095165	020821
3/4″	095235	095264	095948	020822
7/8″	095237	095264	095948	020822

Figure 4-19. Tie rods are used to brace across long distances.

Steel bars are indicated on erection plans with the letters "BAR", with a standard designation of bar width and thickness. For example, a notation of BAR 2½ × ¼ indicates a 2½″ wide bar that is ¼″ thick.

FASTENING SYSTEMS

Steel erection is the process of setting the structural framework in place. Construction is not complete until the structural steel members are properly braced and all connections are secured with bolts or by welding. Traditionally, rivets were used to secure together structural steel joints. However, due to developments in bolt manufacturing and welding technology, rivets are rarely used in modern structural steel construction.

The proper bolt, washer, and nut assemblies are required to secure connections in structural steel construction.

Bolts

The specified bolt, washer, and nut assembly must be utilized to properly secure connections in structural steel construction. Variations in bolt design and materials affect the amount of tensile stress the bolted connection can withstand. The process of tightening a nut on a bolt provides a friction connection. Some types of bolts are used in initial steel erection, some are designed for light loads only, and high-strength bolts are designed to withstand significant loads.

Structural steel drawings may contain tables concerning the amount of torque to be applied to bolts and the nut rotation for connections. Variable-torque impact wrenches are used to ensure all connections meet torque specifications. Details for bolted connections are a source of information on bolts, washers, and nuts. **See Figure 4-20.**

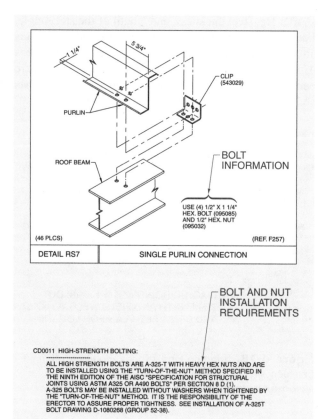

Figure 4-20. Proper bolt and nut installation is essential for proper structural fastener performance. Erection plan details are the primary reference for bolted connections.

For round-head ribbed or heavy hex bolts, the length of the bolt is the distance from the bearing surface of the head to the end of the bolt. For countersunk heads, the length is the distance from the top of the head to the end of the bolt. Three standard thread designs are coarse (UNC), fine (UNF), and extra fine (UNEF). The thread design is stated as the number of threads per inch. Coarse threads are the most common thread design for fasteners in structural steel construction.

Types and Applications. Machine bolts are used for temporary connections and for low-stress connections. Machine bolts are referred to as erection bolts when they are used for temporary connections. Erection bolts are removed when the joint is finished by welding or are replaced with high-strength bolts.

Structural ribbed bolts are used to make high-strength steel connections. **See Figure 4-21.** Ribbed bolts are driven into holes in adjoining members and tightened to the proper amount of torque. Low-carbon steel bolts, designated as A307, can be used for light framework and low-stress applications for column splices and beam and girder connections. High-strength hex head bolts are divided into two basic categories—A325 and A490—by ASTM International. A325 bolts are made by heat-treating medium-carbon steel bolts. A490 bolts are made from alloy steel. A325 bolts have a lower shear capacity than A490 bolts. High-strength hex head bolts contain an A325 or A490 marking on their heads. Additional markings include a manufacturer identification symbol.

Welding and Cutting

Many structural steel construction operations, including the installation of braces, decking, and final connection of structural members, require the use of welding and cutting equipment. Many connections between steel members are permanently joined using shielded metal arc welding (SMAW). Erection plans provide information concerning the type of weld joint, location of weld, welding process, size of weld bead, and surface finish of the weld. **See Figure 4-22.**

Welding Types and Applications. The most common welding process used in structural steel construction is shielded metal arc welding. *Shielded metal arc welding (SMAW)* is an arc welding process in which the arc is shielded by the decomposition of the coating on an electrode that has a similar metallic composition to the steel being joined. Electric current flows through a circuit created by the electrode and the steel members. An electric arc melts the base metal and the electrode, joining the steel members. The arcing process is shielded from the atmosphere and from impurities by a gas created through the decomposition of the electrode coating.

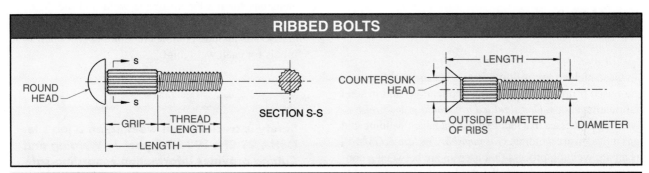

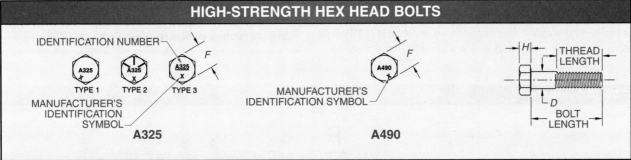

Nominal Bolt Size (D)	Width across Flats (F)	Height (H)	Thread Length
½	⅞	5/16	1
⅝	1 1/16	25/64	1¼
¾	1¼	15/32	1⅜
⅞	1 7/16	35/64	1½
1	1⅝	39/64	1¾
1⅛	1 13/16	11/16	2
1¼	2	25/32	2
1⅜	2 3/16	27/32	2¼
1½	2⅜	15/16	2¼

* in in.

Figure 4-21. Ribbed bolts may be used where steel members are drawn tightly together before fastening. High-strength hex head bolts are identified with markings on the bolt head.

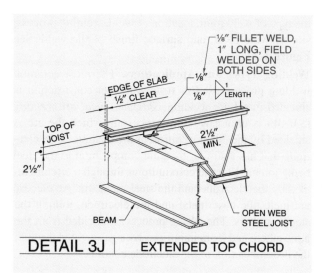

Figure 4-22. Details provide information regarding welded connections.

Gas shielded arc welding is another arc welding process used primarily in industrial applications and in steel fabrication shops. *Gas shielded arc welding* is a group of welding processes that includes gas metal arc welding and gas tungsten arc welding. *Gas metal arc welding (GMAW)* is an arc welding process that uses an arc between a continuous wire electrode and the weld pool. Carbon dioxide or carbon dioxide/argon mixtures are used as shielding gas when steel is welded. *Gas tungsten arc welding (GTAW)* is an arc welding process in which a shielding gas protects the arc between a nonconsumable tungsten electrode and the weld pool. Helium or argon is used as a shielding gas. GTAW produces a very high-quality weld.

The *American Welding Society (AWS)* is a trade association that is devoted to promoting welding and related processes. The AWS develops and distributes material, testing, and certification standards for welding processes. In many instances, tradesworkers performing welding operations on a job site must be certified either by the AWS or by a local governmental agency. Prints or specifications typically include information regarding certification requirements for welders.

Weld Symbols. Erection plans include standard weld symbols specifying the required welds. **See Figure 4-23.** The basic components of a weld symbol include the reference line, the arrow indicating the location of the weld, dimensions for the length of the weld and the depth of penetration, and symbols that describe the type of weld joint. Details included on the erection plans contain weld symbols for each weld joint.

TECH TIP

Safety is crucial when welding on a job site. OSHA 29 CFR 1926 Subpart J—*Welding and Cutting* provides information regarding safe welding and cutting practices on a job site, including having proper ventilation and using personal protective equipment (PPE).

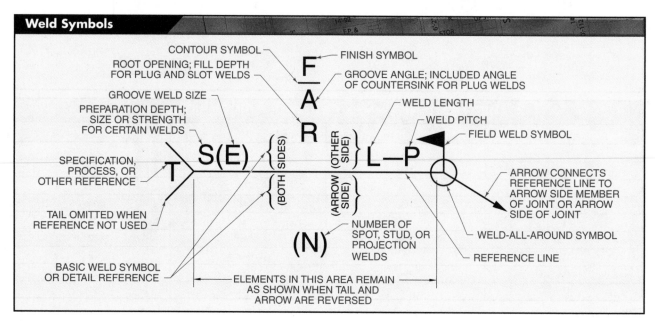

Figure 4-23. The standard format used for weld symbols provides information regarding location, length, and type of welded joint.

Cutting Materials and Equipment. While not specified on erection plans, cutting operations may be required to fabricate steel members on the job site. Structural steel members are cut at the job site using oxyacetylene or plasma arc cutting equipment. *Oxyacetylene cutting* is an oxygen cutting process in which heat is generated by an oxygen and acetylene flame to sever and remove the metal. **See Figure 4-24.** *Plasma arc cutting* is an arc cutting process that uses a constricted arc to heat the metal and removes the molten metal with a high-velocity jet of ionized gas. Plasma arc cutting produces a cleanly cut edge on many types of metals.

Figure 4-24. Steel members may be cut to length using an oxyacetylene cutting torch.

PANEL MEMBERS

After the structural steel framework is erected, the connections secured, and the structure braced, various panel members are attached to the frame to create floors, walls, and roofs. Erection plans provide information concerning the various panel members to be installed.

> **TECH TIP**
>
> OSHA 29 CFR 1926.106—*Working over or near Water* requires that tradesworkers working in conditions where the danger of drowning exists be provided with U.S. Coast Guard-approved life jackets or buoyant work vests.

Floor Decking

Metal floor decking provides a work platform during construction and a base for finish flooring materials. Metal floor decking is commonly covered with concrete to form a floor slab but may also be covered with noncementitious material. Corrugated metal decking is secured to the tops of the joists using self-tapping screws or by welding. Various types of metal floor decking are available. **See Figure 4-25.**

A plan view on the erection plans indicates the decking panel layout and opening locations. **See Figure 4-26.** Details indicate methods for framing openings, bracing around columns, and edge treatments of the decking.

Buildings. Cellular decking provides openings for additional insulation or for the installation of electrical cables or ductwork. For heavy commercial buildings, metal decking is commonly topped with several inches of concrete. The concrete thickness varies depending on the loads to be supported. The concrete thickness and depth of the completed deck is indicated on erection plans or details.

> **TECH TIP**
>
> A controlled decking zone (CDZ) may be established in an area of a structure over 15' and up to 30' above a lower level where metal decking is being installed and forms the leading edge of the work area. The CDZ boundaries must be designated and clearly marked. The CDZ must not be more than 90' wide and 90' deep from any leading edge. Access to a CDZ is limited to only those tradesworkers engaged in leading edge work.

Metal floor decking is commonly covered with concrete to form a floor slab.

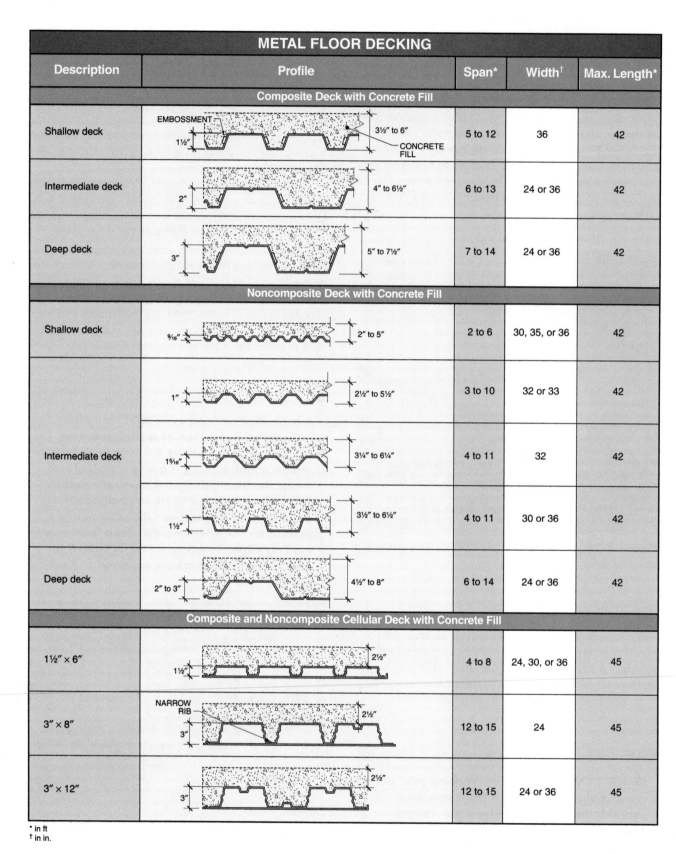

METAL FLOOR DECKING				
Description	**Profile**	**Span***	**Width†**	**Max. Length***
Composite Deck with Concrete Fill				
Shallow deck	EMBOSSMENT — 1½" — 3½" to 6" — CONCRETE FILL	5 to 12	36	42
Intermediate deck	2" — 4" to 6½"	6 to 13	24 or 36	42
Deep deck	3" — 5" to 7½"	7 to 14	24 or 36	42
Noncomposite Deck with Concrete Fill				
Shallow deck	9/16" — 2" to 5"	2 to 6	30, 35, or 36	42
	1" — 2½" to 5½"	3 to 10	32 or 33	42
Intermediate deck	1⅝6" — 3¼" to 6¼"	4 to 11	32	42
	1½" — 3½" to 6½"	4 to 11	30 or 36	42
Deep deck	2" to 3" — 4½" to 8"	6 to 14	24 or 36	42
Composite and Noncomposite Cellular Deck with Concrete Fill				
1½" × 6"	1½" — 2½"	4 to 8	24, 30, or 36	45
3" × 8"	NARROW RIB — 3" — 2½"	12 to 15	24	45
3" × 12"	3" — 2½"	12 to 15	24 or 36	45

* in ft
† in in.

Figure 4-25. Metal floor decking, manufactured in a variety of designs and dimensions, is attached to the top of open web steel joists to create a floor platform.

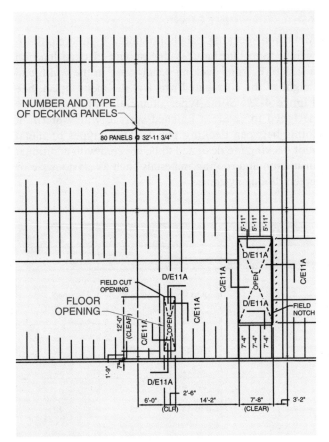

Figure 4-26. Erection plans provide information regarding metal decking installation.

Bridges. The most common bridge decking material is reinforced cast-in-place concrete. Formwork is suspended from the girders and beams and may be supported from below by shoring or falsework. Concrete is placed on top of the formwork. When the concrete has reached a specified strength, the shores and formwork are removed and structural steel or concrete structural members support the completed deck. For some applications, metal decking may be used as bridge deck forms. **See Figure 4-27.** The metal decking remains in place after the shores and falsework are removed.

Portland Cement Association

Figure 4-27. Metal decking may be used as bridge deck forms. The decking remains in place after the shores and falsework are removed.

Wall Panels

Structural steel buildings may be finished with a variety of exterior finish materials. Corrugated metal panels are a common finish material for small buildings. On large commercial buildings, the steel framework is covered with exterior wall finish materials such as glass, stone, precast concrete, or masonry.

Metal Panels. Erection plans and elevations provide information about fastening metal panels to the steel framework. Information may include the type of panel, direction of application, finish trim members at corners and other intersections, and manufacturer name, product numbers, and panel color. Letters representing the panel types are shown on the elevations and keyed to a schedule. **See Figure 4-28.** Wall panels are attached to purlins or girts with self-tapping and self-sealing screws. Metal panels may also be used for soffit and canopy coverings.

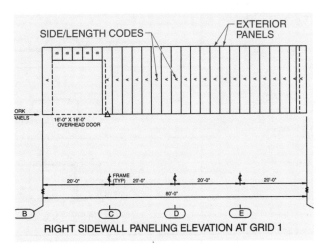

Figure 4-28. The exteriors of many light-gauge metal buildings are covered with prefinished metal panels.

Exterior Wall Panels. Many types of prefabricated exterior wall panel materials are available including glass, stone, precast concrete, and masonry. Prefabricated panels are attached to the framework using shielded metal arc welding (SMAW). Weld plates embedded in or fastened to stone and precast panels provide a connection point for welding to columns and spandrel beams. Exterior metal trim members that support glass or other panel materials are attached to structural steel members with weld clips. Information about attaching panels to steel framework is included in architectural details and elevations.

Roof Decking Panels

Metal roof decking panels provide a high strength-to-weight ratio that reduces the amount of dead load on a structure. Roof decking panels vary in the width and height of the corrugated ribs and the metal finish. **See Figure 4-29.** Some types of metal roof decking are designed to create a finished surface with watertight joints between decking panels. Other types of metal roof decking are designed to be covered with additional insulation and roofing materials such as elastomeric or green roofing systems.

METAL ROOF DECKING				
Description	Profile	Span*	Width†	Max. Length*
Economy		2 to 8	32 to 33	42
Narrow rib	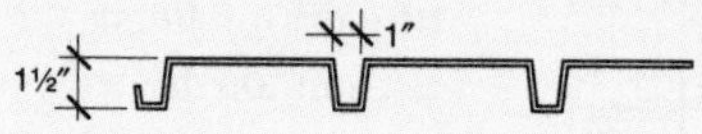	4 to 11	36	42
Intermediate rib	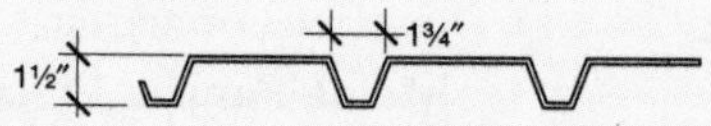	4 to 11	36	42
Wide rib	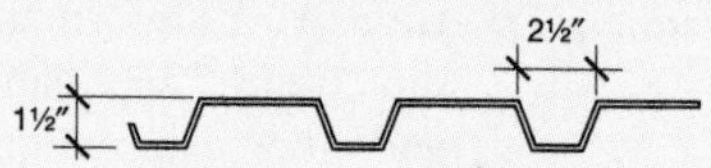	5 to 12	36	42
Acoustical deck	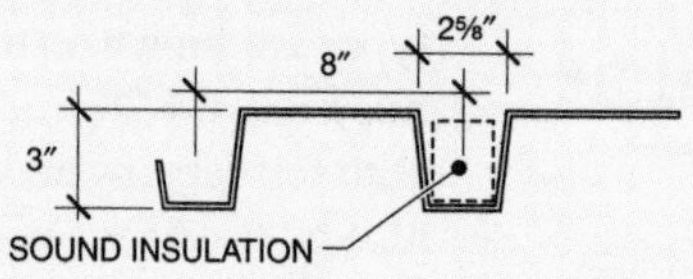	10 to 20	24	42
Cellular	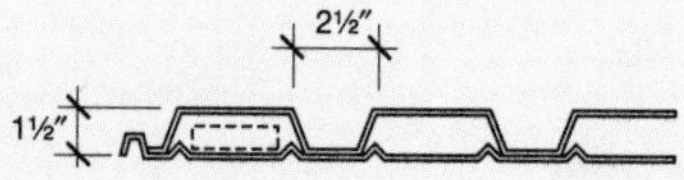	9 to 12	24	40
	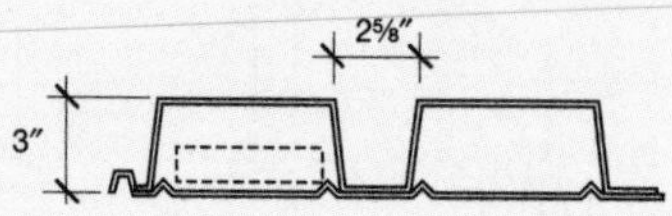	10 to 13	24	40
	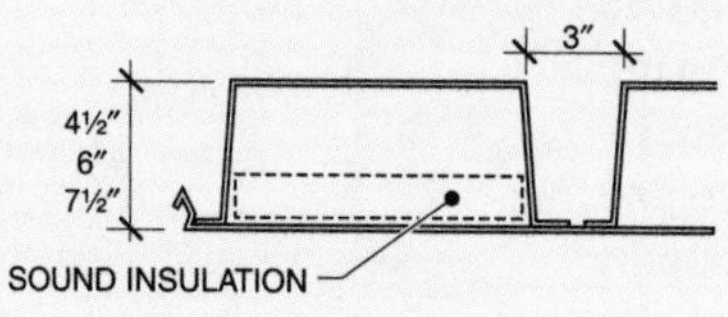	20 to 30	24	40

* in ft
† in in.

Figure 4-29. Metal roof decking is available in a variety of designs and dimensions.

A plan view of a steel roof panel layout is included in the erection plans. **See Figure 4-30.** Letters representing the panel types are shown on the elevations and keyed to a schedule. The schedule includes information about the types and sizes of the panels so that the panels can be placed properly and efficiently. Roof panels are attached to purlins or ceiling joists with self-sealing screws or weld clips. Elevations and details provide additional information about insulation or roofing materials for waterproofing.

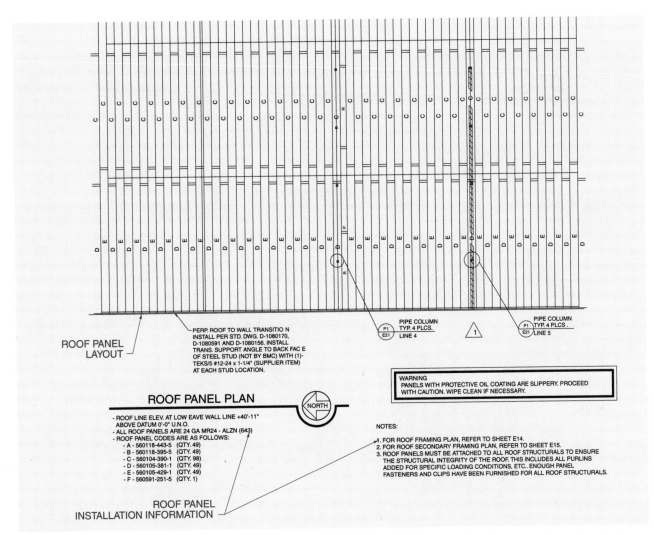

Figure 4-30. Roof decks may be formed of metal decking only or covered with waterproofing materials and insulation.

Refer to the CD-ROM in the back of the book for Chapter 4 Quick Quiz® and related printreading and reference material.

Name _Paul Sellers_   Date _____

Refer to the Johnson Construction project (Sheets 9 and 10).

T   (F)   **1.** Materials for temporary braces are supplied by the owner.

_____   **2.** All field welding must comply with American Welding Society specifications as set forth in ANSI/AWS ___.

(T)   F   **3.** Washers are required on all bolts with slotted connections.

T   F   **4.** The abbreviation "FLL" refers to foundation load line.

_____ 56 _____   **5.** A 1⅛" nominal bolt requires fastener tension of ___ kips for a slip critical connection.

_____   **6.** The maximum shim thickness recommended by the metal building manufacturer is ___".
          A. 0.01
          B. 0.025
          C. 0.05
          D. 0.075

_____ 40 _____   **7.** The column at grid point B7 is ___ lb/ft.

T   F   **8.** Spacing between column grid lines A through D is equal.

_____ 112 _____   **9.** The total number of column anchor bolts is ___.

T   F   **10.** Webs of the four columns along grid line 4 are set parallel with grid line 4.

_____ 5" _____   **11.** The center-to-center spacing for column anchor bolts is ___".

T   F   **12.** Columns C2 and C3 are spaced the same as columns C5 and C6.

_____   **13.** A column designation of AB2 relates to ___.
          A. web size
          B. weight per foot
          C. anchor bolt spacing
          D. grid line letter and number spacing

T   F   **14.** Anchor bolt spacing for columns B2 and C2 is the same.

_____   **15.** Anchor bolts must project ___" above the concrete.

_____   **16.** The overall size of the column base plate for column D7 is ___.

T   F   **17.** Grout thickness for columns with attached bracing members is less than for columns without attached bracing members.

_____ **18.** The minimum distance from the centerline of the column to the edge of the concrete pocket is ___ where gusset plates for column bracing are attached in line with column webs.

_____ **19.** The web depth for the beam between grid points A5 and A6 is ___".

_____ **20.** The weight of the beam between grid points D4 and D5 is ___ lb/ft.
         A. 16
         B. 26
         C. 31
         D. 36

_____ **21.** The distance from grid line 5 to the centerline of the intermediate beam located at the left edge of the open area and spanning the full distance between grid lines A and B is ___.

_____ **22.** The heaviest beam used on the project weighs ___ lb/ft.

_____ **23.** The lightest beam used on the project weighs ___ lb/ft.

_____ **24.** The tube attached to the top of the beam on detail S1 is ___ long.

_____ **25.** Open web steel joist spacing between grid lines B and C is ___.
         A. 1'-1"
         B. 2'-4"
         C. 3'-7"
         D. 3'-8"

T   F   **26.** The notation 28K9 on the open web steel joist layout plan refers to open web steel joists.

_____ **27.** The typical spacing for open web steel joists between grid lines C and D is ___.

T   F   **28.** Three Type J6 open web steel joists are indicated on the layout plan.

T   F   **29.** Open web steel joists are placed parallel with the grid lines 1 through 7.

_____ **30.** ___ rows of bridging are indicated for the open web steel joists between grid lines 3 and 4.

_____ **31.** As shown on Details 1J, 2J, 3J, and 5J, open web steel joists must bear on the steel beam at least ___".

T   F   **32.** The distance from the top of the beams to the top of the open web steel joists is 2".

T   F   **33.** Open web steel joists are attached to supporting beams with two fillet welds, each measuring 1" long.

_____ **34.** Typical diagonal bracing is provided by ___.
         A. 1" × 1" × ⅛" steel angles
         B. 1" × 1" × ⅛" reinforcing bars
         C. 28K9 steel joists
         D. W12 × 24 beams

T	F	**35.** Open web steel joist bundles are lifted by their strapping to avoid damage to the joists when delivered to the job site.
T	F	**36.** The bottom bridging between open web steel joists is continuous between beams and fastened to the bottom web of each perimeter and opening beam.
T	F	**37.** The long dimensions of composite deck panels are set parallel to grid lines 1 through 7.
_____		**38.** The clear space for the largest floor deck opening measures ___.
_____		**39.** The length of all composite deck panels between grid lines B and C is ___.
T	F	**40.** The composite decking should overhang the open web steel joists a minimum of 11″.
T	F	**41.** Openings in the composite deck panels are cut at the job site.

42. The distance from the edge of the concrete floor deck to the edge of the largest floor opening is ___.
   A. 1′-9″
   B. 7′-4″
   C. 7′-10″
   D. 7′-17″

T	F	**43.** The structural steel member installed at the perimeter of the composite deck to form an edge for the concrete is a 4″ × 4″ × ⅜″ thick steel edge angle.
_____		**44.** A(n) ___″ diameter weld is used to attach steel decking to supporting members.
T	F	**45.** The minimum bearing for decking panels on supporting members is 2½″.
T	F	**46.** In composite deck panels, reinforcement is provided by 15 ga sheet metal for single holes less than 12″ diameter.
T	F	**47.** Specifications for the concrete deck indicate embedded aluminum items are not allowed.
_____		**48.** The distance from the centerline of perimeter floor beams to the edge of the concrete floor slab is ___″.
T	F	**49.** The weight of the beam between columns A1 and A2 is 14 lb/ft.
_____		**50.** The diameter of the bracing rods for the ceiling beams is ___″.
T	F	**51.** The roof slopes downward toward grid line A.
_____		**52.** The abbreviation T.O.S. on Section B of Sheet E14 indicates ___.
T	F	**53.** The total difference in roof elevation from the high side to the low side is 2′-1″.
T	F	**54.** Roof beam bracing rods are fastened in place with fillet welds.
T	F	**55.** All roof Z purlins are placed facing toward the downslope side of the roof.

_____ **56.** A total of ___ roof perimeter flange braces are installed along grid lines 1 and 7.

_____ **57.** Flange braces and Z purlins are attached to roof beams with bolts through ___" diameter holes.

T   F **58.** All roof Z purlins are placed in parallel with the lettered grid lines.

_____ **59.** Steel roof members identified as FE3270, FE3272, FE3273, and FE3274 are ___.
    A. Z purlins
    B. rake channels
    C. perimeter beams
    D. flange braces

_____ **60.** All roof Z purlins between grid lines 4 and 5 are manufacturer product code ___.

T   F **61.** Roof purlins are spaced 4' OC between grid lines 1 and 3 at grid lines C and D.

_____ **62.** The minimum clear vertical distance from the third floor finished floor to the underside of the roof beams is ___ at grid line D.

_____ **63.** The total roof slope is ___" in 20'.

T   F **64.** Information concerning the roof screen between grid lines C and D is included on sheet E20.

T   F **65.** Roof purlins are spaced 3' OC between grid lines 3 and 5 at grid lines C and D.

T   F **66.** The closest Z purlin to the outside roofline is 11 1/16" along grid line A.

_____ **67.** Typical spacing of the first-floor girts is ___ OC.

T   F **68.** All horizontal girts along grid lines A and D weigh 22 lb/ft.

_____ **69.** The sag rods that hang from perimeter beams to support girts are ___" diameter.
    A. 1/4
    B. 3/8
    C. 1/2
    D. 5/8

_____ **70.** ___ are used for angular bracing between grid lines 4 and 5 at grid line D.
    A. Slotted tubes
    B. Angles
    C. Channel
    D. Girts

_____ **71.** The sag rods along grid line A between grid lines 1 and 2 are spaced ___ OC.

T   F **72.** The vertical distance from the centerline of the lowest level girt to the top of the second floor perimeter structural steel beams is 4'-6".

T   F **73.** Sag rods are attached to roof beams with a double nut at each end.

T   F   **74.** Where angular braces pass through horizontal girts, they are welded to T clips with a ³⁄₁₆″ fillet weld.

_____ **75.** The horizontal distance from grid line A to the face of the girts is ___″.

T   F   **76.** Sag rods are installed along grid line 1 between grid lines C and D.

_____ **77.** The height of the third floor perimeter floor beam along grid line 7 between grid lines A and B is ___″.
           A. 12
           B. 22
           C. 24
           D. 62

_____ **78.** The distance from grid line C to the centerline of the closest tube column of the roof screen is ___.

T   F   **79.** The platform steel of the roof screen area is level.

_____ **80.** The roof panel designated as "J" is located along grid line ___.

_____ **81.** A(n) ___-colored guttering is to be used around the building.

T   F   **82.** Roof panel installation begins at grid line 7 and proceeds toward grid line 1.

T   F   **83.** The roof panels are 26 ga steel.

T   F   **84.** Detail P1 of drawing E21 is a view looking down from above.

_____ **85.** Bolts attaching overlapping purlins to beams are spaced ___″ OC.

chapter 5

Reinforced Concrete Construction

Reinforced concrete is concrete containing tensile reinforcement in order to resist forces exerted on a structure. The structural properties of concrete and reinforcement are combined to achieve exceptional overall tensile and compressive strength. Reinforced concrete may be cast-in-place or precast. Cast-in-place concrete is concrete that is deposited in the place where it hardens as part of the structure. Precast concrete is concrete cast at a job site and lifted into place, or formed and cast at a casting yard, transported to the job site, and lifted into place.

REINFORCED CONCRETE CONSTRUCTION

The combination of concrete and reinforcing steel creates an integrated construction system that combines the best structural properties of each material. Concrete has a high compressive strength while reinforcing steel has a high tensile strength. When properly prepared and bonded together, the structural properties of each material are realized to achieve exceptional overall strength.

Reinforced concrete is used in large commercial structures to create foundations, floors, columns, beams, roadways, walls, roof decks, and other structural members. **See Figure 5-1.** Reinforced concrete members may be cast-in-place, precast on the job site, or precast off the job site. Reinforced concrete is commonly used with other construction materials, such as masonry and structural steel.

Specifications provide information regarding concrete ingredients, placement, curing, and finishing. Specifications also include information about reinforcing steel requirements and properties. Where noted on architectural drawings and specifications, various types of insulation materials may be cast into concrete walls, floor slabs, or roof panels to improve the insulation qualities of concrete construction members.

Architectural prints contain most of the print-related information for reinforced concrete construction. Information regarding reinforced concrete is shown on foundation plans, floor plans, structural plans, elevations, and various details.

CAST-IN-PLACE CONCRETE

A variety of methods are used to construct the formwork that allows for the placement of cast-in-place concrete. *Formwork* is the total system for supporting fresh concrete, including the sheathing that contacts the concrete, supporting members, hardware, and braces. The formwork used for a particular application depends on the overall size of the concrete structure.

Portland Cement Association

Figure 5-1. Reinforced concrete is a durable and versatile building material that is used in a variety of construction applications.

A *form* is a temporary structure or mold used to support concrete while it is setting and gaining adequate strength to be self-supporting. Forms may be set on or in the ground for footings and slabs-on-grade. Forms may be set on concrete slabs or footings to support walls, columns, and above-grade slabs. Each type or application of cast-in-place concrete requires specific print information regarding dimensions of the finished concrete, reinforcement requirements, inserts, and concrete properties.

Forming

Architectural prints provide the finished dimensions for concrete structures. **See Figure 5-2.** The dimensions include the depth, width, and height for cast-in-place concrete as well as information regarding items placed in or through the concrete.

Information regarding the type and design of forming system is usually not included on the structural prints. For large commercial cast-in-place concrete structures, a separate set of formwork drawings is provided by the concrete form supplier. **See Figure 5-3.** Formwork drawings are developed by manufacturer specialists in formwork design. Formwork design is based on concrete dimensions shown on the architectural prints. Formwork drawings indicate manufacturer form identification numbers and type, placement, formwork fastening systems, form tie systems, and shoring and bracing information. Concrete formwork design must take into consideration all forces to be placed on the form during concrete placement.

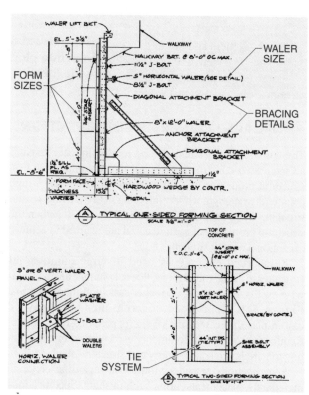

Symons Corporation

Figure 5-3. A separate set of formwork drawings may be provided by form suppliers and manufacturers to ensure safe and proper construction of concrete formwork.

TECH TIP

Per OSHA 29 CFR Subpart Q—*Concrete and Masonry Construction,* formwork must be designed, fabricated, erected, supported, braced, and maintained so that it is capable of supporting all reasonably anticipated vertical and lateral loads that may be applied to the formwork without failure.

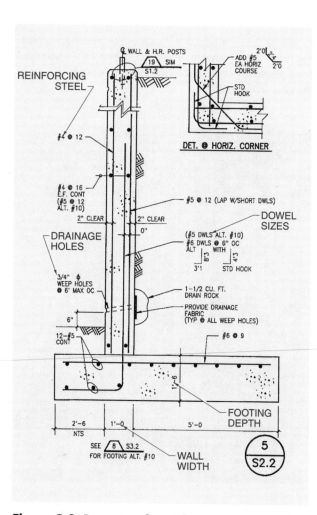

Figure 5-2. Dimensions for reinforced concrete members and reinforcement are indicated on architectural prints.

The amount of concrete placed in a form creates varying load requirements. Hydrostatic pressure within the form creates great pressures that increase as the amount of fresh concrete in the form increases. Formwork design variables include form width and height, concrete properties, and the rate at which concrete is placed in the form. Information about the minimum concrete strength required prior to formwork removal (stripping) may also be included.

Piles. Piles may be required for deep foundations where poor load-bearing capacity of the soil is encountered. A *pile* is a slender concrete, steel, or timber structural member driven or otherwise embedded on end into the ground to support a load or compact the soil. A *caisson pile,* or caisson, is a cast-in-place concrete pile made by driving a steel tube into the ground, excavating it, inserting reinforcement, and filling the cavity with concrete.

A grid of letters and numbers is provided at regular intervals on a plan view. The grid intersection points provide references for pile placement. Piles are shown on foundation plans with information that includes the depth of the lower tip of the pile, elevation at the top of the pile, and elevation at the top of the pile cap. **See Figure 5-4.**

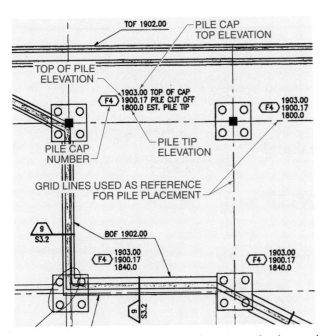

Figure 5-4. Plan views indicate locations, depths, and heights for reinforced concrete piles.

Cast-in-place concrete piles are formed by the surrounding soil, a casing that is removed as concrete is placed, or steel pipe. When piles are formed using the surrounding soil, holes are drilled to a specified depth, reinforcing steel is set in place, and the hole is filled with concrete. The bottom of the pile may be belled out to create a larger bearing surface if very poor load-bearing soil conditions exist. When

the sides of the drilled holes are not stable enough to remain in place until the concrete is placed, a metal casing (sleeve) is set into the hole as drilling proceeds. Reinforcing steel is then set in place and the casing is removed immediately after the concrete is deposited in the hole. A *pipe pile* is a steel cylinder with open ends that is driven into the ground with a pile-driving rig and then excavated and filled with concrete. **See Figure 5-5.** Pipe piles range in size from 10″ to 24″ in diameter. A conical steel shoe is often placed on the tip of a pipe pile to prevent it from filling with soil during driving. Details show the pile diameter, reinforcing steel, and means used to join the pile and pile cap.

Portland Cement Association

Figure 5-5. Pipe piles are driven into the ground and a pile cap is formed over the piles to distribute and transmit imposed loads to the piles.

TECH TIP

Concrete piles are common in heavy commercial construction since they provide excellent compressive strength. Concrete piles are available in 2½″ to 120″ diameters and can be driven up to 270′ deep.

Pile Caps and Footings. A *pile cap* is a structural member placed on top of, and usually fastened to, the top of a pile or group of piles and used to transmit imposed loads to the pile or group of piles. Dimensions for pile caps on structural prints include the width, length, depth, and reinforcing steel requirements. **See Figure 5-6.** Pile caps may be very large and require a large amount of concrete to be placed at one time. Proper staging of concrete delivery is required. Proper forming and bracing of the forms is essential to prevent formwork failure.

Continuous footings are placed on piles, pile caps, or the surface of the ground when adequate load-bearing capacity of the soil exists. Dimensions on structural foundation plan views provide the footing locations, widths, depths, and elevations. **See Figure 5-7.** The proper forming for each section of the footing is determined from the sections. Sections also provide additional dimensions and information about reinforcing steel, keyways, finishes, and projecting dowels or other inserts required to create a bond between the footing and other cast-in-place concrete members such as slabs or walls.

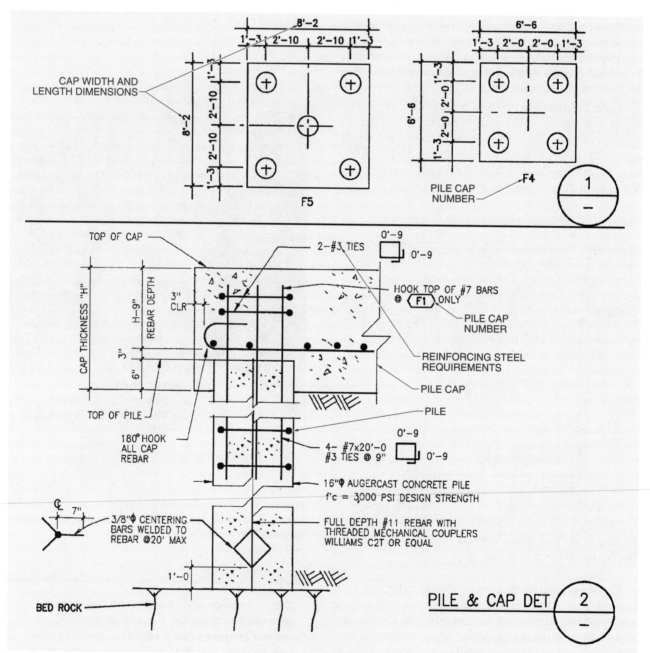

Figure 5-6. Sections through the footings provide specific information. Reinforcing steel in piles is tied to reinforcing steel in the pile cap.

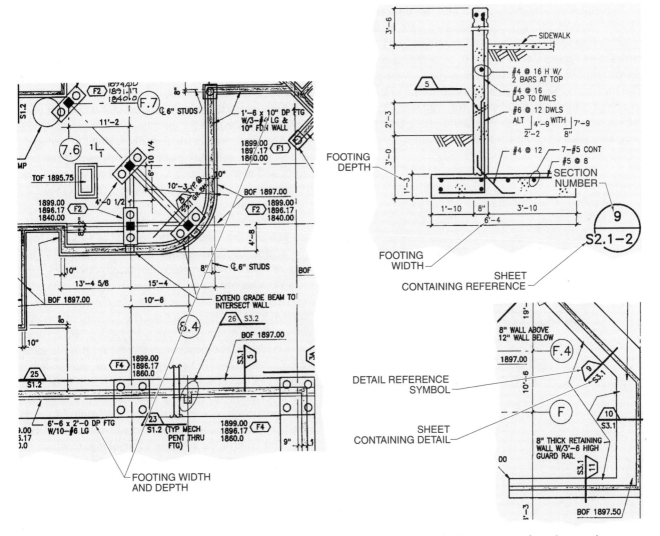

Figure 5-7. Footing dimensions are noted on a plan view as width followed by depth. Sections referred to in plan views must be carefully cross-referenced to ensure proper construction of reinforced concrete footing and foundation walls.

Slabs-on-Grade. Flat or sloping concrete slabs placed on the surface of the ground may be used as a foundation system in light load-bearing applications. Slabs-on-grade are also used for basement and ground-level floors and roadways. Information obtained from the plans regarding slabs-on-grade includes treatment of the area below the slab, items embedded in the slab, and elevations and finishes.

TECH TIP

Industrial floor construction design must take into account flatness requirements of a floor and placement of joints on the floor. Industrial floors typically require a high degree of flatness and require careful joint placement to ensure smooth forklift traffic flow.

Portland Cement Association

A self-propelled laser screed is a vibratory screed that is guided by a laser to obtain a high degree of flatness for slabs.

The edges of a slab-on-grade may be thickened to provide support for light- to medium-weight loads and protection against movement during freezing and thawing cycles. **See Figure 5-8.** A trench is excavated around the perimeter of the area where the floor slab is to be placed. The trench is typically shown on architectural prints. The trench provides adequate room for formwork and the thickened edge of the slab. Spread footings or piles are necessary where heavier loads are anticipated.

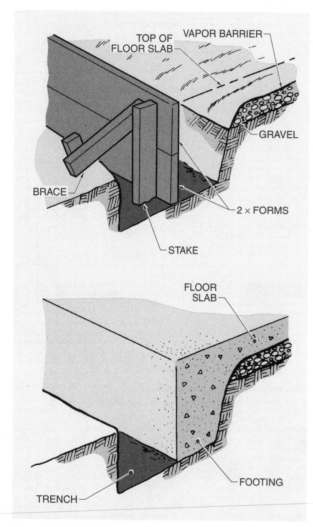

Figure 5-8. A wide trench is excavated to accommodate the formwork and thickened edge of a slab-on-grade.

Formwork may be constructed for perimeter footings and slabs where they project above the surface of the soil. Other sections of a slab may be thickened to support additional loads. Locations for the thickened concrete sections are commonly indicated with dashed lines on plan views and with additional sections.

Various materials may be placed below the slab prior to concrete placement for moisture protection and insulation. Sections and details describe subgrade materials such as gravel, moisture protection such as polyethylene sheets, and insulation such as expanded polystyrene. The compaction requirements and thickness of the subsurface gravel are indicated.

Architectural plan views and details indicate the thickness of a concrete slab and the method of isolation from or attachment to surrounding members. Forming is not required where slabs-on-grade are placed between existing walls or other structural members. However, expansion joints may be needed at these points. An *expansion joint,* or isolation joint, is a joint that separates sections of concrete to allow for movement caused by expansion and contraction of concrete. The locations and types of expansion joints are indicated on details. **See Figure 5-9.** The edge design of a keyway or dowel is shown if the slab must join to other members. Contraction joints in slabs are also shown on plan views and details. A *contraction joint,* or control joint, is a groove made in a concrete surface to create a weakened plane and control the location of cracking. Contraction joints are formed, tooled, or sawed in the concrete.

Portland Cement Association
A piece of preformed asphalt-impregnated fiber material, generally ¼" to ½" thick, is used in an expansion joint to provide space for concrete to expand.

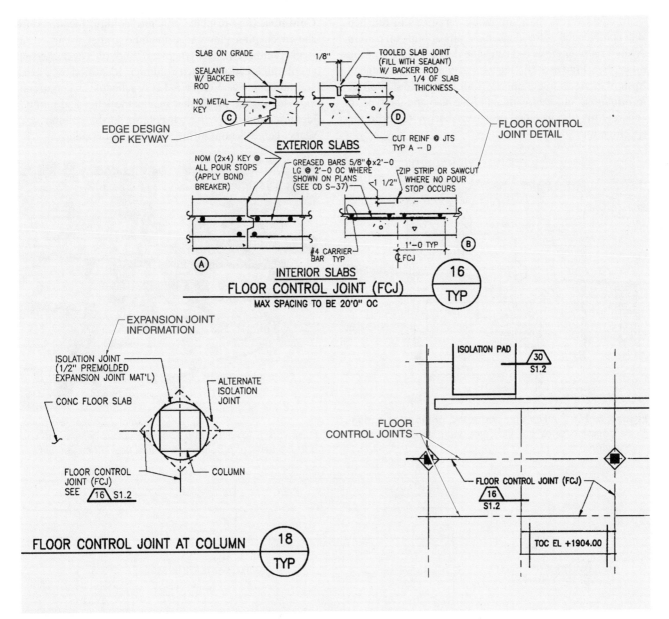

Figure 5-9. Expansion joints around structural members allow for movement of the concrete caused by expansion and contraction and help to prevent cracks in slabs-on-grade. Contraction joints create a weakened plane in the concrete and control the location of cracking.

Electrical conduit, plumbing pipe, and other mechanical systems may be installed in cast-in-place concrete floor slabs. The conduit and pipe is bent and stubbed out of the slab where electrical or plumbing service is required in walls and other locations. The ends of the conduit and pipe are capped or covered with protective tape before placement of the concrete for the slab to prevent concrete or other debris from falling into and blocking the conduit and pipe.

Conduit sizes and locations are shown on the electrical prints. Pipe sizes and locations and information regarding other mechanical systems are shown on the mechanical prints. It may also be necessary to stub reinforcing steel out of the slab to create attachment points for column and wall reinforcing steel. Anchor bolts may also be set in the slab for attaching structural steel columns or mechanical equipment. Locations for reinforcing steel and anchor bolts are found on the architectural floor plans or structural plans.

Elevations to the top of the concrete slab for slabs-on-grade and above-grade slabs are indicated on elevations or provided as a note on plan views. The amount of slope for a floor, roadway, or ramp is also indicated.

In cases where there are many variations in finished slab height, such as for mechanical pits or built-up pads, a symbol on plan views indicates the depth or height of the offset. **See Figure 5-10.** Sections may also be referenced from plan views to obtain side wall and other slab offset information. Surface treatment of the slab and the type of concrete may be included on cover sheet notes or in the specifications.

Columns. Cast-in-place concrete column locations are shown on plan views in a manner similar to structural steel columns and concrete piers. Lines on a grid intersect at column locations. Each column is identified by a letter and number. **See Figure 5-11.** A schedule is commonly included on the prints that relates the letter and number identification code of each column to the column size, shape, height, and reinforcing steel requirements.

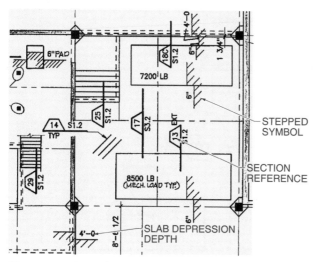

Figure 5-10. Built-up and depressed portions of cast-in-place concrete slabs are indicated with stepped symbols and sections.

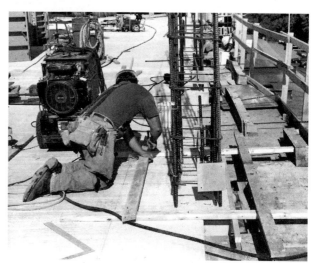

Portland Cement Association
Falsework is fastened to a concrete slab to secure and support the base of a column form and prevent movement.

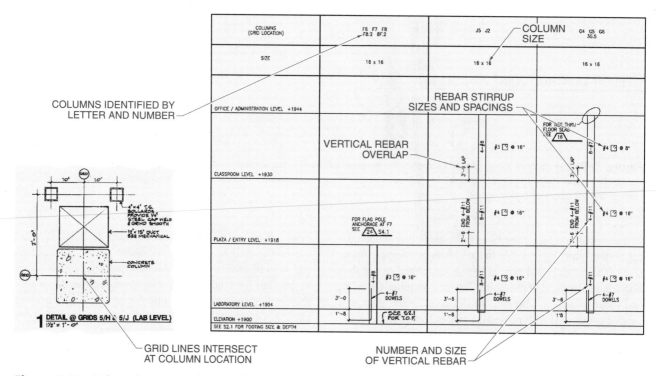

Figure 5-11. Column details are referenced from a plan view using grid lines that represent letters and numbers and intersect at column locations. Column schedules provide elevation, dimensions, and reinforcing steel information.

Formwork and concrete for constructing columns are commonly placed one level at a time. For example, the columns for the first level of a three-story building are formed first and the concrete is placed. The second floor deck is then formed, placed, and finished. The columns between the second and third floor are formed and the concrete is placed after the second floor deck is completed.

Structural steel columns may be encased in concrete to improve their strength and protect them from fire. Forms are set around the structural steel columns and concrete is placed.

Centerlines for columns are established after slabs, footings, and foundations are complete. Reinforcing steel for columns is set in place. This column reinforcing steel may be built in place or prefabricated at another location on the jobsite and lifted into place as a single unit. Formwork for the columns is then constructed around the steel, plumbed, and properly braced. Column forms include fiber forms, wood forms, patented panel forms, steel forms, and fiberglass spring forms. Concrete for the columns is placed with a concrete pump or a concrete bucket attached to a crane. **See Figure 5-12.** Formwork is removed after the concrete has obtained the required strength.

Portland Cement Association

Figure 5-12. Column forms are firmly braced and concrete for the columns is placed with a concrete bucket or concrete pump.

Column Caps. A cap is placed on top of columns to support loads not directly above the column. Column caps, also referred to as pier caps, are used primarily in bridge construction. A column cap may be placed directly over one column or span between columns and carry beams, girders, and joists. **See Figure 5-13.** Dimensions for overall width, thickness, and height for column caps are shown on plan views, sections, and elevations.

Column cap formwork is typically supported by the columns on which the formwork is placed. One forming method uses through-bolts, which are placed in holes formed in the columns. The bolts support brackets that act as seats for the formwork. Another forming method uses a two-piece friction collar that is secured around the columns. The collars have projecting support brackets to secure column cap forms in position.

Walls. Cast-in-place concrete walls are constructed over footings or slabs. Wall length and width dimensions are shown on foundation plan views, elevations, details, and sections. **See Figure 5-14.** Additional information regarding cast-in-place concrete walls includes elevations to the top and bottom of the wall, thickness of the wall, position of wall surface features, embedded items, and reinforcing steel.

For any form that is to be removed after concrete placement, a form-release agent is applied to the area of the form that will come in contact with the fresh concrete prior to its placement. One side of the wall form is set in place and braced. Wall forms are made of wood or metal faces reinforced with wood or steel frames. Wall forms are aligned and reinforced with walers, strongbacks, and braces. Reinforcing steel, blockouts for doors and windows, bulkheads, form facing materials, and waterstops are installed before the opposing side of the wall form is set into place. Wall ties extending between the opposing wall forms maintain the proper distance between the wall forms and distribute concrete loads during placement. Wall ties used in heavy commercial construction include snap ties, wire ties, she bolts, and coil ties. The type of wall ties used for a particular application depends on the wall width and design. For heavy commercial projects with complex forming requirements, form manufacturers provide detailed drawings of wall form, tie, and bracing to ensure proper formwork construction.

For some applications, such as abutment walls, slurry walls, or retaining walls, only one side of a wall is formed and the opposing side is supported by soil or rock. Reinforcing steel is set in place and the forms are set and braced. Additional braces are required for one-sided wall forms to support the hydrostatic pressure that is not offset by another side of a wall as with a conventional wall tie system.

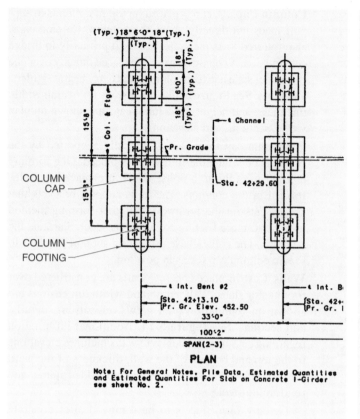

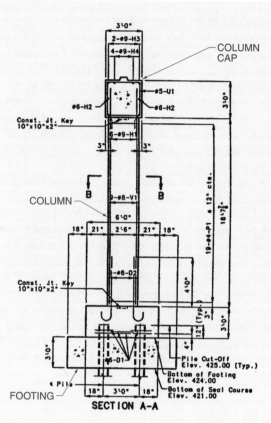

PLAN

Note: For General Notes, Pile Data, Estimated Quantities and Estimated Quantities For Slab on Concrete I-Girder see sheet No. 2.

SECTION A-A

Symons Corporation

Figure 5-13. Column caps, shown in plan views and sections, are formed by clamping or bolting formwork to supporting columns.

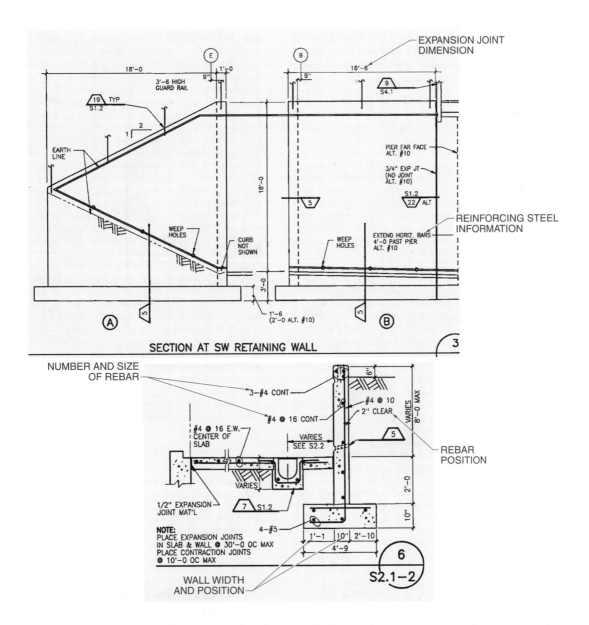

Figure 5-14. Sections through cast-in-place concrete foundation walls show surface treatments, reinforcement requirements, and dimensions.

Gang forms may be used for large wall sections. A *gang form,* or ganged panel form, is a series of smaller prefabricated form panels fastened together to create a larger form. Gang forms allow for some prefabrication of forming systems and can be set in place more quickly than many individual form components. **See Figure 5-15.** Gang forms are set in place with cranes and may be bolted to existing concrete walls to hold them in place. Concrete is placed after wall forms are set and properly braced. The forms are stripped when the required concrete strength is obtained.

Above-Grade Slabs. Various methods are used to construct above-grade concrete slabs, including a monolithic method with joists and the above-grade slab being placed at the same time; supporting the above-grade slab with precast concrete or structural steel beams; or supporting the slab with steel beams and corrugated decking. Openings in above-grade slabs for elevator shafts, stairwells, mechanical chases, and similar openings are shown on plan views with two solid diagonal lines forming an "X" inside a solid shape, such as a square or rectangle. The dimensions of the finished opening in the slab are indicated on the plan views.

The two monolithic slab systems are the one-way joist system and the two-way joist system. The type of system to be used is indicated on plan views and details. **See Figure 5-16.** Hidden lines indicate beam direction and location. Two-way joist systems are also referred to as waffle beam and slab systems.

APA —The Engineered Wood Association

Figure 5-15. A gang form consists of a series of smaller prefabricated form panels set in place and braced as a larger single form.

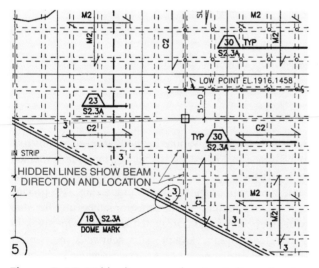

Figure 5-16. Hidden lines on a plan view indicate direction and location for beams in a two-way cast-in-place concrete monolithic slab and beam floor.

Portland Cement Association

Compressed air can be used to separate dome forms from the hardened concrete.

For both monolithic slab systems, decking is used to form a flat surface at the elevation of the bottom of the cast-in-place concrete beams. Steel, wood, or panelized shores support the beams and joists over which the decking is installed. The spacing and width of the beams are laid out on the deck. Pans or domes are used to form voids in the concrete slab to create beams in either one or two directions. A *pan* is a prefabricated form unit used in one-way concrete joist systems. A *dome* is a square prefabricated pan form used in two-way concrete joist systems. **See Figure 5-17.** Spandrel beams are formed around the outside edge of the deck to provide structural support. Perimeter edge forms are constructed to retain the concrete and maintain the proper elevation at the top of the slab.

Reinforcing steel, electrical conduit and raceways, and mechanical piping and other utilities that are designed to be integrated into the slab are installed prior to concrete placement. Forming systems are designed by specialists to ensure the proper number of shores is in place and the proper amount of camber (slight curvature) has been designed into the form. The shores and forms must support the weight of the forms, the fresh concrete, all reinforcement, electrical and mechanical utilities, and the tradesworkers placing and finishing the concrete. **See Figure 5-18.** The shores and formwork are removed after the concrete has been placed and finished and has reached the required strength.

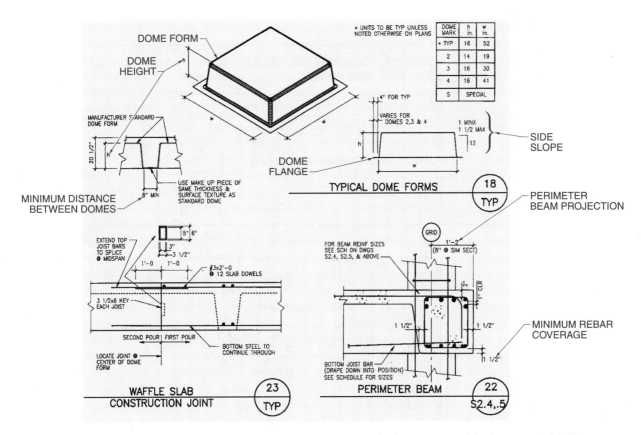

Figure 5-17. Steel domes are available in standard heights and widths for forming monolithic beam and slab floor systems. Perimeter beams support the edges of an above-grade slab.

Figure 5-18. Reinforcing steel is set in place after formwork and domes are installed.

Information for cast-in-place concrete beams, including spandrel beams and interior supporting beams between columns, may be included in a beam schedule on the prints. Cast-in-place concrete beams are identified on a plan view with letters and numbers that correspond to the beam schedule. **See Figure 5-19.**

Precast concrete beams or steel beams may be placed independently of the above-grade concrete slab. The slab thickness and reinforcing steel position is indicated on the details. Formwork is suspended from beams after columns, walls, and beams are in place. Hangers are placed on the beams and allow for temporary joists and bracing brackets to be supported and set to the proper slab height. Lubricated threaded rods are used to support brackets for joists. Lubrication is used to prevent concrete from adhering to the rods. Form decking is then installed over the joists. The slab is tied to the beam by reinforcing steel projecting from a precast concrete beam or welded studs on top of a structural steel beam. An edge form is built around the perimeter of the deck. Electrical conduit, mechanical blockouts, and reinforcing steel are set in place and the concrete is placed and finished. The temporary joists, bracing brackets, and deck forming are removed from below after the concrete has obtained its design strength. The threaded rods are removed and the remainder of the hanging hardware is placed into the concrete deck and remains in place.

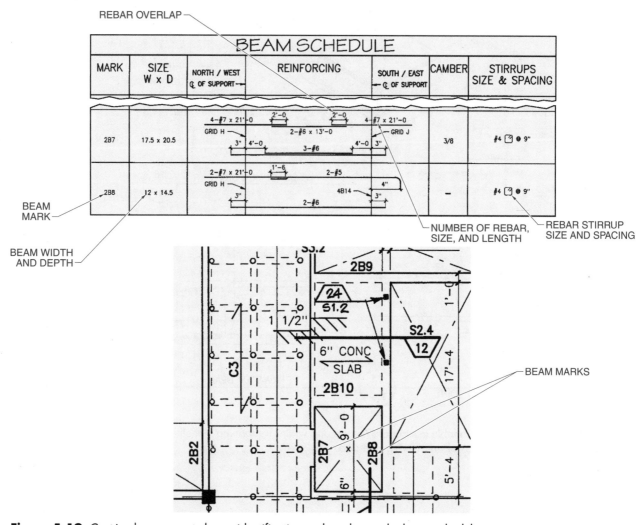

Figure 5-19. Cast-in-place concrete beam identification codes relate to the beam schedule.

Concrete for above-grade slabs is also placed on corrugated decking supported by steel beams or open web steel joists. **See Figure 5-20.** A structural steel C channel is welded around the perimeter of the corrugated decking to retain the concrete. The top of the C channel is set to the top of the slab. Concrete is placed directly on the corrugated decking and finished to the proper elevation and slope.

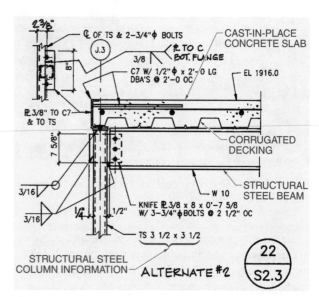

Figure 5-20. Cast-in-place above-grade concrete slabs can be supported by structural steel and corrugated steel decking.

TECH TIP

OSHA 29 CFR Subpart Q—*Concrete and Masonry Construction* provides safety information regarding common equipment used to place and finish concrete. For example, powered and rotating-type concrete trowels must be equipped with a control switch that will automatically shut off the power when operators remove their hands from the handles.

Reinforcing Steel

Cast-in-place concrete is reinforced with a variety of steel materials including steel reinforcing bars (rebar) and welded wire reinforcement. Details, notes, and reinforcing schedules provide specific information regarding the type and position of the reinforcing steel in the concrete. **See Figure 5-21.** In addition, pretensioned and post-tensioned concrete use specialized reinforcement that combines the compressive strength of concrete with the tensile strength of steel.

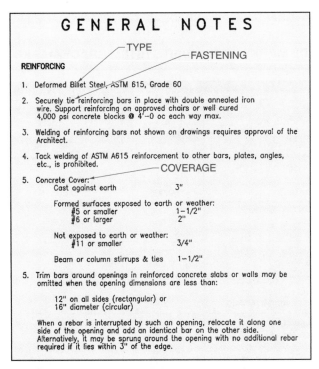

Figure 5-21. General notes on structural drawings provide information about reinforcing steel, fastening, and coverage.

Materials. Rebar is the most common material for reinforcing piles, roadways, footings, columns, walls, and beams. *Rebar,* or steel reinforcing bar, is steel bar containing lugs (protrusions) that form an irregular surface to increase the ability of the bars to interlock and adhere to the surrounding concrete. Rebar is tied together with wire ties in various configurations to reinforce concrete structures.

Rebar size is indicated in increments of ⅛″ diameter. For example, a #4 rebar is ⁴⁄₈″, or ½″, diameter. Several grades of steel are used for rebar that vary according to their tensile strength. Grade 40 rebar has a minimum tensile yield strength of 40,000 psi. Grades 50 and 60 have higher yield strengths. **See Figure 5-22.** Fewer reinforcing bars with higher yield strength are required to provide the same level of tensile strength compared to lower yield bars in a given area. An epoxy coating may be applied to rebar subjected to the weather. Epoxy-coated rebar are commonly used in road and bridge construction to minimize the effects of corrosion on the steel. Rebar is shown on sections with solid circles for end views and solid lines for profile views.

Welded wire reinforcement is also used to reinforce concrete slabs. *Welded wire reinforcement* is heavy-gauge wire joined in a grid and used to reinforce and increase the tensile strength of concrete. Sheets or rolls of the reinforcement are stretched out and set in place to reinforce lightweight slabs such as sidewalks.

Portland Cement Association

Straight and bent rebar for a road slab and edge barrier are positioned and tied together before concrete placement.

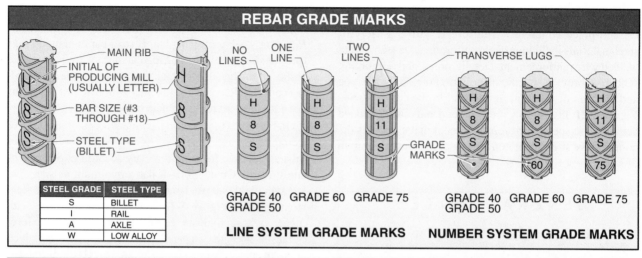

REBAR GRADE MARKS

STEEL GRADE	STEEL TYPE
S	BILLET
I	RAIL
A	AXLE
W	LOW ALLOY

LINE SYSTEM GRADE MARKS — GRADE 40 / GRADE 50, GRADE 60, GRADE 75

NUMBER SYSTEM GRADE MARKS — GRADE 40 / GRADE 50, GRADE 60, GRADE 75

STANDARD REBAR SIZES

Bar Size Designation	Weight per Foot*	Diameter†	Cross-Sectional Area2†
#3	0.376	0.375	0.11
#4	0.668	0.500	0.20
#5	1.043	0.625	0.31
#6	1.502	0.750	0.44
#7	2.044	0.875	0.60
#8	2.670	1.000	0.79
#9	3.400	1.128	1.00
#10	4.303	1.270	1.27
#11	5.313	1.410	1.56
#14	7.650	1.693	2.25
#18	13.600	2.257	4.00

* in lb
† in in.

REINFORCING STEEL STRENGTH AND GRADE

Deformed Billet	Minimum Yield Strength*	Ultimate Strength*
ASTM A615 Grade 40	40,000	70,000
Grade 50	50,000	90,000
Rail Steel ASTM A996 Grade 40	50,000	80,000
Grade 50	60,000	90,000
Axle Steel ASTM A996 Grade 40	40,000	70,000
Grade 50	50,000	90,000
Deformed Welded Wire Reinforcement ASTM A496	70,000	80,000
Plain Welded Wire Reinforcement ASTM A82 Less than W1.2	56,000	70,000
Equal to or greater than W1.2	65,000	75,000

* in psi

Figure 5-22. Rebar is identified using a standard system of lines or numbers. Various types and grades of steel provide different yield strengths.

Portland Cement Association
Reinforcing steel is formed at a fabricating shop when many shapes of similar sizes are required for a construction project.

Welded wire reinforcement size is designated by the wire spacing and either the cross-sectional area of the wire or the wire gauge. **See Figure 5-23.** Wires may be either smooth or deformed. Welded wire reinforcement has yield strengths between 50,000 psi and 70,000 psi. Welded wire reinforcement is shown on details as a dashed line with an "X" at each break in the line.

TECH TIP

Unrolled welded wire reinforcement must be prevented from recoiling by securing each end or turning over the roll.

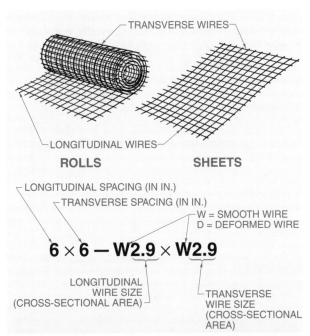

COMMON STOCK SIZES OF WELDED WIRE REINFORCEMENT			
New Designation (W-Number)	Old Designation (Wire Gauge)	Diameter*	Weight†
6 × 6 — W1.4 × W1.4	6 × 6 — 10 × 10	1/8	21
6 × 6 — W2.0 × W2.0	6 × 6 — 8 × 8	5/32	29
6 × 6 — W2.9 × W2.9	6 × 6 — 6 × 6	3/16	42
6 × 6 — W4.0 × W4.0	6 × 6 — 4 × 4	1/4	58
4 × 4 — W1.4 × W1.4	4 × 4 — 10 × 10	1/8	31
4 × 4 — W2.0 × W2.0	4 × 4 — 8 × 8	5/32	43
4 × 4 — W2.9 × W2.9	4 × 4 — 6 × 6	3/16	62
4 × 4 — W4.0 × W4.0	4 × 4 — 4 × 4	1/4	85

* in in.
† in lb per 100 sq ft

Figure 5-23. Welded wire reinforcement size is designated by wire spacing and cross-sectional area or wire spacing and wire gauge.

Installation. Print details for cast-in-place concrete structures provide specific information regarding installation of reinforcing steel. **See Figure 5-24.** Details include the size and length of the rebar, the position of the steel within the finished concrete, and any overlap requirements. Sleeves and other fastening devices are available where reinforcing steel is joined end to end. Per OSHA 29 CFR Subpart Q–*Concrete and Masonry Construction,* all protruding reinforcing steel must be guarded to eliminate the hazard of impalement.

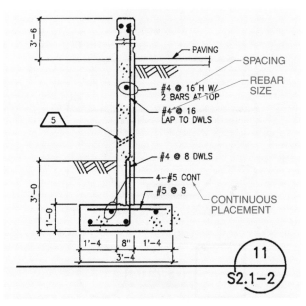

Figure 5-24. Sections through cast-in-place concrete foundation walls and footings indicate rebar size, type, and placement.

Each portion of the architectural and structural prints and details for concrete structures provides reinforcing steel information. For example, cages of vertical rebar with intermediate stirrup ties are formed for columns. The size of the upright and tying rebar for stirrups and the spacing of the rebar in both directions are indicated on the column schedule. **See Figure 5-25.** Various sizes and shapes of bars are required for beams. Reinforcing steel sizes, shapes, and lengths are also provided in the beam schedule. The various shapes are created by heating and bending the bars at a fabricating shop or on the job site. Schedules for the placement of reinforcing steel in cast-in-place concrete joists and slabs are also included on the structural prints.

Portland Cement Association

Tie wires are used to secure vertical rebar in position before concrete placement.

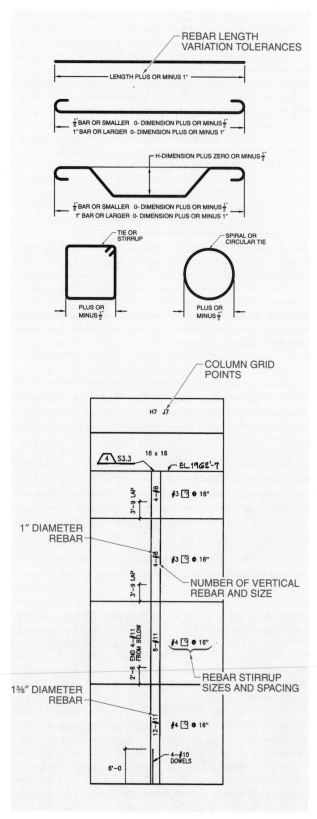

Figure 5-25. Rebar is bent into different shapes for concrete beam reinforcement. A column schedule indicates sizes and spacing of rebar and stirrups.

Cages of reinforcing steel for columns and beams as well as mats of reinforcing steel for walls and slabs are formed by tying rebar together with wire in the configurations as indicated on the prints. When possible, large cages or mats are prefabricated and tied together at a location away from the concrete forms for greater accessibility by tradesworkers. The large cages or mats are then lifted into the forms with cranes and set at their final location before concrete placement occurs. **See Figure 5-26.** Reinforcing steel must be positioned within the forms in a way that ensures that it will be properly embedded in the concrete. The cages or mats are secured and braced to prevent movement during concrete placement. Some rebar making up the cages or mats are commonly designed to project out of the edges of the forms to create an attachment point for adjoining cast-in-place concrete members. The amount of reinforcing steel projecting out of the concrete is noted on the details and schedules.

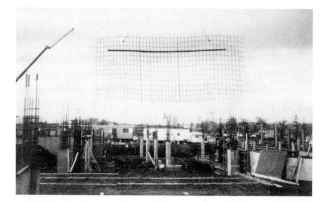

Figure 5-26. A series of individual steel reinforcing bars may be tied together into a mat and set in place as a single unit using a crane.

For flat slabs and roadways, the rebar mat or welded wire reinforcement must be positioned a specific distance above the bottom of the slab using chairs or bolsters. A *chair,* or bar support, is an individual supporting device used to support reinforcing steel in the proper position before and during concrete placement. **See Figure 5-27.** A *bolster* is a continuous supporting device used to support reinforcing steel in the bottoms of slabs. The top wire of the bolster is corrugated at 1-inch centers to maintain the position of the reinforcement. Chairs and bolsters are set in position, and reinforcing steel is set on top of the chairs or bolsters. The reinforcing steel may simply rest on the chairs and bolsters or be tied in place. Chairs and bolsters are also used in other horizontal concrete members such as beams.

CHAIRS

Symbol	Bar Support Illustration	Support	Standard Sizes
SB	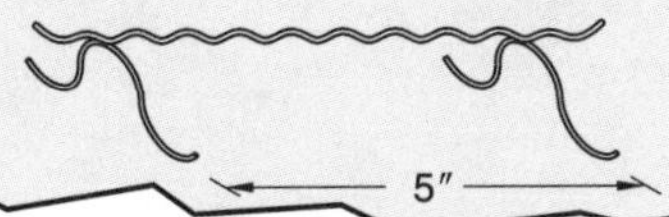 5″	Slab bolster	$\frac{3}{4}″$, 1″, $1\frac{1}{2}″$, and 2″ heights in 5′ and 10′ lengths
CHCM*	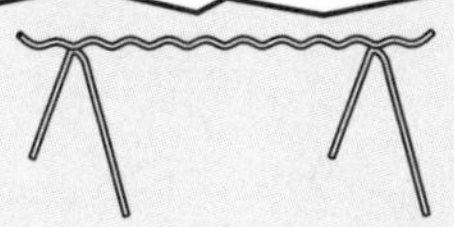	Continuous high chair for metal deck	Up to 5″ heights in $\frac{1}{4}″$ increments
JCU†	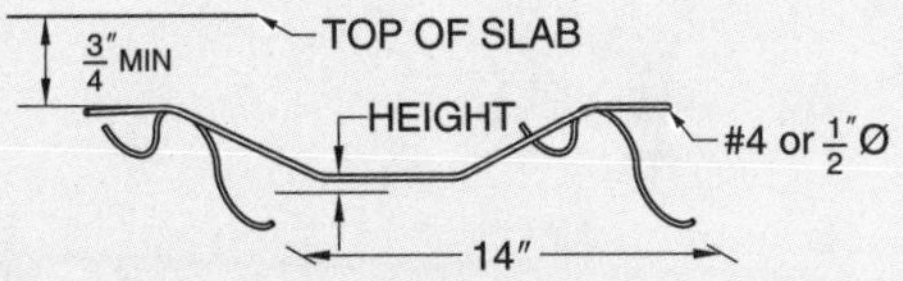 $\frac{3}{4}″$ MIN — TOP OF SLAB — HEIGHT — #4 or $\frac{1}{2}″$ Ø — 14″	Joist chair, upper	14″ span. 1″ through 3″ heights $\frac{1}{4}″$ in increments

* Available in Class 3 only, except on special order.
† Available in Class 3 only, with upturned or end bearing legs.

Figure 5-27. Bolsters and chairs support reinforcing steel in the proper position before and during concrete placement.

Prestressed Concrete. Additional concrete strength is gained by utilizing steel tendons (cables) in tension in a reinforced concrete assembly. *Prestressed concrete* is concrete in which internal stresses are introduced to such a degree that tensile stresses resulting from service loads are counteracted to the desired degree. Two methods of creating prestressed concrete are pretensioning and post-tensioning.

Pretensioning is a method of prestressing concrete in which steel tendons are tensioned before concrete is placed. The tendons are set between the concrete forms and stretched to a specified amount using powerful pretensioning jacks. Plates and anchors at the ends of the pretensioned member secure the tendons in position. Concrete is then placed in the forms around the tendons. The concrete bonds to the tendons as it hardens. When the concrete has achieved its specified strength, tension on the tendons is released and the pull of the tendons places the concrete under compression to minimize deflection. Pretensioning is most commonly used for precast members such as beams or piles. **See Figure 5-28.**

Post-tensioning is a method of prestressing concrete in which steel tendons are tensioned after concrete is placed. In post-tensioning, the tendons are encased in a sheath to prevent them from adhering to the concrete. The sheathed tendons are set in the concrete forms and concrete is placed around them. After the concrete has obtained the specified strength, the tendons are stretched and placed in tension to transfer the load from the tendons to the concrete. Plates and anchors at the ends of the post-tensioned member secure the tendons in position. Post-tensioning is most commonly used for cast-in-place concrete members.

Portland Cement Association

Figure 5-28. Tensioned steel tendons embedded in concrete members provide tensile strength. Pretensioning is most commonly used for precast members.

TECH TIP

Per OSHA 29 CFR Subpart Q—*Concrete and Masonry Construction,* no tradesworkers, except those essential to post-tensioning operations, are permitted to be behind a post-tensioning jack during tensioning operations. In addition, signs and barriers must be erected to limit access to the post-tensioning area.

For pretensioning and post-tensioning, the proper number of tendons of the appropriate diameter must be installed, the proper spacing of the tendons must be maintained during concrete placement, the specified amount of tension must be applied to the tendons, and the specified plates and anchors must be installed.

PRECAST CONCRETE

Precast concrete members are commonly used in many types of heavy commercial construction. Precast concrete members are used for walls, beams, slabs, piles, and a variety of other members. Concrete members may be precast at the job site and lifted into place or precast at a casting yard and transported to the job site where they are lifted into position. Precast concrete members require special reinforcement to ensure they can withstand the stresses of transportation and installation as well as the final structural loads.

Tilt-Up Construction

Tilt-up construction is a method of concrete construction in which wall panels are cast horizontally at the job site adjacent to their eventual position and tilted up into position after removal of the forms. **See Figure 5-29.** Tilt-up construction is commonly used for construction of concrete structures of one to three stories such as warehouses. Information contained on prints regarding tilt-up panels includes wall dimensions, opening dimensions, placement and types of lift anchors and reinforcement, and interior and exterior wall finish.

Portland Cement Association

Precast segmental box girders are lifted by crane and secured to the tops of cast-in-place piers or columns.

Figure 5-29. Tilt-up wall panels are cast on a floor slab and lifted into place using a crane after the concrete has reached the desired strength.

A planning schedule for casting, lifting, and bracing of tilt-up panels is completed prior to the beginning of the construction project to ensure that walls are placed and lifted in an orderly manner. A planning schedule ensures that walls that are cast, raised, and braced prior to other walls do not obstruct casting and placement of the walls that follow. Special care is taken by the jobsite superintendent to insure that the lifting equipment, for example a crane, has the proper load rating to safely lift and place the weight of the precast members.

After tilt-up wall panels are in position and properly braced, horizontal structural members that support floors and roofs are connected to the walls by welding plates or bolted inserts or supported by ledges formed on the walls. Precast concrete, cast-in-place concrete, or structural steel decking and roofs may be used with tilt-up panels.

TECH TIP

Per OSHA 29 CFR Subpart Q—*Concrete and Masonry Construction*, lifting inserts that are embedded or otherwise attached to tilt-up members must be capable of supporting at least two times the maximum intended load applied or transmitted to them.

Forming. Before forming tilt-up panels, a horizontal casting bed must be created. A casting bed is most commonly a cast-in-place concrete slab, but wood platforms or compacted fill can also be used as a casting bed. Low edge forms, equal in height to the tilt-up panel thickness to be cast, are set and attached to the casting bed. The inside edges of the forms are set to the dimensions of the finished panel. Provisions are made around the perimeter of the form to allow for the projection of reinforcing steel as shown on the prints. The forms may have drilled holes or be constructed of two adjoining pieces that allow for placement of reinforcing steel yet still retain the concrete.

When a concrete floor slab is used as a casting bed, a bond breaker is applied to the slab surface before concrete placement. A *bond breaker* is a blend of organic chemicals, inorganic chemicals, or polymers that prevents fresh concrete from adhering to previously set concrete. Prior to applying a bond breaker, the floor slab is blown clean to remove all debris. The bond breaker is applied directly to the surface of the floor slab by spraying or mopping. Bond breakers contain a coloring agent that does not adhere to the concrete but is used to indicate that all areas of the floor slab are properly coated. A bond breaker should not be applied to reinforcing steel since it will inhibit adhesion of the steel to the concrete and reduce the effectiveness of the reinforcing steel.

Blockouts for tilt-up panels are set in place for door and window openings and beam pockets. A *blockout* is a form that creates an open space within a concrete structure under construction in which fresh concrete is not to be placed. Door and window jambs are also set in the wall form. Blockouts and jambs are properly braced in the casting bed to prevent movement during concrete placement.

For some applications, a decorative surface finish may be required on the inside of the wall that is on the lower side of the forms. Form liners or other materials used to create the desired surface finish are placed on the floor slab and secured in position to prevent movement during concrete placement. In addition, welding plates for connecting adjoining wall panels and other structural attachments for interior floors or roof members are installed.

In some applications, electrical and mechanical fixtures may be placed at this time. Junction boxes, switch and receptacle boxes, and conduit may be set in place and secured for embedded electrical work. Pipe and blockouts for ductwork or additional piping may be set for plumbing and HVAC systems. Electrical, plumbing, and HVAC installations are shown on the electrical and mechanical prints. Electrical and mechanical fixtures placed in a tilt-up wall are secured to the reinforcing steel to ensure that they remain in the proper position during concrete placement.

Reinforcing. Details regarding reinforcing steel for tilt-up wall panels are similar to details provided for cast-in-place concrete structures. Additional reinforcement may be required to hold the wall rigid during the lifting process. Lift anchors and brace inserts are set within the forms to provide attachment points for lifting hardware and crane cables. Lift anchors and brace inserts are installed at specific locations and tied to the reinforcing steel. Proper location of the lift anchors ensures that the wall can be lifted safely and will not crack or break during the lifting operation. **See Figure 5-30.** Brace inserts allow the wall panels to be properly braced after being lifted into place and prior to the installation of interior floors or the roof structure. Threads in lift anchors and brace inserts are protected during the placement of the fresh concrete to ensure that threads are not blocked and bolts can be properly installed later. Symbols on details and manufacturer manuals describe the types of lift anchors and brace inserts placed in the tilt-up wall forms. Dimensions on elevations for tilt-up wall panels show proper locations for lift anchors and brace inserts.

Portland Cement Association

A decorative surface finish is used to simulate the appearance of natural stone on precast panels.

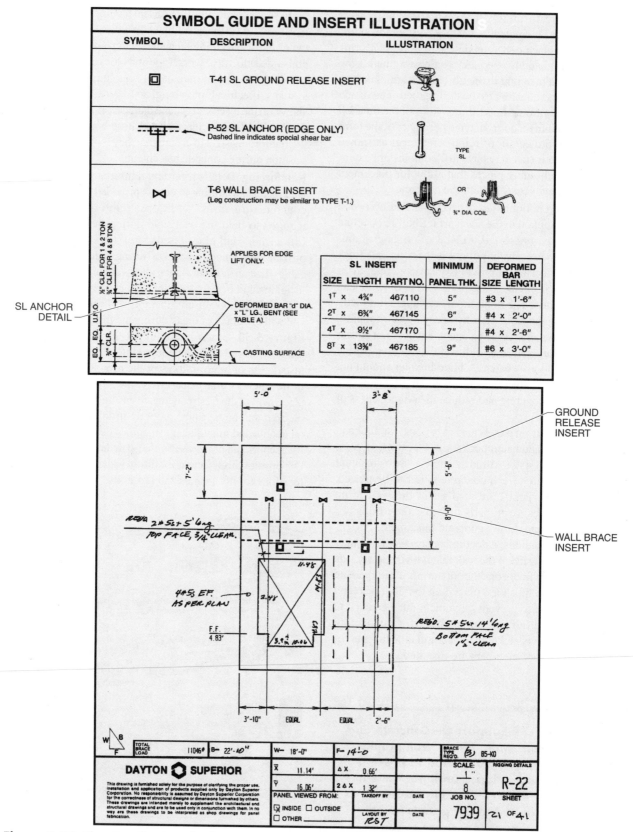

Figure 5-30. Elevations drawings for tilt-up wall panels indicate locations for lift anchors and brace inserts that are set in place and fastened to the reinforcing steel.

Finishing. Concrete is carefully placed in the forms to avoid disturbing the bond breaker, blockouts, jambs, and other previously installed items. Concrete is commonly placed onto a slanted panel that allows the concrete to slide easily onto the floor slab rather than dropping directly onto the slab. The interior surface of the wall is the surface that is in direct contact with the casting bed. The exterior surface of the wall is flush with the top of the edge forms. After the concrete is placed and leveled, exterior wall surface finishes can be applied. One or more tilt-up wall panels may be formed, reinforced, and placed at the same time on the job site, depending on the size of the building and accessibility for concrete placement.

Lifting. Prior to lifting, tilt-up panels with several blockouts and other voids may require additional bracing. Steel beams or C channels may be temporarily fastened to the wall surface for temporary reinforcement during the lifting operation. A crane with a spreader bar is used to lift and set the finished tilt-up wall panels in position after the concrete has obtained the specified strength. Cables are fastened to the lift anchors embedded in the wall with bolts and reinforcing plates as required. Braces are bolted to the brace inserts embedded in the wall and then fastened to the floor or other members designed to hold the wall in position. The wall panel must be properly braced prior to removing the crane rigging. The lifting and bracing process is performed for each wall panel until all wall panels are in their final position.

Fastening. Tilt-up wall panels are fastened together using several methods including flush cast-in-place pilasters, cast-in-place columns, precast columns, steel columns, and flush steel plates. **See Figure 5-31.** Concrete pilasters or columns may be formed between the braced wall panels. The pilasters or columns are cast-in-place and tie the tilt-up panels together. Reinforcing steel projecting from the edges of adjoining tilt-up wall panels extends into the pilasters or columns to tie the panels and pilasters or columns together. Precast concrete columns with steel inserts may be set in place between the tilt-up panels and welded to the tilt-up panels to secure them. A structural steel column may be set between adjoining tilt-up panels, and steel angles embedded in the edges of the tilt-up panels may be welded to the column. Steel inserts in adjoining tilt-up wall panels may be welded to a plate that spans between them and joins them together. Details on architectural or structural prints indicate the types of panel-joining systems used. For many applications, interior floors, walls, and the roof further tie the individual tilt-up wall panels together.

Regardless of the methods used to tie the wall panels together, braces for tilt-up wall panels must remain in place until the panels are fastened together and the floors and roofs are in place as required to ensure the structural integrity of the building. Interior walls, floors, and roofs tie the tilt-up wall panels together and provide structural stability.

Tilt-up wall panels are secured to the concrete floor of the structure by being set in a keyway formed in the slab or footing or by rebar placed in the floor near the wall sections and secured by ties and grout or concrete.

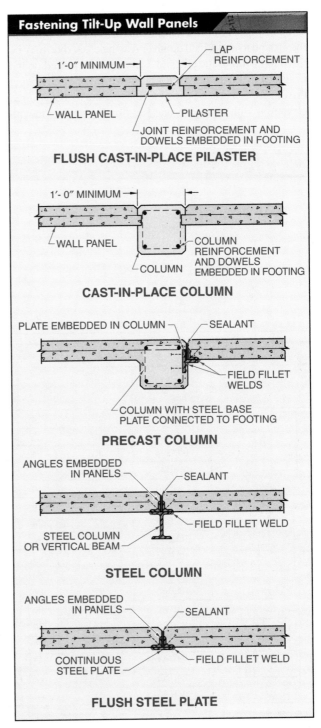

Figure 5-31. Tilt-up wall panels are fastened together using several methods.

Off-Site Precast Members

Concrete members that are precast and transported to the job site include piles, beams, slabs, and wall panels. The primary considerations in construction with large precast concrete members are proper handling, rigging, and installation to prevent cracking and failure of the units.

Piles. Precast concrete piles are driven in a manner similar to other piles. A pile-driving hammer forces the pile into the ground to a specified depth or resistance. A shoe may be fitted to the tip of the pile and a driving head placed on the head of the pile to prevent damage while the pile is being driven. Tops of precast concrete piles are cut to the required elevations with a concrete saw. Reinforcing steel is cut with oxyacetylene cutting equipment to provide the proper amount of projection for attachment of other building members.

Beams and Slabs. Precast concrete beams and girders are used for building and bridge construction. Beams and girders are formed in stemmed, rectangular, L-, or T-shapes according to print details. **See Figure 5-32.** The details provide information regarding the beam or girder length, height, and width. A schedule and structural prints may also be provided to indicate the types and placement for precast concrete beams. Structural engineers design the placement of reinforcing steel based on the loads to be imposed on the beam. Beams are commonly reinforced by pretensioning or post-tensioning.

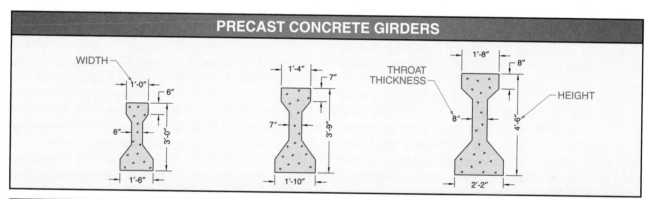

PRECAST CONCRETE GIRDERS

SAFE SUPERIMPOSED SERVICE LOADS* (PLF)

Beam	Designation	Number of Strands	H†	H1/H2†	Span‡								
					18	22	26	30	34	38	42	46	50
RECTANGULAR (B = 12" OR 16")	12RB24	10	24		6726	4413	3083	2248	1684	1288	1000		
	12RB32	13	32			7858	5524	4059	3080	2394	1894	1519	1230
	16RB24	13	24		8847	5803	4052	2954	2220	1705	1330		
	16RB32	18	32			7434	5464	4147	3224	2549	2036	1642
	16RB40	22	40				8647	6599	5163	4117	3332	2728
L-SHAPED	18LB20	9	20	12/8	5068	3303	2288	1650	1218				
	18LB28	12	28	16/12		6578	4600	3360	2531	1949	1524	1200	
	18LB36	16	36	24/12			7903	5807	4405	3422	2706	2168	1755
	18LB44	19	44	28/16			8729	6666	5219	4166	3370	2454
	18LB52	23	52	36/16				9538	7486	5992	4871	4007
	18LB60	27	60	44/16					8116	6630	5481
INVERTED TEE	24IT20	9	20	12/8	5376	3494	2412	1726	1266				
	24IT28	13	28	16/12		6951	4848	3529	2648	2030			
	24IT36	16	36	24/12			8337	6127	4644	3598	2836	2265	1825
	24IT44	20	44	28/16			9300	7075	5514	4378	3525	2868
	24IT52	24	52	36/16				7916	6326	5132	4213
	24IT60	28	60	44/16					8616	7025	5800

* safe loads shown indicate 50% dead load and 50% live load; 800 psi top tension has been allowed, therefore additional top reinforcement is required
† in in.
‡ in ft

Figure 5-32. Precast concrete girders and beams are designed to span long distances.

T-shape beams incorporate beams and slabs into a single unit and are used to span between column caps, beams, or girders in bridge construction. The slab section of the T may be from 8' to 10' wide. Double T-shape beams with two vertical sections are also available. Precast T-shape beams may be joined side by side to create an integrated unit for bridge beams and road surfaces.

After precast concrete beams are set in place, forms may be set perpendicular between them to provide for placement of cast-in-place concrete. The concrete ties the beams together, prevents tipping, and distributes imposed loads. Beams tied together with cast-in-place concrete are commonly placed directly above columns and column caps for bridge construction.

Precast floor slabs are available in a variety of thicknesses and designs and are used for above-grade floor slabs and roof panels. The types and sizes of precast floor slabs are noted on plan views and elevations. The thickness of precast floor slabs is noted in inches, followed by "FS" for solid flat slab or "HC" for hollow-core slabs. For example, a slab designated as 6FS is a 6" thick solid flat slab. The slab type and size may also be noted with a manufacturer product code or name. Methods for fastening the slabs to each other and to supporting members are shown on details and sections. Fastening may be accomplished by bolting or welding steel inserts.

Exterior Panels. Many designs of decorative precast concrete panels are used for wall finish and paving. Decorative concrete panels can be cast with various concrete colors, aggregate sizes and types, and surface finishes. Steel angles or plates are embedded into the panels to allow them to be welded to other members, such as structural steel columns and beams or reinforced concrete members with adjoining steel inserts. Precast wall panels are lifted into place with a crane with the proper load rating, braced, and secured in place. Information included on prints regarding decorative wall panels includes surface finish, the location of the panels, and fastening details.

Precast concrete is also used for ornamental stone, coping, windowsills, lintels, headers, and other applications on exterior walls of concrete or masonry. **See Figure 5-33.** Details provide dimensions and installation information.

Concrete

Prints related to concrete structures and components generally provide information regarding the finished dimensions for the shape and location of the structures and components and information regarding reinforcement. Minimal information is included on architectural and structural prints relating to properties and placement of concrete. The symbol used to represent concrete on sections is a series of small triangles in a random arrangement. Information regarding the composition, placement, and finishing of concrete is typically included in the specifications.

Properties. Architectural prints may include notes concerning specific concrete information such as compressive strength, aggregate size, slump, water-cement ratio, and admixtures. **See Figure 5-34.**

Compressive strength is the measured maximum resistance of concrete to axial loading and is expressed as force per cross-sectional area (typically pounds per square inch). A *compression test* is a quality control test that is used to determine the compressive strength of concrete. A 6" diameter by 12" high test cylindrical mold is filled with concrete from the same batch of concrete that is placed in the forms. The test cylinder is cured under the same basic conditions as the building concrete for 28 days. After 28 days, the cylinder is tested at a laboratory to determine the compressive strength. The mold is removed from the concrete and force is applied using a hydraulic press until the concrete breaks. A pressure gauge is used to indicate the force applied to the concrete. Multiple test cylinders are taken on large concrete projects to fully test all the concrete used at the job site.

Aggregate represents approximately 70% of the overall volume of concrete. Aggregate size and type, typically included in the general notes, affects final concrete strength and finishing properties. General notes may describe the type and maximum size of aggregate to be used. Smaller aggregate is typically used in concrete mixtures when reinforcing steel is placed closely together or for thin panels and slabs.

Slump is the measure of consistency of freshly mixed concrete. A slump test is performed on the job site to determine the slump of the concrete being placed into the forms. Fresh concrete is collected and placed into a 12" high slump cone that is placed on a flat surface.

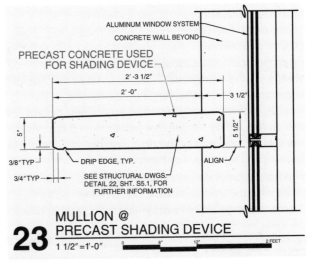

Figure 5-33. Precast concrete may be used for nonstructural purposes, such as sunshades and trim elements.

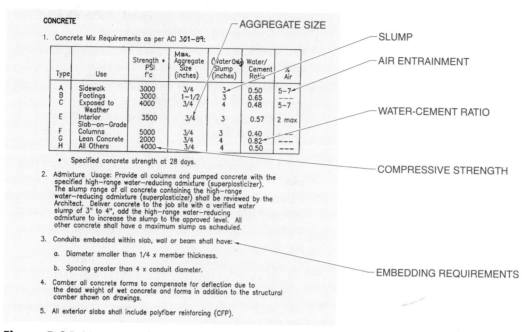

Figure 5-34. Notes regarding concrete mixture requirements, including admixtures and embedded items, are included on the prints.

When the slump cone is full, the cone is carefully lifted from the concrete and placed next to it. A straightedge is placed across the top of the cone and a measurement is taken from the straightedge to the top of the concrete. The distance from the top of the cone to the top of the concrete, in ¼″ increments, is the amount of concrete slump.

The *water-cement ratio* is the ratio of the weight of water (in lb) to the weight of cement (in lb) per cubic yard of concrete. The proper water-cement ratio of concrete is crucial to its final strength. Typically, the lower the water-cement ratio, the stronger the concrete. An adequate amount of water is required to fully hydrate the cement. Too much water weakens the concrete mixture, while too little water results in incomplete hydration.

An *admixture* is a substance other than water, aggregate, or portland cement that is added to concrete to affect its properties. Various admixtures are available that can modify concrete properties, such as setting time, amount of required water, workability, flowability, and air entrainment. *Air entrainment* is the occlusion of minute air bubbles in concrete during mixing. Air entrainment increases the workability of concrete and the resistance of concrete to freezing and thawing. **See Figure 5-35.**

Placement. Information regarding concrete placement is included in the specifications. General rules to be followed during concrete placement include placing concrete as close as possible to its final location to limit segregation of the mixture; placing concrete at a rate that does not place undue stresses on the formwork; avoiding the addition of excessive water at the job site; properly vibrating concrete to ensure proper consolidation; and properly curing concrete to ensure it conforms to final strength requirements.

RECOMMENDED AIR CONTENT PERCENTAGE

Nominal Maximum Size of Coarse Aggregate*	Exposure†	
	Mild	Extreme
⅜ (10 mm)	4.5	7.5
½ (13 mm)	4.0	6.0
¾ (19 mm)	3.5	6.0
1 (25 mm)	3.0	6.0
1½ (40 mm)	2.5	5.5
2 (50 mm)	2.0	5.0
3 (75 mm)	1.5	4.5

* in in.
† in percent

Figure 5-35. The desirable percent of air entrainment in concrete varies according to aggregate size and concrete exposure conditions.

Refer to the CD-ROM in the back of the book for Chapter 5 Quick Quiz® and related printreading and reference material.

Name _Paul._ Date _____

X (T)   F   **1.** Concrete has high tensile strength.

____ D ____   **2.** A slump test measures ___.
   A. formwork stability
   B. compressive strength of fresh concrete
   C. on-grade slab compaction
   D. consistency of freshly mixed concrete

(T)   F   **3.** Hydrostatic pressure within concrete forms increases as the amount of concrete in the forms increases.

~~tilt up~~
~~construction~~
tilt up   **4.** ___ construction is a method of concrete construction in which wall panels are cast horizontally at the job site adjacent to their eventual position and tilted up into position after removal of the forms.

(T)   F   **5.** A pipe pile is a steel cylinder with open ends that is driven into the ground with a pile-driving rig and then excavated and filled with concrete.

X   C   **6.** The water-cement ratio that results in the highest concrete strength is ___.
   (A) 0.25
   B. 0.50
   C. 0.75
   D. 1.0

(T)   F   **7.** Additional braces are typically required for one-sided wall forms.

gang form   **8.** A(n) ___ is a series of smaller prefabricated form panels fastened together to create a larger form.

(T)   F   **9.** Domes form voids in concrete decks in a two-way concrete joist system.

____ B ____   **10.** Solid diagonal lines on an above-grade slab plan view indicate ___.
   A. cross-bracing
   B. a floor opening
   C. shoring
   D. dome placement

pan   **11.** A(n) ___ is a prefabricated form unit used in one-way concrete joist systems.

T   (F)   **12.** Rebar must not project out of the face or edges of reinforced concrete.

X (T)   (F)   **13.** Tilt-up walls are commonly used for high-rise structures.

___formwork___ **14.** ___ is the total system for supporting fresh concrete, including the sheathing that contacts the concrete, supporting members, hardware, and braces.

(T)   F   **15.** A #5 rebar is ⅝″ in diameter.

___C___ **16.** A ___ is a continuous supporting device used to support reinforcing steel in the bottoms of slabs.
    A. chair
    B. rack
    C. bolster
    D. none of the above

(T)   F   **17.** Precast T-shape beams are used for bridge construction.

X (T)   F   **18.** Precast concrete wall finish panels are fastened to structural steel members with drilled holes and through-bolts.

___B___ **19.** A pile cap is located ___.
    A. below-grade to support piles
    B. on top of piles
    C. between footings and foundations
    D. between columns and column caps

___D___ **20.** A ___ is a space within a concrete structure under construction in which fresh concrete is not to be placed.
    A. buck
    B. form
    C. brace
    D. blockout

T   (F)   **21.** Aggregate makes up approximately 20% of concrete volume.

___⅜″___ **22.** The diameter of #3 rebar is ___″.
    A. ³⁄₁₆
    B. ⅜
    C. ¾
    D. 3

X (T)   F   **23.** Lugs on steel rebar are designed to inhibit corrosion.

(T)   F   **24.** Grade 60 rebar has greater yield strength than a grade 40 rebar.

___Caisson Pile___ **25.** A(n) ___ is a cast-in-place concrete pile made by driving a steel tube into the ground, excavating it, inserting reinforcement, and filling the cavity with concrete.

___A___ **26.** Chairs are used to support rebar for ___.
    A. slabs
    B. walls
    C. columns
    D. abutments

T   (F)   **27.** Rebar for columns is set after formwork is in place.

___Form___ **28.** A(n) ___ is the temporary structure or mold used to support concrete while it is setting and gaining adequate strength to be self-supporting.

_X_

_____Column_____ **29.** Friction collars are used to support formwork for ___.

X ⓣ F **30.** Locations for electrical conduit are shown on slab-on-grade plan views.

_____Steel_____ X **31.** For an above-grade slab, a perimeter form that is welded around the perim-
C Channel Steel   eter of corrugated steel decking is made from ___.

T Ⓕ **32.** Contraction joints for concrete slabs act to provide thermal insulation.

_____28_____ **33.** Concrete cylinders are tested for compressive strength after ___ days.

Ⓣ F **34.** Tilt-up wall sections may be joined together after being lifted into place by
cast-in-place concrete columns.

X T Ⓕ **35.** Bond breakers are most commonly used with cast-in-place concrete.

_____ **36.** ___ is the measured maximum resistance of concrete to axial loading.

Ⓣ F **37.** Electrical conduit is placed in tilt-up wall panels prior to concrete placement.

_____pile_____ **38.** A(n) ___ is a slender concrete, steel, or timber structural member driven or
otherwise embedded on end into the ground to support a load or compact
the soil.

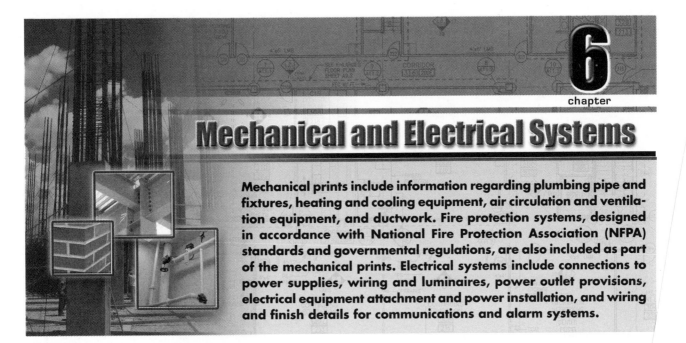

6 chapter

Mechanical and Electrical Systems

Mechanical prints include information regarding plumbing pipe and fixtures, heating and cooling equipment, air circulation and ventilation equipment, and ductwork. Fire protection systems, designed in accordance with National Fire Protection Association (NFPA) standards and governmental regulations, are also included as part of the mechanical prints. Electrical systems include connections to power supplies, wiring and luminaires, power outlet provisions, electrical equipment attachment and power installation, and wiring and finish details for communications and alarm systems.

MECHANICAL AND ELECTRICAL SYSTEMS

Mechanical systems are used to circulate water, gas, and other fluids and provide heated, cooled, and conditioned air for a building. Mechanical prints include information related to plumbing pipe and fixtures, ductwork, and heating, ventilating, and air conditioning (HVAC) equipment. Mechanical prints are identified with the prefix "M" followed by the sheet number.

Electrical systems include connections to power supplies, wiring and luminaires (lighting fixtures), provision of power outlets, installation of required attachments and power for electrical equipment, and wiring and finish for communication, security, and fire alarm systems. Electrical prints are identified with the prefix "E" followed by the sheet number.

Mechanical and electrical prints are used in conjunction with other prints included in the set. Allowances for placement of mechanical equipment, pipe and ductwork, panelboards, luminaires, switches, and electrical conduit must be made when constructing structural and finish members. Wall, floor, and roof openings must be properly constructed to allow for installation and access to pipes, ductwork, and conduit. Anchorage and supports may be embedded or attached to structural members to support mechanical and electrical equipment. BIM systems that integrate structural, architectural, mechanical, and electrical prints into a single system greatly reduce the conflicts or clashes between structural, mechanical, and electrical components. Properly utilized BIM techniques also allow for additional prefabrication of mechanical and electrical installations.

A table of contents is typically included on the cover sheet of a large set of mechanical and electrical prints indicate the sheet numbers and their contents. A variety specialized symbols and linetypes are used on mechanical and electrical prints. A list of the symbols and linetypes are often included on cover sheets along with general notes relating to the construction.

PLUMBING

Mechanical prints for plumbing installations include information relating to the stormwater drainage system, sanitary drainage and vent piping system, cold and hot potable water supply, piping for compressed air and other gases, and miscellaneous piping for items such as vacuums and irrigation systems. Each of the plumbing systems must be constructed according to the plumbing code adopted for the particular jurisdiction. A variety of pipes and fittings are used and are represented on mechanical prints with specialized abbreviations, symbols, and linetypes. **See Figure 6-1.** Where plumbing systems are complex and require detailed description, portions of the floor plan may be drawn to a larger scale. The large-scale drawings, often provided on separate sheets, allow for better understanding of the plumbing systems in a specific portion of a structure.

Pipes are installed before the wall, ceiling, and floor finishes are applied. Pipes are stubbed and capped as necessary where plumbing fixtures are to be attached. Fixtures, such as water closets or lavatories, are installed after finishes are applied.

PLUMBING/PIPING ABBREVIATIONS AND ANNOTATION

ABBREVIATIONS

AFF	ABOVE FINISH FLOOR
AFG	ABOVE FINISH GRADE
AH	AIR HANDLING UNIT
AT	ATTENUATOR
BLR	BOILER
CLG	CEILING
CO	CLEAN-OUT
COIW	CLEAN-OUT IN WALL
COTF	CLEAN-OUT TO FLOOR
COTG	CLEAN-OUT TO GRADE
DIV	DIVISION
DWG	DRAWING
EF	EXHAUST FAN
ET	EXPANSION TANK
EXH	EXHAUST
FD	FLOOR DRAIN
FS	FLOOR SINK
GTV	GAS TANK VENT
IE	INVERT ELEVATION
LAB	LABORATORY
MFG	MANUFACTURER
NC	NORMALLY CLOSED
UON	UNLESS OTHERWISE NOTED

ANNOTATION

FRWH	FREEZE RESISTANT WALL HYDRANT
NO	NORMALLY OPEN
OSA	OUTSIDE AIR
PRV	PRESSURE REDUCING VALVE
RA	RETURN AIR
RC	ROOF COWL
RD	ROOF DRAIN
TYP	TYPICAL
UH	UNIT HEATER
VTR	VENT THRU ROOF
WHA	WATER HAMMER ARRESTER
W/	WITH
@	AT
Ø	DIAMETER
~	FLAT OVAL
ARV	ACID RESISTANT VENT
ARW	ACID RESISTANT WASTE
ARVTR	ACID RESISTANT VENT THRU ROOF
CW	COLD WATER
HW	HOT WATER
HWC	HOT WATER CIRC.
SV	SUMP PUMP VENT
AR	ACID RESISTANT

PLUMBING/PIPING SYMBOLS

GENERAL SYMBOLS

ELBOW UP	
ELBOW DOWN	
TEE UP	
TEE DOWN	
CONCENTRIC REDUCER/INCREASER	
ECCENTRIC REDUCER/INCREASER	
UNION	
RISE/DROP IN PIPE	
VENT THRU ROOF	
PIPE SLEEVE	
EXPANSION JOINT	
FLEXIBLE CONNECTION	
PIPE ANCHOR	
PIPE GUIDE	
CAP	
ROOF DRAIN	
ROOF DRAIN (ABOVE)	
CLEAN-OUT (EXPOSED)	
CLEAN-OUT (FLUSH TO FLOOR OR GRADE)	
FLOOR DRAIN	
FLOOR SINK	
WATER HAMMER ARRESTER	
STRAINER	
THERMOMETER	
FLOOR/WALL PENETRATION	
AUTOMATIC BALL DRIP	
FLOW SWITCH	

CIRCULATING PUMP (POINTS IN DIRECTION OF FLOW)	
PRESSURE GAUGE	
VALVE (AS INDICATED OR SPECIFIED)	
GLOBE VALVE	
VALVE IN RISER	
ANGLE VALVE	
CHECK VALVE	
PRESSURE & TEMPERATURE RELIEF VALVE	
PRESSURE REDUCING VALVE (POINTS TOWARDS LOW PRESSURE)	
GAS VALVE	
SOLENOID VALVE	
VALVE W/TAMPER SWITCH	
HOSE BIBB/WALL HYDRANT	
HOSE BIBB (ELEVATION)	
DOWNSPOUT NOZZLE	
FIRE DEPARTMENT CONNECTION	
CIRCUIT SETTER	
TRAP SET (AS SPECIFIED) # INDICATES TRAP TYPE	
INDICATES SCREWED FITTINGS	
INDICATES WELDED FITTINGS	
INDICATES FLANGED FITTINGS	
POINT OF CONNECTION BETWEEN NEW & EXISTING WORK #" INDICATES EXISTING PIPE SIZE	

Figure 6-1. Abbreviations and symbols are used on mechanical prints to represent plumbing fixtures, equipment, and pipes.

TECH TIP

The Facility Services Subgroup in *MasterFormat 2004* includes Division 21—Fire Suppression, Division 22—Plumbing, Division 23—Heating, Ventilating, and Air Conditioning, Division 25—Integrated Automation, Division 26—Electrical, and Division 27—Communications.

Stormwater Drainage and Sanitary Drainage Systems

A *stormwater drainage system* is the piping system used to convey rainwater or other surface water from landscaped areas, paved areas, and roofs to a storm sewer or other places of disposal or storage, such as catch basins, retention ponds, or detention ponds. Captured stormwater may be reused at the building site for irrigation or other purposes. Information regarding the stormwater drainage system is typically included on civil prints and may also appear on plumbing prints. Large pipe made of precast concrete or various plastic compounds that are placed during excavation of the site and connected to the main storm sewer are shown on civil prints.

A *sanitary drainage system* is a system of sanitary drainage pipes and fittings that convey wastewater and waterborne waste from pluming fixtures and appliances, such as water closets and sinks, to the sanitary sewer system. Sanitary waste, or sewage, is kept separate from the stormwater drainage system since sewage is a hazardous waste that must be treated and processed. Some older urban areas still utilize a combined sewer system. A *combined sewer system* is a drainage system that utilizes one set of waste piping for both stormwater and sanitary drainage systems. Due to environmental concerns, combined sewer systems are no longer used for new installations. However, drainage systems for new structures in areas with combined sewer systems are still connected to existing combined sewer systems.

A sanitary drainage system must be properly vented. A *vent piping system* is a system of vent pipes and fittings that provides circulation of air to or from a sanitary drainage system to protect trap seals from siphonage or back pressure. A mechanical site plan may be included in the mechanical prints showing information regarding the location of waste piping and connections to storm and sanitary sewers. **See Figure 6-2.** Plan views and details indicate locations for drain piping in the immediate vicinity outside of the building and within the building.

Piping. In a stormwater drainage system, roof drains or storm drains in an impervious area, such as a roadway, collect rainwater and other surface water and discharge it into leaders and building storm drains. Several storm drains in a building or structure may feed into a building storm sewer, which then connects to a municipality or city storm sewer. In some situations, surface water may be discharged directly into retention or detention ponds. Water from retention or detention ponds is held temporarily for later discharge into the stormwater system or is filtered at the site and recycled for irrigation or gray water uses. Piping for these recycled water systems is shown on civil plans or plumbing plans.

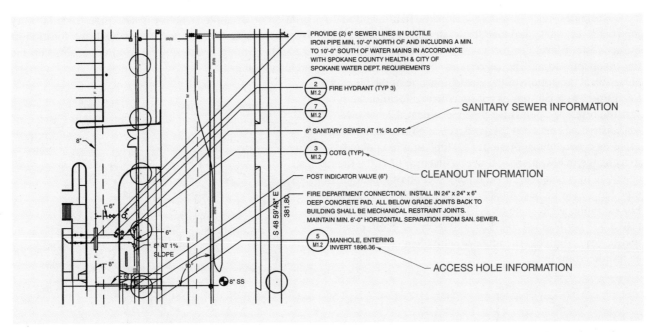

PROVIDE (2) 6" SEWER LINES IN DUCTILE
IRON PIPE MIN. 10'-0" NORTH OF AND INCLUDING A MIN.
TO 10'-0" SOUTH OF WATER MAINS IN ACCORDANCE
WITH SPOKANE COUNTY HEALTH & CITY OF
SPOKANE WATER DEPT. REQUIREMENTS

FIRE HYDRANT (TYP 3)

6" SANITARY SEWER AT 1% SLOPE — SANITARY SEWER INFORMATION

COTG (TYP)

POST INDICATOR VALVE (6") — CLEANOUT INFORMATION

FIRE DEPARTMENT CONNECTION. INSTALL IN 24" x 24" x 6"
DEEP CONCRETE PAD. ALL BELOW GRADE JOINTS BACK TO
BUILDING SHALL BE MECHANICAL RESTRAINT JOINTS.
MAINTAIN MIN. 6'-0" HORIZONTAL SEPARATION FROM SAN. SEWER.

MANHOLE, ENTERING
INVERT 1896.36

ACCESS HOLE INFORMATION

Figure 6-2. Plumbing installations, such as sanitary sewers, cleanouts, and access holes, are shown on mechanical site plans.

In a sanitary drainage and vent piping system, wastewater and waterborne waste exits a fixture, such as a lavatory or urinal, and flows through a fixture trap and into a fixture drain. Several fixture drains may then feed into a horizontal branch. Horizontal branches on several levels in the same vicinity then feed into a waste stack, which leads to the building sewer and into the sanitary sewer.

Plumbing prints indicate the size, purpose, type, and elevation of each pipe through the use of symbols and abbreviations. **See Figure 6-3.** For example, a broken line with an "RDL" abbreviation on a plan view indicates a roof drain line. Cross-referencing the plan views with the plumbing abbreviations and symbols indicates the use for each pipe. Plan views and, in some cases, isometric details are used to show piping arrangements. Detail sections related to the plan views provide reference points for the isometric details.

Pipe sizes are expressed as the inside diameter (ID) of the pipe in inches. Outside diameters of pipes vary depending on the material used for the pipe and the schedule (wall thickness) of the pipe.

Polyvinyl chloride (PVC), acrylonitrile butadiene styrene (ABS), polyethylene (PE), and cast iron (CI) pipe are commonly used as waste pipe. Specifications provide information regarding the type of pipe to be used for various applications. General notes on the mechanical prints may also provide information about the material used for waste pipes.

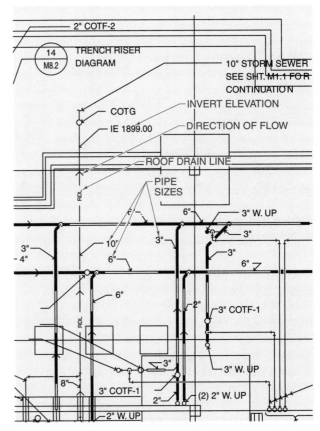

Figure 6-3. Plan views on plumbing prints provide an overall view of plumbing installations and indicate pipe sizes and flow directions.

Waste pipes must be properly sloped when installed to provide adequate drainage for wastewater and water-borne waste. Elevations for waste pipes are specified to the invert elevation. The *invert elevation* is the inside elevation at the bottom of the inside flow line of the pipe. **See Figure 6-4.** Elevations are specific to a location and provide reference points for pipe placement. Elevations for waste pipes between these reference points are indicated as a percentage of slope. A slope of 1% is equal to a change in elevation of 1′ over a distance of 100′. To calculate the change in elevation of a section of pipe, the pipe length (in inches) must be multiplied by the slope. For example, a 2% slope on a 10′ section of pipe is 2.4″ (10′ × 12″ per foot = 120″; 120″ × 0.02 = 2.4″). On plan views, vertical pipes are indicated with open circles and an architectural note.

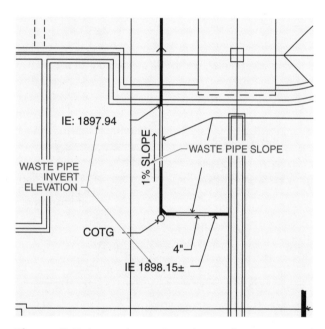

Figure 6-4. Invert elevations and pipe slopes are noted on mechanical prints.

Odors and gases are removed from sanitary drainage piping and conveyed away from inhabited areas by vent pipes. Vent pipes branch off of sanitary drainage piping and commonly terminate above the roofline. Vent pipes are shown on plan views and isometrics using hidden (dashed) lines. **See Figure 6-5.** Mechanical prints also indicate locations for cleanouts that are part of a sanitary drainage system.

Location dimensions for waste pipes are typically not included on mechanical prints. Grid lines and major structural members are usually shown on the prints as reference points. Exact waste pipe positions are determined from dimensions included on architectural and structural prints. For example, where a waste pipe for a lavatory is to be placed through an interior wall, the location of the wall opening is determined from dimensions on the architectural prints.

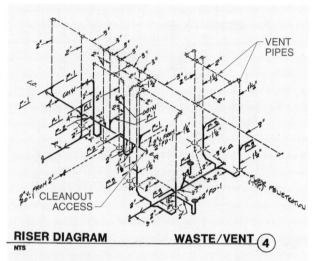

RISER DIAGRAM NTS   **WASTE/VENT** 4

Figure 6-5. Hidden (dashed) lines on isometric drawings indicate vent pipes. Plumbers install pipes, fittings, and hangers for waste pipes according to plumbing codes and local building code requirements.

For cast-in-place concrete floors, walls, and roofs, pipe may be set in place prior to concrete placement, or sleeves or chases may be provided so that pipe can pass through the floor, wall, or roof. In some cases a core drill may be used to drill through the concrete wall, floor, or roof to create an opening for the pipe. Details for installing sleeves are provided in the mechanical prints. **See Figure 6-6.** Chases, sleeves, and other openings for mechanical pipes are permitted only in accordance with the structural prints. Excessive holes in structural supporting members could create structural failures. Any penetrations or openings must be properly sealed after pipe installation to prevent airflow in accordance with local fire code requirements.

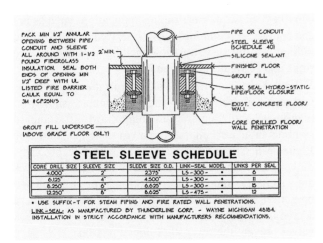

Figure 6-6. Details are commonly used to provide information regarding pipe and conduit sleeve installation. Sleeves allow for connections between levels of a building without losing structural integrity.

Fittings, Fixtures, and Terminations. A variety of fittings and fixtures are combined to make up a plumbing system. Fittings, such as elbows, tees, and unions, are rarely identified on mechanical prints except when shown in some detail. Fixtures, such as lavatories or water heaters, are typically identified on mechanical prints and are keyed to a schedule or the written specifications for additional information.

The technique used to join pipe and fittings depends on the type of pipe and fittings specified for the system. Plastic pipe and fittings, such as ABS, PVC, and PE, are commonly solvent-welded together, while no-hub cast iron pipe is typically joined using rubber gaskets and stainless steel sleeves.

Sanitary drainage and vent piping systems and stormwater drainage systems must be properly supported at the specified elevations. For underground installations, engineered fill may be required to ensure piping is kept at the proper elevation after backfilling of pipe trenches. *Engineered fill* is material such as sand or crushed stone that meets certain physical requirements and placement methods. The underground support requirements are listed in the specifications or shown on civil prints. Pipes throughout the structure are supported from structural members using hangers and other supports. The type of pipe support is not indicated on the mechanical prints. A general note regarding pipe support may be included on the mechanical prints or in the specifications.

Roof and floor drains are indicated on mechanical prints with notations and numbers that are keyed to a schedule or with general notes on the prints or in specifications. Roof and floor drain diameters are shown on plan views and isometric drawings.

The ends of waste and vent pipes may be stubbed and capped where fixtures are to be connected after room finishes are applied. Prior to approval by a building inspector, waste and vent systems may be subjected to a pressure test in which a certain amount of water or air pressure is built up inside the piping system for a certain period of time. Leaks are detected and fixed as required before final approval is granted.

Gray Water Systems. *Gray water* is water that is collected at a building site and reused for irrigation or other nonpotable water applications. The use of gray water minimizes the need for treated water use at a site. Various methods of reusing gray water within a structure may be utilized. Water from various locations of the structure such as roof areas and any impervious paved areas are common sources of gray water.

Gray water may be treated in several ways. Rainfall or snowmelt from a roof may be captured by the gutter and downspout system and piped into a holding tank. Water from these tanks is pumped through a basic filtration system to remove large impurities. The resulting gray water can be used for lavatories or other water needs not requiring drinkability standards. Surface water from paved areas may be captured in retention or detention ponds. This water may then be piped and pumped for irrigation. Each of these systems minimizes the need for treated water and reduces the surge on stormwater systems during times of heavy rain or melting snow.

Reed Manufacturing Co.
A direct tapping machine drills and taps potable water mains while the mains are under pressure.

Potable Water Supply System

A *potable water supply system* is a system of water service pipe, water distribution pipe, fittings, and valves inside or outside a building (but within the property lines) that supplies and distributes potable water to points of use within a building. *Potable water* is water that is free from impurities that could cause disease or harmful physiological effects. Common points of use for potable water are restrooms, kitchens, water fountains, and laboratory areas. Mechanical prints provide information regarding pipes, connections, valves, and fixtures.

Piping. Mechanical prints indicate the sizes and general locations of pipe conveying hot and cold potable water. **See Figure 6-7.** Pipe sizes are expressed as the inside diameter (ID) of the pipe. Additional pipe information is indicated on plan views and isometric drawings in a manner similar to waste piping diagrams. Information about the pipe to be used is provided in the written specifications.

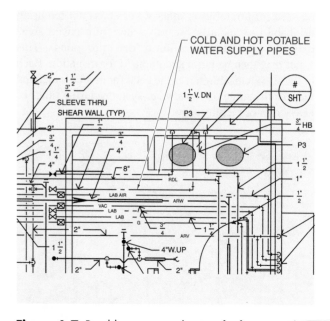

Figure 6-7. Potable water supply pipes for fixtures such as sinks are shown on mechanical prints.

Copper, PVC, and cross-linked polyethylene (PEX) are common materials used for water supply piping. Copper tube is typically joined using soldered fittings. However, copper tube may also be joined using rolled groove and flared connections. PVC pipe and fittings are solvent-cemented together. PEX piping systems are comprised of flexible tubing that are joined with a variety of crimp, clamp, push-fit, or compression fittings. PEX tubing is suitable for low pressure piping applications such as hot and cold water plumbing supplies, service lines, hydronic radiant heating systems, snow melting applications, ice rinks and refrigeration warehouses.

Valves. A *valve* is a device that controls the flow of fluids within pipes. Valve types and locations are shown on plan views and isometric plumbing drawings. Standard symbols indicate the types of valves to be installed. **See Figure 6-8.** Common valve types include gate, butterfly, globe, and check valves. Gate valves and butterfly valves are primarily intended for fully open or fully closed operation. Globe valves are designed to be used where a graduated throttling of fluid flow is required. Check valves prevent the reversal of flow direction.

Fixtures. A plumbing fixture schedule is usually included as part of the mechanical prints. Notes on plan views and isometric plumbing drawings relate to the fixture schedule. **See Figure 6-9.** Schedule information may include the fixture manufacturer names and model numbers, sizes, fittings, methods for attaching the fixture to the structure, waste pipe sizes, vent pipe sizes, water supply pipe sizes, and general notes. Details for connections to fixtures or equipment, such as hot water tanks, are provided where necessary.

Other Piping Systems. Piping systems for compressed air, vacuums, gases, or fuels are required for certain commercial structures, such as manufacturing facilities or laboratories. Pipe used in commercial applications is described in the same manner as drainage and water supply pipe. Special abbreviations and symbols may be used on plan views to describe the pipe. Details are often included to describe connections to fuel supply tanks or vacuum units. Plan views and isometrics of the piping and valve configurations are also provided.

FIRE PROTECTION

Municipalities and governmental agencies have adopted regulations and requirements for fireproofing, firestopping, and fire protection equipment and systems. The regulations and requirements are based on the intended use and potential occupancy of the structure. The National Fire Protection Association (NFPA) has developed standards to assist local governments in developing fire protection requirements for local building codes. Mechanical prints for fire protection systems are designed in accordance with these standards and governmental regulations. Mechanical prints may contain a schedule outlining the fire protection devices and equipment required for each occupied space of a building. **See Figure 6-10.** The type of system, sprinkler type, and hazard level may be indicated on the schedule.

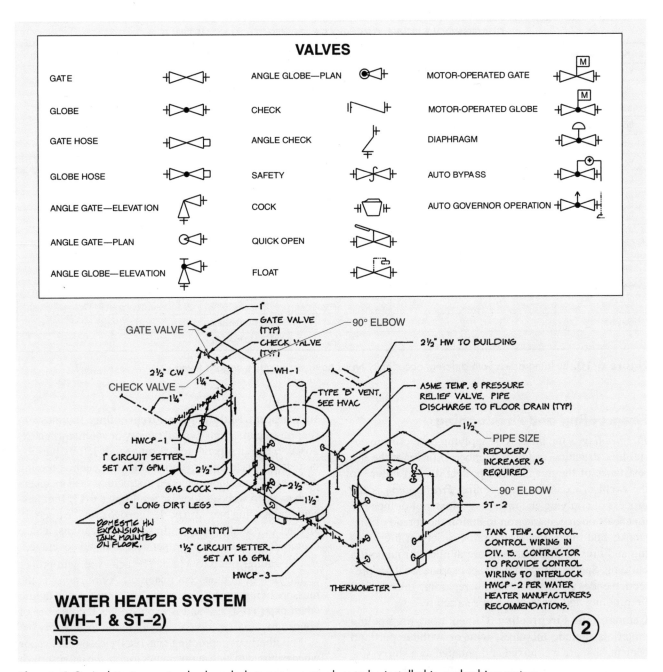

Figure 6-8. Architects use standard symbols to represent valves to be installed in a plumbing system.

PLUMBING FIXTURE SCHEDULE													
Symbol	Fixture	Mfr.	Model No.	Mounting	Type	Material	Size	Drain	Trap	W	V	HW	CW
P1	Water closet	American Standard "Afwall"	2477.016	Wall	Siphon Jet	White vitreous china	Elongated bowl	—	—	4"	2"	—	1"
P2	Urinal	American Standard "Jetbrook"	6570.022	Wall	Blowout	White vitreous china	21" × 14½" × 15⅛"	—	—	2"	2"	—	1"
P3	Lavatory	American Standard "Aqualyn"	0476.028	Counter	Self-rimming	White vitreous china	Oval 20" × 17"	American Standard #7723.018	1½" × 17GA	1½"	1½"	³/₈"	³/₈"

Figure 6-9. Plumbing fixture descriptions are provided by a schedule in mechanical prints.

CLASSROOM LEVEL SPRINKLER PROTECTION SCHEDULE					
Room Name	Occupancy Hazard	System	Suspended Ceiling	Sprinkler Type	Temperature Classification
Elevator Lobby	LH	WP	Yes	Pendant	Ordinary
Elect.	OH-1	WP	Yes	Pendant	Intermediate
Telephone	OH-1	WP	Yes	Pendant	Ordinary
Info. Net. Res. Ctr.	OH-2	WP	Yes	Pendant	Ordinary
Student Lounge	OH-2	WP	Yes	Pendant	Ordinary
Studio	OH-2	WP	Yes	Pendant	Ordinary
Cont. Education	OH-2	WP	Yes	Pendant	Ordinary
Cont. Education	OH-2	WP	Yes	Pendant	Ordinary
Control Room	OH-2	PA	Yes	Pendant	Ordinary
Audio/Visual	OH-2	PA	Yes	Pendant	Ordinary
Tele Seminar	OH-2	WP	Yes	Pendant	Ordinary
TV Seminar	OH-2	WP	Yes	Pendant	Ordinary
Stair	OH-2	WP	Yes	Pendant	Ordinary
Corridor	LH	WP	Yes	Pendant	Ordinary
Corridor	LH	WP	Yes	Pendant	Ordinary
Corridor	LH	WP	Yes	Pendant	Ordinary

Figure 6-10. Building areas with different occupancy types have varying fire hazard levels.

Fireproofing and Firestopping

Fireproofing is the process of applying protective material to a structural or finish member to increase the fire resistance of the member and protect it from failure in the event of excessive heat or fire. *Firestopping* is the process of applying and installing a material or member that seals open construction to inhibit the spread of fire, smoke, and fumes in a structure. A variety of fireproofing and firestopping materials are applied throughout a building. Specifications and the appropriate local fire protection codes contain information regarding the required fireproofing and firestopping in a structure.

Cementitious Fireproofing. Cementitious fireproofing materials include mixtures, with or without portland cement, that are wet mixed and pumped as a slurry. Cementitious fireproofing materials completely encase the structural or finish member to prevent direct contact with flames and inhibit the spread of heat to the member. A variety of cementitious fireproofing materials are available, each with different material bases and fireproofing qualities. For example, a fireproofing mixture may be gypsum or portland cement based with vermiculite.

Cementitious fireproofing materials are typically sprayed onto a structural or finish member. The base surface must be clean and free of any material that would prevent proper adhesion. Manufacturer information related to the proper mixture and application must be followed carefully to achieve the required fire rating.

Intumescent/Endothermic Fireproofing. Intumescent fireproofing materials expand rapidly in volume when exposed to high temperatures, closing voids left by burning or melting construction materials. Intumescent fireproofing materials are commonly applied to structural steel members such as open web steel joists. **See Figure 6-11.** Intumescent fireproofing may also be designed to provide a certain amount of sound insulation. In some applications, a base coat may be required for proper adhesion and coverage.

Fire ratings of up to 3 hr can be achieved with proper application of intumescent materials. A *fire rating* is the measure of resistance of a material or component to failure when exposed to fire and is expressed as the number of hours a material or component will retain its integrity.

Endothermic fireproofing materials release water vapor when exposed to high temperatures to cool the outer surface of the member inhibiting the spread of fire in a building. For some applications, endothermic mats are wrapped around electrical raceways or electrical cable trays containing critical circuit conductors.

Caulk/Putty. Fire-rated caulk or putty may be applied as a firestopping material around conduit and other piping where it penetrates walls and floors. Airtight seals are created by the caulk or putty to inhibit the spread of fire, smoke, and fumes in a structure. Some types of firestopping caulk expand when heated to ensure an airtight seal. Firestopping caulk is available in tubes or bulk for application with a putty knife. Firestopping putty can be applied by hand and pressed into place.

FIREPROOFING

FIRESTOPPING

Figure 6-11. Fireproofing material is applied to structural steel members to inhibit structural failure due to heat. Firestopping caulk that seals walls and floors may be applied around pipes.

Sleeves/Flashing. Sleeves or sheet metal flashing may also be used as firestopping for wall and floor penetrations. Cast-in-place sleeves are attached to the formwork before concrete is placed to provide an opening for pipes. Some sleeves are intumescent devices and do not require firestopping foam or caulk. Other sleeves are filled with foam or caulk to create an airtight seal after pipes have been positioned. Expandable firestopping sleeves may also be fitted onto piping and conduit and slid into position at wall and floor penetrations.

Sheet metal flashing may also be used as a firestopping material. Thin-gauge sheet metal is cut and fastened in place around pipes and conduit as needed to stop airflow. Caulk or putty may also be applied to ensure an airtight seal around the flashing.

Fire Protection Systems

Fire protection systems include detection and/or fire suppression systems. Fire detection systems include equipment and devices that sense heat, smoke, or flame and activate a local or remote fire alarm or fire suppression system. Fire suppression systems disperse water, gas, or chemicals into fire hazard areas. Mechanical prints indicate locations of detection and fire suppression equipment and devices, such as pipes, sprinkler heads, valves, and alarm boxes, and indicate areas to be protected. **See Figure 6-12.**

Fire detection systems operate by sensing one or more products of fire. Heat, smoke, and flame detectors are the most common types of fire detection systems installed in commercial buildings. Heat detectors are typically used in dirty environments or where dense smoke is produced. The most common heat detectors either react to a broad temperature change or a predetermined fixed temperature. Smoke detectors detect visible or invisible particles produced by combustion. Flame detectors are electronic devices that scan an area for specific types of infrared, visible, and ultraviolet light emitted by flames during combustion. When a flame detector recognizes the light from a fire, the detector sends a signal to activate an alarm.

Three types of fire suppression systems are wet, dry, and gaseous. Each system has different applications and installation and operation requirements. The system specified for installation is based on the usage of the area requiring fire protection. For example, a dry fire suppression system is often specified for a room where freezing temperatures are anticipated. The NFPA publishes a wide variety of codes and standards related to fire protection systems. Alarm attachments notify fire protection districts when the fire protection system of a building is activated. Wet, dry, and gaseous systems are attached to an alarm valve that initiates notification of the fire department when water, air, or gas begins passing through the system.

Wet Pipe Systems. Pipes of a wet fire suppression system are connected to a main water supply source. Exterior fire hydrants are shown on civil prints or on site plans included on mechanical prints. Sprinkler system pipes convey water throughout a building to various locations where fire suppression is required. Water is stored in the pipe until it is released in the event of a fire. Wet pipe system pipes are typically shown on plan views using a solid line with the abbreviation "WP". Isometric fire protection piping drawings may also be provided.

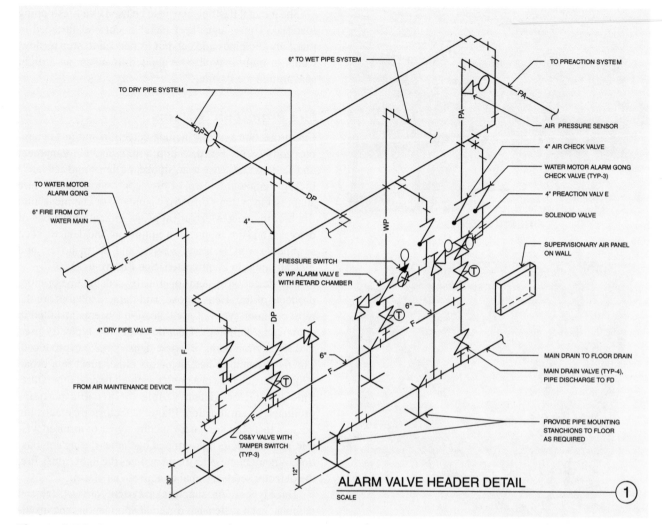

6" TO WET PIPE SYSTEM

TO PREACTION SYSTEM

TO DRY PIPE SYSTEM

PA

PA

AIR PRESSURE SENSOR

4" AIR CHECK VALVE

WATER MOTOR ALARM GONG
CHECK VALVE (TYP-3)

TO WATER MOTOR
ALARM GONG

DP

4" PREACTION VALVE

6" FIRE FROM CITY
WATER MAIN

SOLENOID VALVE

4"

WP

SUPERVISIONARY AIR PANEL
ON WALL

F

PRESSURE SWITCH

6" WP ALARM VALVE
WITH RETARD CHAMBER

DP

T

6"

T

4" DRY PIPE VALVE

F

MAIN DRAIN TO FLOOR DRAIN

MAIN DRAIN VALVE (TYP-4),
PIPE DISCHARGE TO FD

FROM AIR MAINTENANCE DEVICE

T

6"

PROVIDE PIPE MOUNTING
STANCHIONS TO FLOOR
AS REQUIRED

F

F

OS&Y VALVE WITH
TAMPER SWITCH
(TYP-3)

30"

12"

ALARM VALVE HEADER DETAIL

SCALE

①

Figure 6-12. Fire suppression systems disperse water, gas, or chemicals into fire hazard areas. Alarms that are part of a fire detection system typically notify the fire protection district automatically when the fire suppression system is activated.

TECH TIP

A fire standpipe (FSP) system is a strategically placed network of water-filled pipes and hose valves that allows firefighters to directly supply water to a fire.

Sprinkler pipes are manufactured from steel, cast iron, or other materials that will not fail due to heat from a fire. Valves are installed throughout the piping system to allow for periodic testing and to provide for maximum safety. A sprinkler head is installed at the termination point of each pipe in the areas to be protected from fire. Mechanical prints indicate pipe and sprinkler head locations. **See Figure 6-13.**

Additional information regarding pipes, valves, and sprinkler heads is included in the specifications. Sprinkler heads in a wet pipe system are designed to open and release water when activated by heat. Sprinkler heads are installed after ceiling tile is in place in areas where suspended ceiling systems are installed.

Dry Pipe Systems. Dry pipe systems are commonly used in areas subjected to freezing temperatures to eliminate the possibility of water in the pipes freezing and rupturing the pipes. Pipes for a dry pipe system are connected to the same water supply used for a wet pipe system. However, in a dry pipe system, the sprinkler pipes are filled with pressurized air or nitrogen. The pressurized air or nitrogen holds a remote dry pipe valve in a closed position. In the event of a fire, the valve opens, allowing water to flow to the sprinkler heads and be distributed to the fire location.

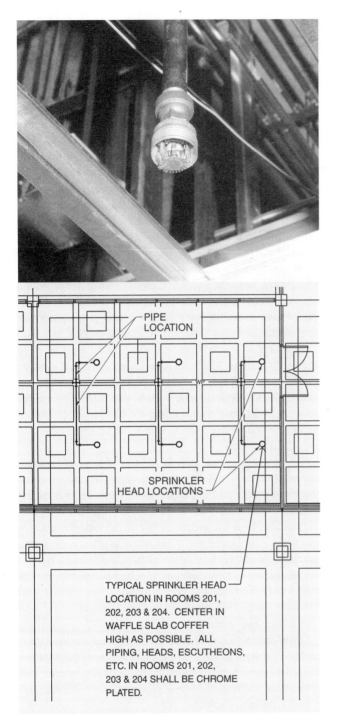

Figure 6-13. Placement of sprinkler heads in accordance with the plan views ensures proper dispersion of fire suppression materials.

A *preaction fire protection system* is a type of dry pipe system in which two separate events—smoke detection and heat development—must occur for the sprinkler system to activate. Preaction fire protection systems are commonly used in locations where there is a concern of accidental discharge of water, such as in computer rooms or control rooms. Pipes for a preaction system are connected to the same water supply used for a wet pipe system. A preaction valve is installed in the water supply piping. Smoke detection allows water to flow into the sprinkler pipes, but the sprinkler heads do not open and discharge water until they are activated by heat from the fire.

Gaseous Dry Pipe Systems. Dry pipe systems may also be designed to deliver a fire-suppressing gas instead of water to a fire location to suppress the fire. A fire-suppressing gas, such as halon, carbon dioxide, nitrogen, or argon, is stored in a tank connected to the sprinkler pipes. The pipes are not connected to the water supply. In the event of a fire, the gas is released from the tank and conveyed to the fire location. The properties of the gas smother the fire. Gaseous dry pipe systems may be installed to provide fire protection to areas where electronic equipment, such as critical computer equipment, is installed and cannot be exposed to water.

Areas in which dry pipe systems are to be installed are indicated on plan views. **See Figure 6-14.** Entire structures may also be protected by dry systems. Dry system pipes are noted on plan views with an abbreviation such as "DP" for dry pipe.

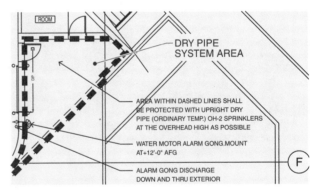

Figure 6-14. Dry pipe fire protection systems prevent pipe failures caused by frozen water in the pipes.

HEATING, VENTILATING, AND AIR CONDITIONING SYSTEMS

Large commercial buildings have a wide variety of heating, ventilating, and air conditioning (HVAC) requirements. Complex systems are required to provide the temperature control and air circulation required for large commercial buildings. Special attention is given to systems where sustainable construction techniques are used. Proper installation of controls and alternative heating and cooling systems can result in highly effective and efficient use of energy. Mechanical prints contain the majority of information about HVAC systems. Plan views, elevations, and details provide fabrication and installation information.

Specifications, general notes, and schedules contain information regarding manufacturer designs and model numbers. Shop drawings may also be produced by a subcontractor or supplier to provide information regarding the proper installation of equipment, pipes, and ductwork. While most of the information regarding HVAC systems is included on mechanical prints, additional information is found by referring to architectural, structural, and electrical prints. A commercial HVAC system may contain a heating, cooling, ventilation, humidification, dehumidification, and/or air filtration system.

Hot water or steam is generated in a boiler and distributed throughout the building.

Heating Systems

A heating system increases the temperature of a building area. Heating systems are classified based on the medium used to convey the heat or heat source used. Common heating systems used in commercial buildings include hot water (hydronic), steam, geothermal, solar thermal, electric heat, heat pump, and gas-fired systems.

Hot Water and Steam Systems. Hot water (hydronic) and steam heating systems use hot water and steam, respectively, as mediums to heat building areas. A hot water heating system is widely used in commercial buildings.

Hot water or steam is generated in a boiler. Water to be heated is pumped through the boiler pipes. **See Figure 6-15.** Expansion tanks allow the water to expand as it is heated without causing a dangerous amount of pressure to build up. The hot water or steam is circulated through pipes in the building spaces using a circulating pump. The water or steam passes through terminal units in the building spaces or through coils in the ductwork. Heat is radiated from the water or steam at terminal units to air passing through the units. Terminal units may be equipped with blowers and ductwork that circulate air through the terminal units. The heated air is distributed to ductwork and building areas. The cooled water is directed through return pipes and back to the boiler where it is reheated and recycled.

Boiler symbols are shown on mechanical plan views and elevations. Specific location dimensions are not provided on mechanical prints; rather, dimensions are obtained from plan views on architectural and structural prints. Information about the boiler type, size, capacity, and heating load is included on a schedule on the mechanical prints or in the specifications. Mechanical and electrical prints indicate the vent and water piping and the power and control wiring connections to boilers. Where applicable, BIM systems can provide additional coordination for placement of piping and prevent on-site conflicts with structural members and electrical systems.

Thermostats connected to the terminal units indicate the need for heat and may activate valves or fans to adjust the temperature to the proper level. Each portion of the hot water or steam system is shown on plan views and elevations. General locations for circulating pumps for hot water supply and hot water return pipes are shown on mechanical prints. Expansion tanks connected to boilers are also shown. Schedules on mechanical prints provide pump information, including manufacturer name and model number, flow rate in gallons per minute (gpm), and motor size and speed.

Pipe connections between the pumps, expansion tanks, and boilers are indicated on mechanical prints. Information regarding pipe to be installed between the pumps, expansion tanks, and boilers includes the pipe and valve size and meter types and locations.

Pipe notations indicate the purpose of the pipe and the direction of water flow within the pipe. **See Figure 6-16.** In a one-pipe system, water passes through each terminal unit in a continuous flow back to the boiler. In a two-pipe system, separate water supply and water return systems are installed. In a four-pipe system, four separate pipe systems are installed for hot and cold water supply and return. Pipes are identified as hot water supply, hot water return, cold water supply, and cold water return. Cold water supply and return pipes are used for cooling operations.

BOILER SCHEDULE (GAS FIRED)   MFR: WEIL-McLAIN

| # | MODEL | CAPACITY (MBH) | | MIN. GAS PRES. ("WC) | HEATING WATER | | | BLWR. (HP) | OPER. WEIGHT (LBS) | REMARKS |
		INPUT (MIN.)	OUTPUT (MIN.) (1)		TEMP. LVG(F)	FLOW GPM	VOL (GAL)			
B-1, 2 (2)	SERIES 88 1488	4474	3550	7"	180	275	339	3	10,675	GAS ONLY GORDON-PIATT BURNER

NOTES
(1) GROSS 1-B-R OUTPUT.
(2) B-1 BASE BID. B-2 IS ALTERNATE NO. 5.

AIR COOLED CONDENSING UNIT SCHEDULE   MFR: LIEBERT

| # | MODEL | UNIT | | | | | | COMPR. MOTORS | | | CONDENSER | | OPERATING WEIGHT (LBS) |
| | | CAPACITY (MBH) | AMB. AIR °F | REFRIG. TYPE | COND. TEMP. °F | MIN. CIRCUIT AMPS | V/PH | NO. | AMPS | | NO. OF FANS | TOTAL KW | |
									FULL LOAD	LOCKED ROTOR			
CU-1	DMC-027A	23.0	95	R-22	—	—	208/1	1	14.1	—	1	—	180

PUMP SCHEDULE

| # | MANUFACTURER | MODEL | TYPE | SERVICE | FLOW (GPM) | HEAD (FT) | MOTOR | | | NOTES |
							HP	RPM	V/φ	
HWP-1, 2	BELL & GOSSETT	1531-2qf BB	CLOSE-COUPLE	HEATING WATER	240	60	7.5	1750	460/3	(1) (11)
BCP-1, 2	"	1531-3 AB	"	"	275	25	3	1750	460/3	(2) (12)
CHWP-1	"	1531-2qf AB	"	SF-1 HW	150	35	3	1750	460/3	(13)
CHWP-2	"	SERIES 60-1qf A	IN-LINE	SF-4 HW	33	35	eg	1770	460/3	
CHWP-3	"	SERIES 60-1qf A	"	SF-5 HW	30	35	qg	1770	460/3	ALT. NO. 2
CHWP-4	"	SERIES 60-1qf A	"	SF-6 HW	27	25	qg	1750	115/1	ALT. NO. 1
CHP-1	"	1531-3 BB	CLOSE-COUPLE	CHILLED WATER	350	65	10	1750	460/3	(10)
CCHWP-1	"	1531-2qf AB	"	SF-1 CHW	170	35	3	1750	460/3	(9)
CCHWP-2	"	SERIES 60-1qf A	IN-LINE	SF-2 CHW	44	35	1	1770	460/3	
CCHWP-3	"	SERIES 60-2 A	"	SF-3 CHW	88	35	1qf	1750	460/3	
CCHWP-4	"	SERIES 60-2 A	"	SF-4 CHW	64	35	1qf	1750	460/3	
CCHWP-5	"	SERIES 60-1qf A	"	SF-5 CHW	20	35	eg	1750	460/3	ALT. NO. 2

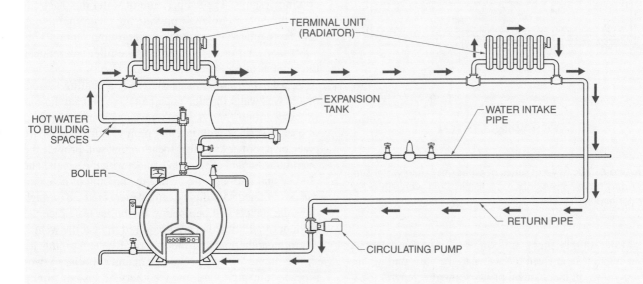

Figure 6-15. Schedules on mechanical prints provide information regarding boilers, condensing units, and pumps. In a hot water or steam system, water or steam is pumped from the boiler to the areas to be heated.

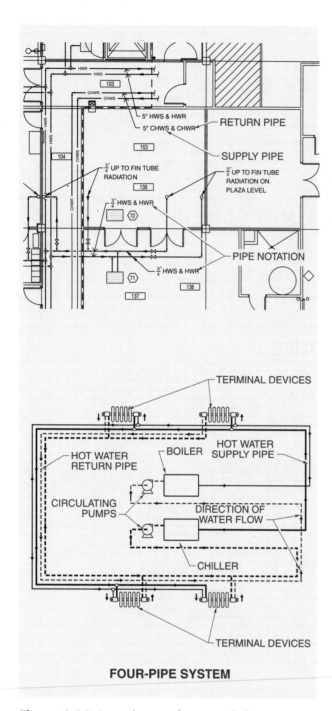

5" HWS & HWR
5" CHWS & CHWR
RETURN PIPE
SUPPLY PIPE
¾" UP TO FIN TUBE RADIATION
¾" UP TO FIN TUBE RADIATION ON PLAZA LEVEL
¾" HWS & HWR
PIPE NOTATION
¾" HWS & HWR

TERMINAL DEVICES
HOT WATER RETURN PIPE
BOILER
HOT WATER SUPPLY PIPE
CIRCULATING PUMPS
DIRECTION OF WATER FLOW
CHILLER
TERMINAL DEVICES

FOUR-PIPE SYSTEM

Figure 6-16. Several series of pipes, including water supply and return pipes, are used to circulate water in a hot water heating system.

Water supply pipes are connected to terminal units that transfer heat from the water to the surrounding air. Plan views on mechanical prints show the general locations of terminal units. A schedule for terminal units describes the water flow requirements (in gpm) and airflow requirements in cubic feet per minute (cfm).

See Figure 6-17. Architects use various geometric shapes, such as hexagons, ovals, and circles, for ease in identifying and cross-referencing the symbols and numbers to the schedules. Details of terminal units are also provided to indicate water and ductwork connections.

Geothermal Heating Systems and Solar Thermal Heating Systems. The use of naturally occurring heat in the ground or energy from the sun provides a substitute or a supplement to hot water heating systems. Either of these heat sources, depending on the geographic location and site conditions, can be utilized in a manner similar to a hot water heating system with water heated by a boiler. Civil plans and mechanical prints indicate the use of a geothermal or solar thermal system.

A *geothermal heating system* is a heating system that uses pipes installed below the surface of the ground to utilize the naturally occurring temperature of the earth to regulate fluid temperature. Fluid is circulated through these pipes by a pumping system. The liquid remains underground for a period of time long enough to achieve the same temperature as the surrounding earth. In cold conditions, this will increase fluid temperature to an acceptable temperature level and the fluid will be returned to a hot water heating system at a temperature then requiring less additional heating energy. For example, fluid enters the geothermal system at 40°F. After the water flows through the underground geothermal piping it is returned to the boiler at 55°F. The additional 15°F of heat produced by the geothermal system require less boiler fuel to bring the fluid to the temperature needed to heat the structure.

Geothermal systems can also be used in water based cooling systems. The temperature of water that is circulated through a chiller to provide air conditioning can also be regulated through the underground piping. This cools the water prior to entering the chiller and reduces the amount of fuel needed.

A *solar thermal heating system* is a heating system that moves water through solar panels to preheat the water before it enters a boiler. Solar panels are comprised of a dark surface with internal piping. Solar panels are place on a roof or any other location that will provide the maximum amount of sunlight contact with the panels. The climate and building site dictate the amount of heat that can be obtained with this system. Water is passed through the solar panels and pumped to the boiler to reduce the amount of additional energy required to heat the water.

Both geothermal and solar thermal system piping are shown on mechanical prints in a manner similar to two-pipe or four-pipe systems. Geothermal system piping may also appear on civil plans. Pumps, solar panels, and storage tanks appear on mechanical and plumbing prints and details.

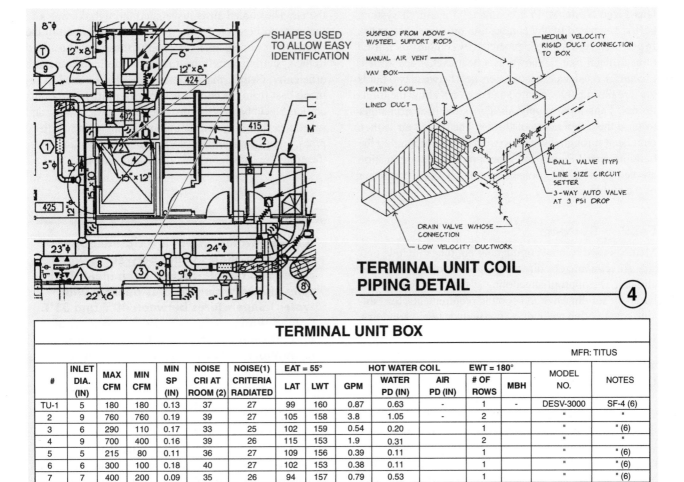

SHAPES USED
TO ALLOW EASY
IDENTIFICATION

SUSPEND FROM ABOVE
W/STEEL SUPPORT RODS

MANUAL AIR VENT

VAV BOX

HEATING COIL

LINED DUCT

MEDIUM VELOCITY
RIGID DUCT CONNECTION
TO BOX

BALL VALVE (TYP)

LINE SIZE CIRCUIT
SETTER

3-WAY AUTO VALVE
AT 3 PSI DROP

DRAIN VALVE W/HOSE
CONNECTION

LOW VELOCITY DUCTWORK

TERMINAL UNIT COIL PIPING DETAIL ④

TERMINAL UNIT BOX

MFR: TITUS

#	INLET DIA. (IN)	MAX CFM	MIN CFM	MIN SP (IN)	NOISE CRI AT ROOM (2)	NOISE(1) CRITERIA RADIATED	EAT = 55°			HOT WATER COIL		EWT = 180°		MODEL NO.	NOTES
							LAT	LWT	GPM	WATER PD (IN)	AIR PD (IN)	# OF ROWS	MBH		
TU-1	5	180	180	0.13	37	27	99	160	0.87	0.63	-	1	-	DESV-3000	SF-4 (6)
2	9	760	760	0.19	39	27	105	158	3.8	1.05	-	2		"	"
3	6	290	110	0.17	33	25	102	159	0.54	0.20		1		"	" (6)
4	9	700	400	0.16	39	26	115	153	1.9	0.31		2		"	"
5	5	215	80	0.11	36	27	109	156	0.39	0.11		1		"	" (6)
6	6	300	100	0.18	40	27	102	153	0.38	0.11		1		"	" (6)
7	7	400	200	0.09	35	26	94	157	0.79	0.53		1		"	" (6)
8	8	750	375	0.23	41	27	117	153	1.7	0.25		2		"	"

Figure 6-17. Plan views on mechanical prints are cross-referenced to a terminal unit schedule. Terminal units transfer heat from hot water systems to air circulated through the terminal unit and ductwork.

Electric Heat Systems. Electric heating systems include in-line electric resistance heating elements, radiant heat panels, and electric baseboard heaters. In-line electric heating systems consist of electric resistance heating elements and a fan that circulates air across the heating elements installed within the ductwork. Heated air is then distributed to building areas through the ductwork. Ductwork for an in-line electric heating system is shown on mechanical prints, while connections for the heating elements are indicated on electrical prints.

Radiant heat panels use an electric resistance heating element embedded in areas such as a ceilings, walls, or floors. Radiant heat panels may also include ceramic panels. The heated panel radiates heat to the surrounding building area. Radiant heat panels provide heat to a localized area only.

Electric baseboard heaters use electric resistance heating elements to generate heat and are typically located along the base of the exterior walls of a building. Electric baseboard heaters use natural air circulation to move air throughout a building area. Information regarding radiant heat panels and electric baseboard heaters is included on electrical prints.

Heat Pumps. A heat pump system is a direct expansion refrigeration system that contains devices and controls that reverse the flow of refrigerant in a system. Reversal of the refrigerant flow enables a heat pump to transfer heat from the outdoors to produce a heating effect and transfer heat from the indoors to produce a cooling effect. Commercial heat pumps commonly use water as the heat source. Heat is transferred from the water to the air in the building areas.

Gas-Fired Systems. In a commercial gas-fired system, heat is generated by the combustion of natural gas or oil. In gas-fired systems, heat generated by combustion is used to heat metal plates, known as a heat exchanger. A *heat exchanger* is a device that transfers heat between two fluids that are physically separated and does not allow the fluids to mix. Fans blow supply air across the heat exchanger to raise the temperature of the air. The heated air is then circulated throughout the ductwork and distributed to the building areas. Fumes and other products of combustion are vented from the combustion chamber to the exterior of the building.

Cooling Systems

Methods used to cool commercial buildings include outside air economizers, direct expansion cooling, and water chillers. The appropriate cooling system for a building is based on building size and cooling requirements. Sustainable construction techniques also include tinted windows, sunshades, operable exterior windows, and interior air circulation systems that provide natural building cooling and reduce the loads on cooling systems.

Outside Air Economizers. An outside air economizer uses outside air to cool building areas. The use of cool outside air for cooling reduces the costs associated with cooling the building. An economizer cycle is an HVAC system cycle in which building areas are cooled by outside air only. However, the use of outside air for cooling is only allowed when the air is at an appropriate temperature and humidity. When the outside air is too warm, too cold, or too humid, the source of the outside air is closed and the mechanical cooling system is activated.

Direct Expansion Cooling. Some commercial air conditioning systems, such as rooftop units or heat pumps, use direct expansion cooling. In a direct expansion cooling system, refrigerant is vaporized in the evaporator coils by heat from the surrounding warm air. The compressor pressurizes the refrigerant and discharges the vapor into a condenser coil. The refrigerant returns to a liquid state and is circulated through an expansion valve. At this point, the pressure and evaporation temperature are reduced. Air is passed over the evaporator coils to cool the air before being distributed to areas to be cooled. Direct expansion cooling systems are commonly used in small- to medium-capacity commercial cooling systems.

Water Chillers. A *water chiller* is a device that cools water and is commonly used in large commercial buildings requiring large quantities of cooled water for cooling purposes. The cooled water is circulated through terminal units located within the building or in coils in an air handling unit. As the cooled water passes through the terminal units or coils, air is circulated through the unit to cool the air. The cooled air is distributed through ductwork to the areas to be cooled. The water is then returned to the chiller to be cooled again. A cooling tower may be added to the system to remove heat from the chilled water more efficiently. Geothermal systems may also be used to remove heat from the chilled water prior to its return to the chiller. Mechanical prints provide plan views, details, and schedules for all system components, including chillers, pipes, pumps, and terminal units. Chiller and cooling tower locations are shown on plan views and site plans.

TECH TIP

A chilled-water system is a closed-loop system that circulates water between a water chiller and remote-cooling equipment. Typically, chilled-water systems operate with water temperatures between 40°F and 55°F. Chilled-water systems provide economical and efficient cooling for a large building or multiple buildings served by a central chilled water plant.

Ventilation Systems

Heated and cooled air is circulated by fans and air handling units through various ductwork configurations, suspended ceilings, or raised flooring systems. Ductwork provides heated and cooled air to diffusers to distribute the conditioned air to building areas. Air is then removed from these areas and returned for reheating, recooling, or exhausting.

Air Handling Units. An *air handling unit,* or air handler, is a prefabricated air distribution assembly that uses fans, ductwork, heating and cooling coils, humidifiers, dehumidifiers, and controls to condition and distribute air throughout a building. In an air handling unit, outside air enters through an inlet and a damper that controls the amount of outside air entering the air handling unit. Return air from the building areas is introduced into another duct. The outside air and return air are then mixed and circulated past heating and cooling coils, humidifiers, and/or dehumidifiers inside the air handling unit. The conditioned air is distributed to the ductwork and building areas.

Mechanical prints provide information about air handling unit and fan requirements on schedules that are cross-referenced to plan views and elevations on mechanical prints. Air handling units and fans are specified by the overall diameter of the fan blades, motor size and speed (in HP and rpm), and the amount of air moved by the fan (in cfm). **See Figure 6-18.**

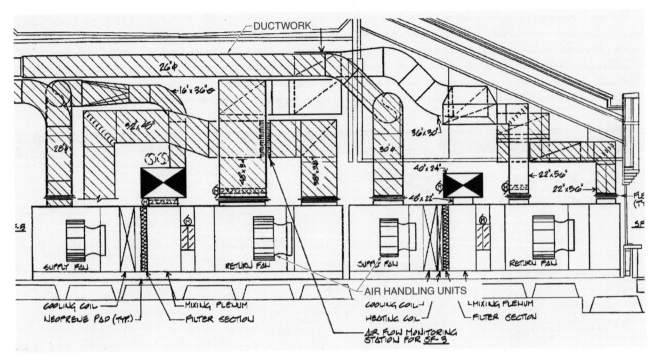

Figure 6-18. Air handling units convey a large volume of air through commercial building ductwork.

Ductwork. Flexible or rigid ductwork is used to circulate heated or cooled air from a source to the building areas. Flexible ductwork makes installation easier, especially when space is limited, and absorbs pulsations and vibrations from mechanical equipment. Flexible ductwork is commonly used for final connections in commercial buildings and is cut to length at the job site.

Galvanized sheet metal or ductboard is commonly used for rigid ductwork. Sheet metal ductwork is typically prefabricated at a shop and transported to the job site for installation. Ductboard ductwork may be fabricated at a shop or on the job site. Ductwork may be insulated to minimize heat loss. Symbols on mechanical prints indicate the type of material used and the shape of the ductwork. **See Figure 6-19.**

For ventilation systems, ductwork is installed to exhaust air from the building to the exterior or to areas within the building for conditioning and recycling. For food preparation areas or manufacturing operations, large quantities of air may need to be removed quickly and replaced by makeup air systems.

Ductwork designs are shown on mechanical prints with plan views and elevations. Ductwork diameters are indicated for round ducts. Width and height are shown for rectangular or square ducts. Methods for attaching ductwork to structural members are not shown on plan views and elevations of mechanical prints. Instead, attachment information is included in the specifications or general notes.

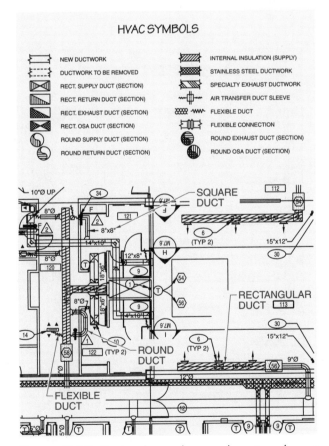

Figure 6-19. Ductwork may be round, rectangular, or square and is fabricated from various materials.

Different types and shapes of air diffusers and grilles are installed in an HVAC system. **See Figure 6-20.** An *air diffuser* is an air distribution outlet used to deflect and mix air and is commonly fitted with vanes or louvers. Standard types and sizes of air diffusers are available from a variety of manufacturers. Each diffuser is identified by type and size using symbols on plan views or schedules on mechanical prints. In many heating and cooling systems, return air is removed from conditioned building areas and returned to the source for reheating, recooling, or reconditioning. A *grille* is a decorative perforated or louvered cover installed at the inlet to the return air ductwork or in a ceiling or floor panel.

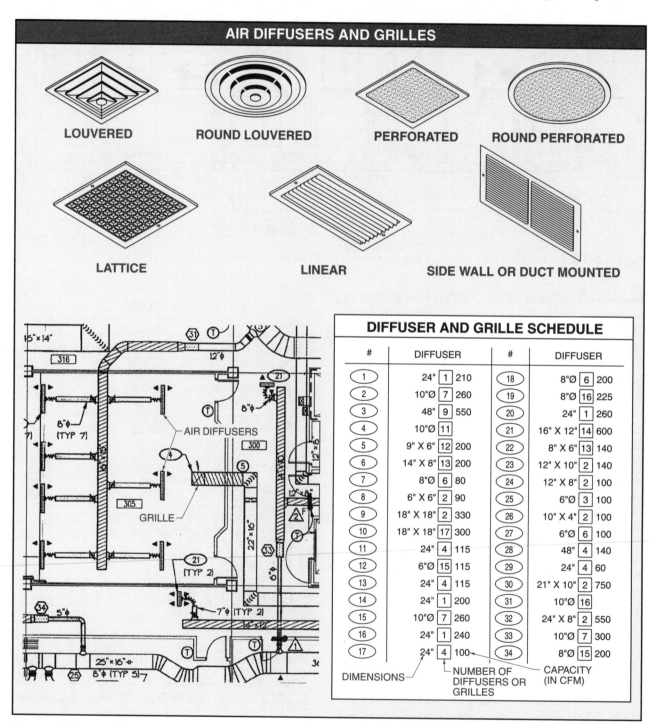

Figure 6-20. Air diffusers and grilles are used at the inlets and outlets of ductwork. Air diffusers and grilles are indicated on plan views and cross-referenced to schedules.

Louvers and vents are installed where ventilation ductwork passes through exterior walls or roofs. Louver or vent size and type are indicated on schedules or shown on elevations and details. Ventilation hoods for collecting fumes or smoke inside buildings are shown with details. Fans and air handling units are installed in the ventilation ductwork to ensure air is moving in the proper direction at the required velocity.

Humidification and Dehumidification Systems. Commercial HVAC systems must be able to regulate the amount of moisture in the treated air. *Humidification* is the process of adding moisture to the air. Humidification is required in cold climates where continuous heating of indoor air causes the humidity level to decrease. Low humidity levels can cause static-electric buildup and discomfort to the building inhabitants. In some situations, proper humidity levels must be maintained to ensure product quality. Moist air is provided to commercial building areas by introducing steam or water into the supply air ductwork through a variety of methods. Information concerning humidification systems is shown on mechanical prints and equipment schedules.

Dehumidification is the process of removing moisture from the air. Dehumidification may be required for building inhabitant comfort or to ensure proper humidity levels in manufacturing environments. Passive or desiccant dehumidification may be used to regulate the amount of moisture in the air in various building areas. In passive dehumidification systems, moisture is removed from the air using the existing cooling coils of the system. As air is cooled, moisture condenses out of the air and decreases the humidity level. In desiccant dehumidification systems, moisture is removed from the air when the air contacts a desiccant material that absorbs moisture. Desiccant dehumidification systems are common in commercial HVAC systems, in manufacturing operations requiring low humidity levels, and in pneumatic systems to prevent corrosion. Information concerning dehumidification equipment is shown on mechanical prints and schedules.

Filtration Systems. An air-filtration system is typically included as a component of an HVAC system. Filtration is provided by screens and filters through which air flows but particles do not. Electrostatic filters remove particles and clean the air as the air passes through electrically charged plates and collector cells. Locations for air-filtration systems within an HVAC system are shown on detail ductwork drawings, and the equipment is listed on mechanical schedules.

ELECTRICAL SYSTEMS

Electrical prints contain information about the wiring, electrical equipment, and electrical finish materials to be installed. **See Figure 6-21.** As with mechanical prints,

specific dimensions for electrical wiring, equipment, and finish materials are not provided on electrical prints. Rather, dimensions are obtained from plan views and elevations on architectural and structural prints. Some electrical information may be included on an electrical site plan, which indicates power sources and exterior lighting. In some instances, additional electrical prints are provided specifically for low-voltage systems, such as security and communication systems.

Figure 6-21. Specific dimensions for location of electrical wiring, equipment, and fixture installation are indicated on plan views and elevations on architectural and structural prints.

Wiring

Electrical wiring is located underground, placed in conduit, left as exposed cable, and installed in walls, above ceilings, and under floors. Electrical loads vary from low-voltage loads for items such as thermostats to high-voltage loads for welding equipment or heavy manufacturing machinery. Each wiring condition is shown on the electrical prints.

Electrical prints may be divided into separate sections for lighting, power supply, and signals such as fire alarms and smoke detectors. Plan views for each application are cross-referenced to ensure all electrical needs are met.

Large standby power generators may be used to supplement the main power supplied to a building in case the main power is taken offline.

A *conductor* is a wire used to control the flow of electrons in an electrical circuit. Conductor size is designated by an American Wire Gauge (AWG) number that indicates the conductor diameter. The higher the gauge number, the smaller the conductor diameter. Where extremely large electrical loads are carried, the load placed on a conductor may be indicated to ensure proper sizing of the conductor. Conductors can be bare, covered, or insulated. Insulation used to encase conductors includes moisture-resistant thermoplastic (TW), heat-resistant thermoplastic (THHN), moisture- and heat-resistant thermoplastic (THW), moisture-resistant thermoset (XHHW), and moisture-, heat-, and oil-resistant thermoplastic (MTW) insulation.

A *cable* is a flexible assembly of two or more conductors with a protective outer sheathing. Common cables include armored cable (AC), metal-clad cable (MC), nonmetallic-sheathed cable, and service-entrance cables.

The number of conductors in an electrical cable is noted on plan views by slash marks along the solid line that indicates the cable run. The number of slash marks is equal to the number of conductors. For example, three slashes along a solid line indicate a three-conductor cable. A cable without any slash marks indicates a two-conductor cable. **See Figure 6-22.**

The cable used and the conduit type or placement are indicated with symbols on plan views. *Conduit* is tube or pipe that supports and protects electrical conductors. Metallic and nonmetallic conduit, such as rigid metal conduit (RMC), rigid nonmetallic conduit (RNC), electrical metallic tubing (EMT), and flexible metal conduit (FMC), are common types of conduit used in commercial construction.

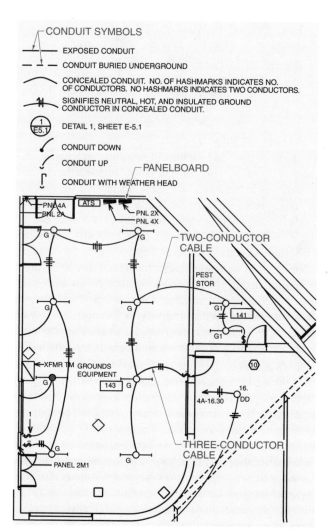

Figure 6-22. Electrical cables and conduit are indicated on electrical prints using electrical symbols. Slash marks on cable runs indicate the number of conductors.

For some situations, conduit must be installed before the placement of structural members such as cast-in-place concrete or masonry. Exact conduit locations are not indicated on plan views or electrical prints. Conduit is placed according to dimensions on architectural and structural prints. During construction, conduit ends are capped to prevent them from filling with concrete, mortar, or other debris. Conduit is connected together using various fittings. Conduit is bent on the job site as required for installation. Upon completion of concrete and masonry work, cables are pulled through the conduit.

Large-diameter cables, such as service-entrance cables that carry electrical service from the main power source to panelboards, are specified by the number of conductors, wire gauge, number and gauge of grounding conductor, and overall cable or conduit diameter. For example, a notation of 4#4, 1#10 GRD, 1¼"C indicates a cable containing four #4 conductors, one #10 grounding conductor, and an overall cable or conduit diameter of 1¼". **See Figure 6-23.**

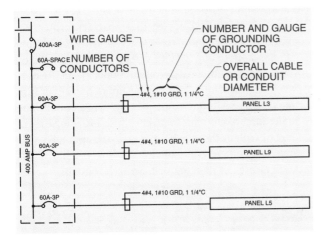

Figure 6-23. Large-diameter cables are specified by the number of conductors, wire gauge, number and gauge of the grounding conductor, and overall cable or conduit diameter.

TECH TIP

The National Electric Code® was first created in 1897. In 1911, the National Electric Code® was transferred to the Electrical Committee of the National Fire Protection Association (NFPA). Since 1959, the NEC® has been updated every three years.

Supports. Electrical cables are fastened to structural members, suspended behind walls and ceilings, placed in cable trays, buried underground, or installed in cast-in-place concrete floors. Clamps are used to fasten conduit and cables to structural steel members. Sections included on electrical prints provide information concerning cable trays and sleeves for electrical cable and conduit. **See Figure 6-24.**

Cable tray locations are shown on plan views. A *cable tray* is an assembly of sections and associated fittings that form a rigid structural system used to support cables and raceways. Support and fastening information for cables installed in walls is not commonly provided on electrical prints. The National Electrical Code® (NEC®) has established requirements for safely and properly fastening cable and conduit to structural members. Per the NEC®, cables and conduit must be secured to wall studs or structural members within a certain distance of switches, receptacles, or other electrical fixtures.

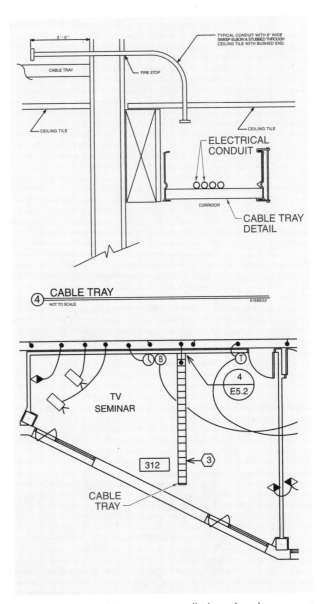

Figure 6-24. Cable trays are installed overhead to support electrical cables and raceways.

Equipment

Electrical prints indicate the general location of transformers, panelboards, junction boxes, fuses, circuit breakers, busways, and switches. Schematic drawings indicate the various electrical loads, circuits, and demands on an electrical system. **See Figure 6-25.**

Transformers are commonly specified for use in commercial buildings. A *transformer* is an electrical device that contains no moving parts and is used to increase or decreases voltage and current ratings of an alternating current (AC) circuit. Transformers are sized by the number of kilovolt-amperes (kVA) they can handle.

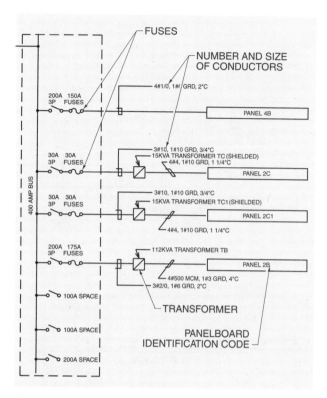

FUSES

NUMBER AND SIZE OF CONDUCTORS

4#1/0, 1#f GRD, 2"C

200A 150A
3P FUSES

PANEL 4B

30A 30A
3P FUSES

3#10, 1#10 GRD, 3/4"C
15KVA TRANSFORMER TC (SHIELDED)
4#4, 1#10 GRD, 1 1/4"C

PANEL 2C

30A 30A
3P FUSES

3#10, 1#10 GRD, 3/4"C
15KVA TRANSFORMER TC1 (SHIELDED)

PANEL 2C1

4#4, 1#10 GRD, 1 1/4"C

200A 175A
3P FUSES

112KVA TRANSFORMER TB

PANEL 2B

4#500 MCM, 1#3 GRD, 4"C
3#2/0, 1#6 GRD, 2"C

TRANSFORMER

100A SPACE

100A SPACE

PANELBOARD IDENTIFICATION CODE

200A SPACE

400 AMP BUS

Figure 6-25. Schematic drawings of power systems indicate cable types, conductor sizes, transformers, and fuses or circuit breakers.

Circuit breakers and fuses open a circuit when an overload condition or short circuit occurs. A *circuit breaker* is an overcurrent protection device with a mechanism that may manually or automatically open a circuit when an overload condition or short circuit occurs. A *fuse* is an overcurrent protection device that provides protection from short circuits and overloads. Circuit breakers and fuses include both current and voltage ratings. The listed current rating (in amperes) of circuit breakers and fuses is the maximum amount of current the circuit breaker or fuse can safely carry without tripping or blowing a circuit. The voltage rating of a circuit breaker or fuse is the maximum amount of voltage that can be applied to the circuit breaker or fuse. The voltage rating of circuit breakers and fuses is equal to or greater than the voltage in the circuit.

TECH TIP

Conductors are sized using the American Wire Gauge (AWG) numbering system. The smaller the AWG number, the larger the diameter of the conductor and the greater the capacity. For example, a #8 AWG conductor is larger and can handle more current than a #12 AWG conductor.

Panelboards. A *panelboard* is a single frame or box, or a group of frames and boxes, with buses and overcurrent protection devices, such as circuit breakers, that may have switches to control light, heat, or power circuits. Panelboards are placed in a cabinet or cutout box that is accessible only from the front. Locations of panelboards are shown on plan views or electrical prints. The electrical plan views relate to the overall electrical service schematic and details. **See Figure 6-26.** Panelboards are represented by an architectural symbol and may have a letter or number code that is keyed to a schedule of electrical equipment. A panelboard schedule indicates the size of the panelboard, mounting directions, and equipment to be connected to each circuit.

Finishes

After electrical panelboards, conduit, and cables are installed, connections are made to finish electrical equipment, including luminaires, switches, receptacles, communication systems, and alarm systems. Care must be taken when wiring a large panelboard with many circuits, finish fixtures, and luminaires to ensure proper circuits are provided for the finish fixtures and luminaires.

Per Article 408.7 of the NEC®, all unused openings in a panelboard for circuit breakers or switches must be covered using approved closures.

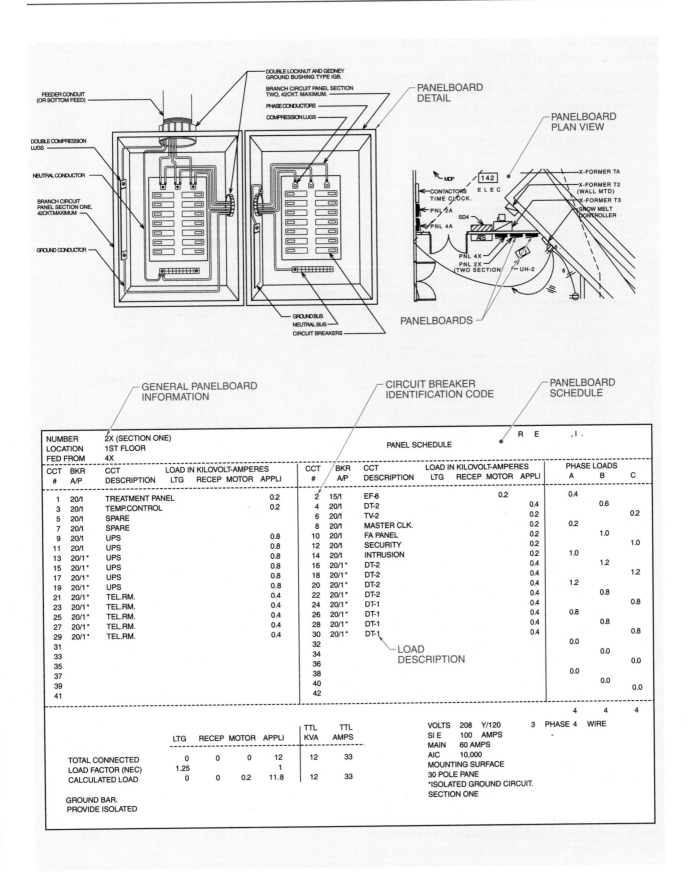

Figure 6-26. Panelboard information is provided on details, plan views, and schedules.

Lighting. Separate plan views and schedules are provided for luminaire installation. **See Figure 6-27.** Symbols or abbreviations indicate the type of luminaire to be installed at each location. At the end of each lighting circuit, a notation indicates the panelboard and circuit numbers for connection to the power source. Types and locations of switches are also shown on electrical plan views.

of receptacle at each location. As on lighting plans, the panelboard number and circuit numbers are indicated at the end of each electrical run.

LUMINAIRE SCHEDULE

TYPE 'A' RECESSED MOUNTING 2' X 4' FLUORESCENT FIXTURE WITH A 3" DEEP, 18 CELL PARABOLIC LOUVER. PROVIDE THREE F40T12 ENERGY SAVING LAMPS PER FIXTURE. LITHONIA 2PM3-GH-340-18-S-277-ES-GLR, DAYBRITE, COLUMBIA, METALUX OR APPROVED SUBSTITUTION.

TYPE 'A1' SAME AS TYPE 'A', EXCEPT WITH PLASTER FRAME FOR RECESSED MOUNTING IN PLASTER CEILING. LITHONIA 2PM3-FH-340-18-S-277-ES-GLR, DAYBRITE, COLUMBIA, METALUX OR APPROVED SUBSTITUTION.

SAME AS TYPE 'A', EXCEPT TWO LAMP FIXTURE WITH 12 CELL PARABOLIC LOUVER. PROVIDE TWO F40T12 ENERGY SAVING LAMPS. LITHONIA 2PM3-GH-240-12-5-277-ES-GLR, DAYBRITE, COLUMBIA, METALUX OR APPROVED SUBSTITUTION.

RECESSED MOUNTING 2' X 4' FLUORESCENT FIXTURE WITH REGRESSED FRAME AND ACRYLIC LENS. PROVIDE TWO F40T12 ENERGY SAVING LAMPS PER FIXTURE. LITHONIA 2SPGH-240-RN-A12.125-277-ES-GLR, DAYBRITE, COLUMBIA, METALUX OR APPROVED SUBSTITUTION.

TYPE 'G' CHAIN HUNG SURFACE MOUNTING TWO LAMP OPEN STRIP FIXTURE WITH WIRE GUARD. PROVIDE TWO F40T12 SUPER SAVER LAMPS PER FIXTURE. LITHONIA C240, DAYBRITE, KEYSTONE, METALUX OR APPROVED SUBSTITUTION.

TYPE 'G1' SAME AS TYPE 'G', EXCEPT WITH COLD WEATHER BALLAST.

TYPE 'H' ROUND RECESSED COMPACT FLUORESCENT DOWNLIGHT WITH CLEAR ALZAK REFLECTOR. PROVIDE TWO 13 WATT PL LAMPS PER FIXTURE. PRESCOLITE CFR813-572-HPF, EDISON PRICE, HALO, CAPRI OR APPROVED SUBSTITUTION.

SYMBOLS AND ABBREVIATIONS INDICATE LUMINAIRE TYPE

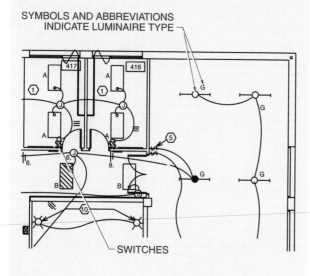

SWITCHES

Figure 6-27. Notations on electrical prints and schedules provide detailed information regarding luminaires.

Power. Symbols indicate fan connections, pump connections, motor controls, standard receptacles, switches, and special receptacles for heavy equipment. **See Figure 6-28.** Electrical prints show receptacle locations and the type

ELECTRICAL SYMBOLS

$	SINGLE POLE SWITCH – P DENOTES PILOT LIGHT
$₃	THREE-WAY SWITCH
$₄	FOUR-WAY SWITCH
$ᴅ	DIMMER SWITCH
$ᴋ	KEY OPERATED SWITCH
⊕	DUPLEX RECEPTACLE
⊕	DUPLEX RECEPTACLE W/GFI
⊕	DUPLEX RECEPTACLE – WEATHERPROOF, GFI
⊕	DOUBLE DUPLEX RECEPTACLE
▲	SPECIAL RECEPTACLE
▥	DUPLEX RECEPTACLE – FLOOR MTD.
▨	SPECIAL RECEPTACLE – FLOOR MTD.
⊙	CLOCK OUTLET
Ⓑ	BELL
◯	MOTOR CONNECTION
Ⓙ	JUNCTION BOX
⊡	PUSHBUTTON

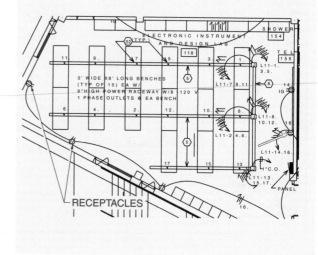

RECEPTACLES

Figure 6-28. Electrical prints indicate the general locations of receptacles and the type of receptacle at each location.

A schedule of all electrically powered equipment and power requirements may be included in the electrical prints indicating each piece of equipment, the amount of power required to operate the equipment, and circuit assignments. **See Figure 6-29.** For example, item CCHWP-1 is a 3 horsepower (HP) pump connected to a 480 volt (V), 3ϕ power supply containing a 3-pole, 30 amp (A) safety switch that is protected by an 8 amp (A) fuse.

MECHANICAL EQUIPMENT CIRCUIT SCHEDULE

ITEM	NAME	VOLT/PHASE	HP/AMP	SAFETY SWITCH	FUSE	CIRCUIT	CIRCUIT No.	MAG. STARTER
AC-1	AIR COMP.	480-3φ	40 HP	3P-100A	80 A	3#4, 1 1/4"C	4M-38.40.42.	DIV.15
	AIR DRYER	480-3φ	40 HP	3P-30A	50 A	3#6, 1"C	4M-43.45.47.	SIZE 1, FVNR
AS-1	AIR SHOWER	460-1φ	2 HP	3P-30A	5.6A	3#12, 3/4"C	4M-32.34.36.	SIZE 0
B1	BOILER	480-3φ	3 HP	3P-30 A	8 A	3#12, 3/4"C	4M-1.3.5	DIV 15
B2	BOILER	480-3φ	3 HP	3P-30 A	8 A	3#12, 3/4"C	4M-7.9.11	DIV 15
BCP-1	CIRC PUMP	480-3φ	3 HP	3P-30 A *	8 A	3#12, 3/4"C	MCC-1A	SIZE 1, FVNR
BCP-2	CIRC PUMP	480-3φ	3 HP	3P-30 A *	8 A	3#12, 3/4"C	MCC-1F	SIZE 1, FVNR
CAB-1	CABINET HEATER	120-1φ	1/60 HP	DIV.15	N/A	2#12, 3/4"C	2M-9.	N/A
CAB-2	CABINET HEATER	120-1φ	1/60 HP	DIV.15	N/A	2#12, 3/4"C	2M-9.	N/A
CAB-3	CABINET HEATER	480-3φ	1/60 HP	3P-30A	N/A	2#12, 3/4"C	4B-22.24.26.	N/A
CAB-4	CABINET HEATER	480-3φ	1/60 HP	3P-30A	N/A	2#12, 3/4"C	4B-28.30.32.	N/A
CCHWP-1	PUMP	480-3φ	3 HP	3P-30 A *	8 A	3#12, 3/4"C	MCC-1B	SIZE 1, FVNR
CCHWP-2	PUMP	480-3φ	1 HP	3P-30 A	2.8 A	3#12, 3/4"C	MCC-1B	SIZE 1, FVNR
CCHWP-3	PUMP	480-3φ	1.5 HP	3P-30 A *	4 A	3#12, 3/4"C	MCC-2A	SIZE 1, FVNR
CCHWP-4	PUMP	480-3φ	1.5 HP	3P-30 A *	4 A	3#12, 3/4"C	MCC-2A	SIZE 1, FVNR
CCHWP-5	PUMP	480-3φ	3/4 HP	3P-30 A *	2.25 A	3#12, 3/4"C	MCC-1A	SIZE 1, FVNR
CCHWP-6	PUMP	120-1φ	1/3 HP	1P-30 A	12 A	2#12, 3/4"C	2M-2	SIZE 00, FVNR
CH-1	CHILLER	480-3φ	309 MCA	3P-400 A	400 A	SEE RISER	M.D.P	DIV.15
CH-2	FUTURE CHILLER	480-3φ	309 MCA			SEE RISER	MDP	
CHP-1	PUMP	480-3φ	10 HP	3P-30 A *	20 A	3#12, 3/4"C	MCC-1F	SIZE 1, FVNR
CHP-2	PUMP	480-3φ	10 HP	-	-	3/4"C ONLY	MCC-1B	
CHWP-1	PUMP	480-3φ	3 HP	3P-30 A *	8 A	3#12, 3/4"C	MCC-1B	DIV 16
CHWP-2	PUMP	480-3φ	3/4 HP	3P-30 A *	2.25 A	3#12, 3/4"C	MCC-2A	DIV 16
CHWP-3	PUMP	480-3φ	3/4 HP	3P-30 A *	2.25 A	3#12, 3/4"C	MCC-1A	DIV 16
CHWP-4	PUMP	120-1φ	1/2 HP	3P-30 A *	15A	2#12, 3/4"C	MCC-1E	SIZE 1, FVNR
CT-1	COOLING TOWER	480-3φ	15 HP	3P-60 A	30 A	3#10, 3/4"C	4M-13,15,17	DIV.15
		480-3φ	10 KW	3P-30 A	N/A	3#12, 3/4"C	4M-20,22,24	N/A
CT-2	COOLING TOWER	480-3φ	15 HP			2) 3/4"CO.		
CU-1	CONDENSER	208-1φ	14.FLA	3P-30A	20A	3#10, 3/4"C	2 X 1-2.4.	SIZE 1, FVNR
	PAINT BOOTH	120-1φ	15A	N/A	N/A	2#12, 3/4"C	2A-30.	
	PAINT BOOTH	480-3φ	3/4 HP	3P-30A	225A	2#12, 3/4"C	4A-23.25.27.	DIV.15
CWP-1	PUMP	480-3φ	20 HP	3P-60 A	40 A	3#8, 1"C	4M-2,4,6	DIV.15
CWP-2	PUMP	480-1φ	20 HP			1"C ONLY	4M	N/A

Figure 6-29. Schedules for electrically powered equipment provide details about each piece of equipment, including switches, fuses, and circuits.

Low-Voltage Systems. Low-voltage power systems are comprised of a step-down transformer, wiring, and electrical fixtures. A *step-down transformer* is an electrical power regulating device with more windings in the primary winding than the secondary winding, resulting in a load voltage that is less than the initial applied voltage. Low-voltage circuits are common for temperature control, communications, security, and voice/data/video (VDV) systems. Light-gauge conductors are used to wire low-voltage circuits. For many commercial buildings, additional electrical prints and specifications are provided for low-voltage systems.

Locations for thermostats that control HVAC systems are shown on plan views of electrical prints. **See Figure 6-30.** Connections from the thermostat to the heating or cooling equipment to be controlled are shown on plan views.

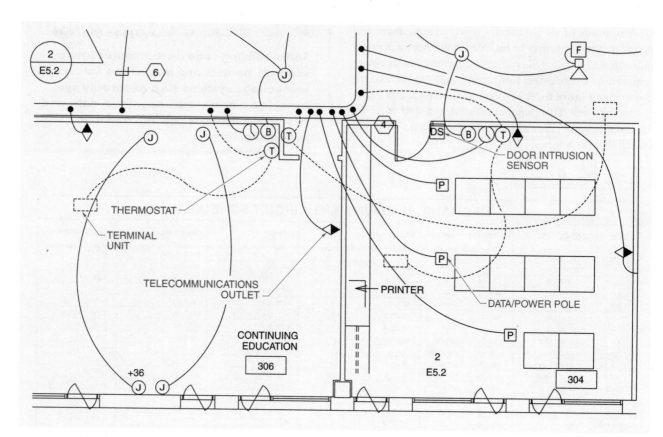

Figure 6-30. Plan views of low-voltage systems indicate the general locations for thermostats, telecommunications outlets, and other low-voltage equipment.

Symbols and abbreviations included on floor plans, electrical prints, or low-voltage plans indicate equipment locations for electrical systems, including television monitors, security cameras, computer equipment, speakers, microphones, satellite dishes, and intercom systems. Configurations of wiring for security and alarm systems for smoke detection, heat detection, air monitoring, and visual security are also shown on electrical prints. In some situations, elevations may provide additional information related to fixture placement.

Fire alarm and security systems are typically tested by the local governmental authorizing agency prior to building occupancy. For example, various types of nonstaining smoke may be released inside a building to ensure that all fire alarm systems are operable prior to issuing an occupancy permit.

Refer to the CD-ROM in the back of the book for Chapter 6 Quick Quiz® and related printreading and reference material.

Name _Paul Sellar_   Date _____

Refer to Riverpoint project (Sheets 17 to 24). Cross-referencing between several sheets may be required to answer questions.

(T)   F   ✗ **1.** Natural gas service is provided to the hot water storage tank in Room 004.

_____   **2.** Immediately after entering the building and passing through the main shut-off valve, the size of the main water supply pipe is ___".
- A. 3
- B. 4
- C. 5
- D. 6

____C____   **3.** The 4" diameter pipe that is parallel with grid line G inside Room 018 is a ___.
- A. gas service pipe
- B. roof drain line
- C. vent pipe
- D. waste stack

T   F   **4.** Elongated sleeves are installed where plumbing pipes penetrate waffle slab stems.

(T)   F   **5.** Air handler 3 has an outside air damper size of 126" × 36".

_____   **6.** The vertical vent pipes for the boilers are ___" in diameter.
- A. 12
- B. 18
- C. 20
- D. 24

(T)   F   **7.** Sprinkler pipe is installed approximately 8" to 12" above the acoustical ceiling.

T   F   **8.** The contractor installing the sprinkler system is responsible for all pipe connected to the system including underground work.

____6"____   **9.** The pipe diameter for condenser supply and return pipes directly at the chiller is ___".

_____   **10.** The electrical chase is located ___.
- A. just east of air handler 3
- B. near the cooling tower
- C. just north of the elevators
- D. in Electrical Room 006

N-M10.2   **11.** Details for gas meter installation and connections are on Sheet ___ of the mechanical prints.

_____ **12.** The size of the gas feed line down to the boiler on the basement floor plan is ___".
  A. 3
  B. 4
  C. 5
  D. 6

_____ **13.** The size for hot water supply and hot water return pipes for air handler 3 is ___" diameter.

_____ **14.** Note P7 in Room 012 refers to ___.
  A. a 7" pipe vent
  B. pipe 7
  C. the plumbing fixture schedule
  D. Sheet 7 for additional details

T   F   **15.** The main water service enters the basement through the floor in Room 018.

T   F   **16.** Exhaust fan 6 is served by the same panelboard as the temperature control panels in Room 018.

T   F   **17.** Fire alarm system power is provided by panelboard P2-1F4-1.

T   F   **18.** Water heater detail 1 notes a pressure-relief valve discharge to a 2" floor drain.

T   F   **19.** Exhaust fan 2, which is south of the chiller in Room 007, intakes air from an 18" square grille.

T   F   **20.** All sprinkler heads on the basement level are chrome.

_____ **21.** The 48" × 104" vertical duct near grid line intersection H5 is connected to ___.
  A. outside air intake to air handler 3
  B. outside air intake to air handler 3A
  C. supply air from air handler 3
  D. supply air from air handler 3A

T   F   **22.** Water supply pipes below the basement floor slab are PVC.

_____ **23.** The diameter of the vent pipe for the sinks in Room 010 is ___".

T   F   **24.** All cleanouts in the unexcavated area of the basement are 4" in diameter.

_____ **25.** Showerheads in Rooms 009 and 010 are fed with a(n) ___" diameter mixing pipe.

T   F   **26.** A water heater is located in Room 012.

_____ **27.** A 14" × 8" air grille accepts 300 cfm and is shown in the southwest corner of Room 003 to ___.
  A. supply air from air handler 3
  B. supply air from air handler 4
  C. exhaust air through air handler 3
  D. exhaust air through air handler 4

_____ **28.** The circles containing a "J" on the east side of Mechanical Room 018 represent ___.

_____ **29.** The 62" × 48" vertical duct between grid lines 4 and 5 and near grid line H is connected to ___.
       A. supply air from air handler 3
       B. supply air from air handler 3A
       C. exhaust air through air handler 3
       D. exhaust air through air handler 3A

_____ **30.** The minimum depth of the main service ductbank from the transformer to the main control closet is ___" below finished grade.

T   F   **31.** Two check valves are to be installed on the water supply for water heater #1.

T   F   **32.** The three sinks in Room 010 are supplied by ½" cold and hot water lines.

_____ **33.** The minimum distance between conduits in the main ductbank section is ___".

_____ **34.** Unit heater 6 is ___.
       A. a gas-fired heat exchanger type furnace
       B. an electric heat exchanger type furnace
       C. connected to the boilers
       D. supplied by air handler 3

_____ **35.** Elevator 2 control panel is connected to circuit ___ in panelboard P2-BF2-3.

_____ **36.** Air separators are installed on ___.
       A. the supply side of the piping
       B. the return side of the piping
       C. both the supply and return sides of the piping
       D. none of the above

_____ **37.** The size of core-drilled holes or sleeve outside diameters through the concrete wall near grid line intersection C2 is ___".

T   F   **38.** The gas service line entering the building is 6" above grade.

_____ **39.** The slope of roof drain lines is ___%.

T   F   **40.** Access for the telephone lines is concealed under the floor slab.

T   F   **41.** The communications panel is located in Room 017.

_____ **42.** The abbreviation "AFF" on basement foundation plan Note 29 indicates ___.

T   F   **43.** Water heater and boiler vent ducts are installed 12" to 18" high.

_____ **44.** The supply rate of the water flow valve into the water heater in Room 004 is ___ gpm.

T   F   **45.** The air flowing through the 24" square duct near grid line intersection H5 is moving in an upward direction.

_____ **46.** The manufacturer model number for the pipe sleeves installed near grid line intersection F4 is ___.

T   F   **47.** The overall size of air handler unit 3 is 18'-6" × 29'-6".

_____ **48.** The fire suppression system shown on the prints is a ___ system.
      A. wet pipe
      B. dry pipe
      C. preaction
      D. gaseous

_____ **49.** The flow rate for the main sprinkler water service is ___ gpm.

_____ **50.** There are a total of ___ sprinkler heads in the two restrooms on the basement level.

T   F   **51.** The Maintenance Office (Room 014) has four electrical receptacles.

_____ **52.** The condensate drain pipe on air handler 4 is ___" in diameter.
      A. 2
      B. 3
      C. 4
      D. 5

_____ **53.** Air handler 3A is connected to ___.
      A. cooling pipes only
      B. heating pipes only
      C. both heating and cooling pipes
      D. none of the above

T   F   **54.** A smoke detector and fire alarm pull station are provided in stairway 000A.

T   F   **55.** Master control closet BF2 contains the controls for both boilers.

T   F   **56.** The size of the vent pipes for the lavatories and urinals in Room 009 is 1½".

T   F   **57.** There are no manual pull station fire alarms on the roof level.

T   F   **58.** All receptacles in hallway 002 are powered from panelboard P2-BF2-4.

_____ **59.** The electrical power for air handler 3 in Room 018 is provided by panelboard number ___.
      A. MCC-BF2
      B. P2-BF2-4
      C. SF-3
      D. BLR-1

_____ **60.** Low-voltage cables serving the chiller have ___.
      A. a series of 8¾" diameter conductors
      B. 8 wires, each #14 gauge
      C. 3 cables, each 4" in diameter
      D. 14 wires, each #8 gauge

T   F   **61.** The emergency wall switch for the ventilation system is located on the outside of the entry to Room 007.

T   F   **62.** Heat recovery coils are installed in the outside air and exhaust ducts for air handler 3.

_____ **63.** Unit heater 1 in Room 015 has a(n) ___-conductor cable and a grounding conductor.
     A. exposed conduit with a two
     B. exposed conduit with a three
     C. concealed conduit with a two
     D. concealed conduit with a three

_____ **64.** The minimum distance between the top of the electrical busway support and the ceiling above is ___".

_____ **65.** The main ductbank ___.
     A. extends under the concrete slab at the north end of Room 018
     B. is in a cable tray through Corridor 002
     C. drops down from the first floor in Service Area 018
     D. location must be obtained from the site plan

_____ **66.** Electrical service from panelboard P2-BF2-2 serves the ___.
     A. alternate bid item #2
     B. cooling tower
     C. first-floor communications room
     D. first-floor electrical room

_____ **67.** A(n) ___ cable is used to ground the transformer and the main electrical room.

T   F   **68.** Pump 1 serves only boiler 1 and pump 2 serves only boiler 2.

T   F   **69.** Pump 5 serves the heat recovery piping system.

_____ **70.** The cooling tower solenoid valve is connected to circuit ___ in panel P2-BF2-3.

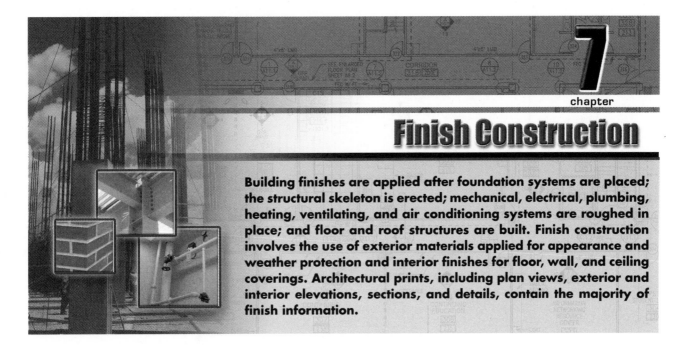

chapter 7

Finish Construction

Building finishes are applied after foundation systems are placed; the structural skeleton is erected; mechanical, electrical, plumbing, heating, ventilating, and air conditioning systems are roughed in place; and floor and roof structures are built. Finish construction involves the use of exterior materials applied for appearance and weather protection and interior finishes for floor, wall, and ceiling coverings. Architectural prints, including plan views, exterior and interior elevations, sections, and details, contain the majority of finish information.

EXTERIOR FINISH

Foundation systems are constructed and structural members are erected with consideration given to the application of exterior finishes. Architectural design may also indicate that structural members remain exposed in various areas on the interior or exterior of a structure. For example, it is common in bridge and road building that structural members are also the exterior finish members. Exterior finish members include bearing and non-load-bearing walls, roof coverings, and exposed structural members.

Exterior elevations show the final appearance of the exterior of a building using an orthographic projection (multiview) drawing. **See Figure 7-1.** Masonry, concrete, metal, glass, exterior insulation and finish systems (EIFS), and other exterior finishes are indicated on exterior elevations. Grid lines on exterior elevations are commonly used to relate the exterior finish materials to structural elements. Elevations are noted for floor levels and rooflines. Typical information is often referenced on details included with the prints.

TECH TIP

Exterior elevations are scaled views that show the shape and size of the roof and exterior walls of a building. Building materials, such as masonry, EIFS, or glass, are specified for the walls through the use of symbols and notations. Door and window openings are shown in their proper locations.

Walls

Weather protection, appearance, security, and thermal insulation are a few of the factors considered when selecting an exterior wall finish for a commercial structure. Commercial exterior wall finishes are commonly designed for low maintenance and high durability.

Commercial wall finish materials, including masonry, concrete, glass, metal, plaster, and EIFS, are used to create an attractive and durable wall finish. Wood siding, commonly used in residential construction, is uncommon on large commercial structures. Heavy-duty door and window systems are installed in commercial buildings because they can stand up to heavy use better, are more secure, and require less maintenance than light-duty door and window systems.

Exterior wall sections provide information about exterior finish materials. **See Figure 7-2.** Exterior finish materials are attached to wood, metal, masonry, structural steel, or concrete structural members.

Exterior gypsum board is a common sheathing material installed over wood or light-gauge metal framing. A variety of types of exterior gypsum board are available. Treated-core gypsum board has a moisture-resistant core and water-resistant surfaces. Treated-core gypsum board may remain exposed to the elements for a maximum of one month after application. Glass mat-faced gypsum board is a paperless gypsum panel with a water-resistant treated core that is surfaced with glass mat facings and a bond-enhancing primer. Glass mats covering the panels create an integrated unit that can withstand moisture exposure. Glass mat-faced gypsum board is commonly used as a substrate for EIFS.

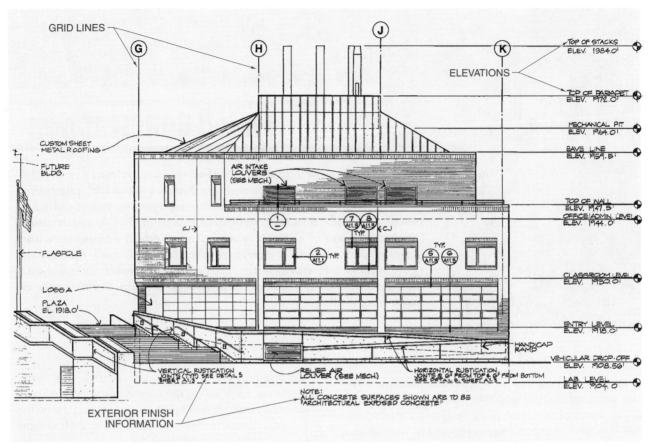

Figure 7-1. Each side of a commercial structure is shown with an exterior elevation that describes exterior finishes and final appearance.

Glass mat-faced gypsum board provides increased moisture protection and is often used under exterior cladding materials such as brick, EIFS, or stucco.

TECH TIP

Gypsum board finishes range from level 0, which does not require tape or finishing, to level 5, which includes a skim coat of joint compound over the entire surface.

Gypsum board may also be installed for interior load-bearing walls and interior partitions. Non-treated-core gypsum board is not intended for long-term exposure to the weather. Type X gypsum board is intended for fire-rated construction. Treated-core Type X gypsum board has various additives that provide fire resistance for use in fire-rated exterior wall assemblies. **See Figure 7-3.**

Masonry. Exterior masonry walls may be brick, concrete masonry units (CMUs), or manufactured or natural stone. Masonry walls are structural bearing walls or veneer facing installed around and between structural members. **See Figure 7-4.**

Masonry units for exterior finish are shown on exterior elevations and wall sections. Special brick, concrete, or stone units may be fabricated specifically for a particular job.

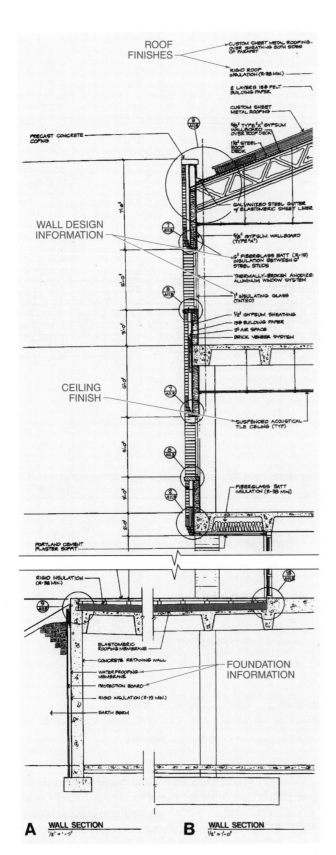

ROOF FINISHES — CUSTOM SHEET METAL ROOFING OVER SHEATHING BOTH SIDES OF PARAPET

RIGID ROOF INSULATION (R-38 MIN.)

2 LAYERS 15# FELT BUILDING PAPER

CUSTOM SHEET METAL ROOFING

PRECAST CONCRETE COPING

5/8" TYPE 'X' GYPSUM WALLBOARD OVER ROOF DECK

1 1/2" STEEL ROOF DECK

GALVANIZED STEEL GUTTER W/ ELASTOMERIC SHEET LINER

WALL DESIGN INFORMATION

5/8" GYPSUM WALLBOARD (TYPE 'X')

6" FIBERGLASS BATT (R-19) INSULATION BETWEEN 6" STEEL STUDS

THERMALLY-BROKEN ANODIZED ALUMINUM WINDOW SYSTEM

1" INSULATING GLASS (TINTED)

1/2" GYPSUM SHEATHING

15# BUILDING PAPER

3" AIR SPACE

BRICK VENEER SYSTEM

CEILING FINISH

SUSPENDED ACOUSTICAL TILE CEILING (TYP)

FIBERGLASS BATT INSULATION (R-38 MIN)

PORTLAND CEMENT PLASTER SOFFIT

RIGID INSULATION (R-38 MIN.)

ELASTOMERIC ROOFING MEMBRANE

CONCRETE RETAINING WALL

WATERPROOFING MEMBRANE

PROTECTION BOARD

RIGID INSULATION (R-19 MIN.)

EARTH BERM

FOUNDATION INFORMATION

A WALL SECTION 1/2" = 1'-0"

B WALL SECTION 1/2" = 1'-0"

Figure 7-2. Overall foundation, wall, floor, ceiling, and roof finishes are detailed on wall sections.

Figure 7-3. Treated-core Type X gypsum board may be used as exterior wall sheathing.

Figure 7-4. Masonry walls are laid as either structural supporting walls or as a veneer facing for structural members.

Where masonry units form the structural wall, the exterior finish, including brick bond, face finish of concrete masonry units, mortar color, lintels, and stone patterns and types, are indicated in the specifications and on elevations. Special brick shapes and coursing information are included on architectural elevations. **See Figure 7-5.**

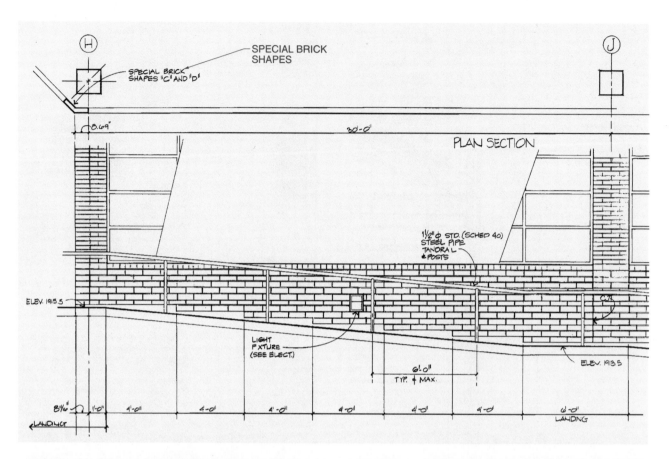

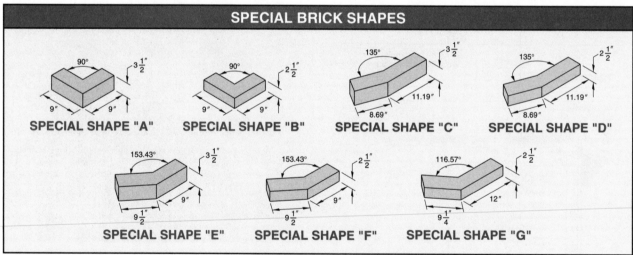

Figure 7-5. Special brick shapes and coursing information are included on the architectural elevations.

Locations and types of masonry control and expansion joints are shown on elevations and details. Wall sections indicate decorative effects and methods for attaching masonry veneer to structural members. Architects may require that a sample section of the masonry wall be laid as a test panel prior to application of masonry to provide a view of the finished product for the owner and architect.

Masonry veneer walls are commonly built in place. However, prefabricated masonry veneer panels may be constructed off site and set in place in a manner similar to precast concrete panels. The prefabricated panels have steel support frames and inserts for attachment to structural members.

Concrete. Architectural concrete is often used as an exterior finish on commercial buildings. *Architectural concrete* is concrete that is exposed to view and requires care in selecting, forming, placing, and finishing of the surface to meet appearance standards. Concrete surface finishing information is provided in the specifications or general notes on the architectural prints. Architectural concrete finishes are created by placing form liners inside the concrete forms prior to concrete placement, sandblasting the surface after concrete has set, or rubbing the surface after it has set to smooth it to the required finish. **See Figure 7-6.** Exterior elevations show exposed concrete wall finishes with architectural notes.

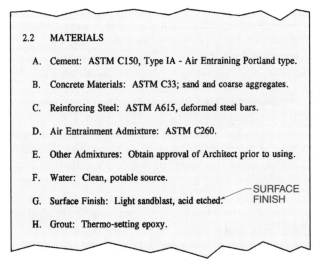

2.2   MATERIALS

A. Cement: ASTM C150, Type IA - Air Entraining Portland type.

B. Concrete Materials: ASTM C33; sand and coarse aggregates.

C. Reinforcing Steel: ASTM A615, deformed steel bars.

D. Air Entrainment Admixture: ASTM C260.

E. Other Admixtures: Obtain approval of Architect prior to using.

F. Water: Clean, potable source.

G. Surface Finish: Light sandblast, acid etched. ——— SURFACE FINISH

H. Grout: Thermo-setting epoxy.

Figure 7-6. Architectural concrete requires additional finishing to create the surface finish indicated by the architect.

Precast concrete panels may also be applied as an exterior finish material. Concrete panels are cast in a variety of designs with many surface finishes and transported to the job site. The panels are lifted into place with a crane and fastened to the structure by welding metal inserts cast into the panels or clips fastened to the structural members.

Glass. Exterior glass finishes are either glass panels set in metal or wood frames or glass block. *Glass block* is an opaque or transparent hollow block made of glass that is used in non-load-bearing walls and partitions. Glass block is set in mortar, similar to brick or concrete masonry units.

Areas where glass block is to be installed are shown on exterior elevations and architectural floor plans. Information concerning glass set in metal or wood frames is obtained from exterior elevations and detail sections. **See Figure 7-7.** Non-load-bearing metal or wood frames are attached to structural members per the manufacturer specifications. Frame and glass panel connections are shown on details.

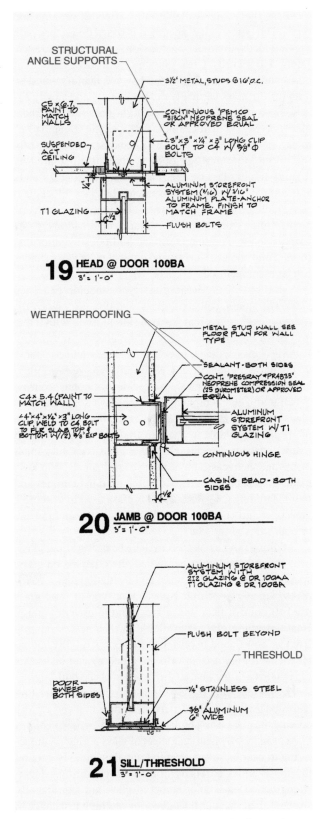

Figure 7-7. Details of head, jamb, and sill attachment and finish show framing members, sealants, frames, and glazing.

Glass sheets are available in a variety of designs and types. The type of glass specified is based on factors such as wind loads, thermal transmission, privacy, appearance, safety, and security. Clear glass includes sheet glass, float glass, and plate glass. *Sheet glass* is glass manufactured by drawing the glass vertically or horizontally and slowly annealing (cooling) it to produce a high-gloss surface. Sheet glass is available in three thicknesses: single strength (³⁄₃₂″ thick), double strength (⅛″ thick), and heavy sheet (from ³⁄₁₆″ to ⁷⁄₁₆″ thick). *Float glass* is glass manufactured by floating liquid glass on a surface of liquid tin and slowly annealing it to produce a transparent, flat glass. Float glass has excellent optical properties and is used in mirrors, architectural windows, and specialty applications. *Plate glass* is sheet glass that is ground and polished after it is formed and cooled. Curved and conical plate glass shapes can be manufactured.

Variations of the basic glass types include patterned glass, wire glass, cathedral glass, and obscure glass. *Patterned glass* is glass that has one side finished with a fine grid or an unpolished surface so the glass is translucent. Patterned glass is used in office partitions and for doors and windows when privacy is desired. *Wire glass* is glass embedded with wire mesh to provide additional security. When wire glass is broken, the wire mesh holds the glass pieces together. *Cathedral glass*, or stained glass, is a type of art glass available in a wide variety of colors and with many surface treatments. Cathedral glass is cut into small pieces and then reassembled using lead channels. Windows made from cathedral glass are often used in churches and public buildings. *Obscure glass* is glass used to obscure a view or create a design by sandblasting and/or etching one or both sides of the glass.

Additional types of glass include heat-absorbing glass, heat-strengthened glass, tempered glass, laminated glass, spandrel glass, security glass, and insulating glass. *Heat-absorbing glass*, or tinted glass, is a type of float glass used to control heat and glare in large areas of glass. Heat-absorbing glass is available in thicknesses ranging from ⅛″ to ½″, and in blue, bronze, green, or gray colors. Heat-absorbing glass absorbs solar energy and dissipates the majority of the heat to the exterior of the building.

Heat-strengthened glass is glass produced by reheating annealed float glass close to its softening point and rapidly cooling the glass using high-velocity air. *Tempered glass* is glass produced by heating a sheet of glass during manufacturing to near its softening point and then quickly cooling the glass under carefully controlled conditions. Heat-strengthened glass and tempered glass are significantly stronger than sheet or float glass, have a high resistance to thermal stress, and have high impact resistance.

Laminated glass is specialty glass produced by placing a clear sheet of polyvinyl butyral (PVB) between two sheets of glass and subjecting the composite to intense heat and pressure. Laminated glass is used in schools and other public buildings where safety is a concern.

Spandrel glass is tinted glass or glass with a polyvinyl fluoride coating. Heat-treated or laminated glass is used to manufacture spandrel glass.

Security glass is glass composed of multiple layers of polycarbonate plastic and/or glass bonded together under intense heat and pressure and coated with a PVB or polycarbonate film. Bullet- and burglar-resistant glass are two types of security glass.

TECH TIP

Self-cleaning glass contains a thin layer of titanium dioxide that allows the glass to break down dirt using UV rays from the sun.

Insulating glass is glass made of two pieces of sheet glass that are separated by a sealed air space. The edges are sealed on all four sides by a glass, metal, or plastic closure. A typical piece of insulating glass measuring ⁷⁄₁₆″ thick is composed of two sheets of double-strength glass (⅛″) with a ³⁄₁₆″ air space and a glass closure. Insulating glass windows may be coated or the air space may be filled to provide additional insulating properties. A *low-emittance (low-E) coating* is a metal or metallic oxide coating that reduces the passage of heat and ultraviolet rays through windows. Argon or krypton gas may be used to fill the air space to reduce the passage of heat from one piece of glass to the other.

Mastics, caulks, sealants, rubber or plastic seals, and metal or wood stops secure glass into frames. Most information regarding glass to be installed in a commercial building is included in the written specifications. Details may refer to a specific type of glass for an identified area. Exterior elevations and floor plans provide dimensional information for the glass panel sizes.

Frames for glass panels are installed and attached to the structural members per manufacturer specifications. Suction cups are attached to the face of the glass to lift panels into place. Sealants and frame mullions and stops are applied to secure the glass in the frame and waterproof the joint between the frame and glass.

Metal/Glass Curtain Walls. Metal and glass panels are both used for curtain walls on commercial buildings. A curtain wall is a non-load-bearing, prefabricated or job-built, glass or metal panel supported by metal frame members or set with various clip systems that are attached to structural members. Insulated or noninsulated metal panels and glass panels are set in frames and attached to structural members to form an integrated, weatherproof unit. **See Figure 7-8.**

Curtain Walls

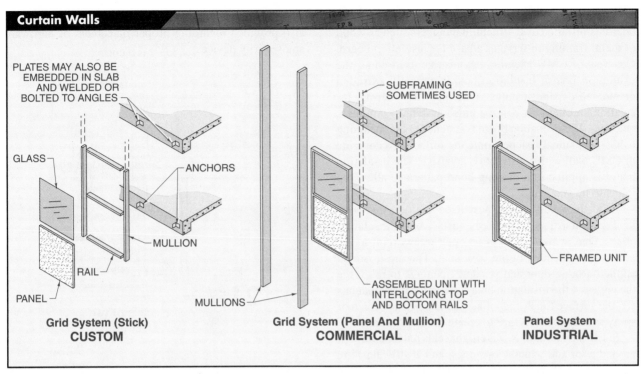

Figure 7-8. Curtain wall systems integrate metal framing members and metal or glass panels into prefabricated or job-built units of various designs.

The design of curtain wall systems must account for the expansion and contraction of components making up the curtain wall panels. Curtain walls are fastened to structural members with welding clips or inserts. Curtain wall panel locations are shown on elevations. Curtain wall manufacturers may provide shop drawings that identify each panel and location by letters and numbers. A *shop drawing* is a detailed drawing or set of drawings, created by the contractor, subcontractor, supplier, fabricator, or manufacturer that shows how building elements will be fabricated or installed and contains other pertinent information such as construction materials, finishes, or dimensions.

Mullions several stories in length are installed vertically to cap the joints between curtain wall panels. Mullions and curtain wall frames are made of aluminum or other metals treated to be corrosion resistant. Methods for sealing joints between curtain wall panels, including rubber or vinyl weather stripping or various caulks and sealants, are shown on details.

Prefinished corrugated metal sheets are used for exterior siding on small structural steel buildings. The types and colors of metal siding are shown on elevations. Wall sections indicate insulation installation.

Exterior Insulation and Finish Systems (EIFS). *Exterior insulation and finish systems (EIFS)* are exterior finish systems consisting of exterior sheathing, insulation board, reinforcing mesh, a base coat of acrylic copolymers, and a finish of acrylic resins. **See Figure 7-9.** EIFS are used on commercial buildings to provide a unique color or appearance.

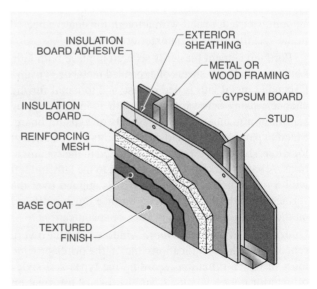

Figure 7-9. Exterior insulation and finish systems (EIFS) incorporate insulation board and polymer-based materials into a weatherproof exterior finish.

Exterior sheathing, such as glass mat-faced gypsum board, is installed over structural members such as wood or metal framing to serve as a base for insulation board. Concrete or CMU backing may also be used under the insulation board. Insulation board, typically extruded or molded expanded polystyrene, provides thermal insulation and flexibility in the finish system to minimize cracking. The insulation board is fastened to the sheathing using mechanical fasteners or to the surface of a concrete or CMU wall with construction adhesive. The insulation board is applied in a running-bond pattern to allow the joints to be staggered.

A ¼″ thick base coat of acrylic copolymers and portland cement is troweled onto the surface of the insulation board. One or two layers of an open-weave reinforcing mesh are embedded in the base coat. The mesh is also applied around door and window openings to reinforce the edges of the insulation board. After the base coat has set, the surface finish coat of acrylic resins is troweled or sprayed onto the structure to create the desired finish.

In EIFS installations, watertight seals must be formed around door and window openings, and at sills, flashing points, and expansion joints. Sealants are applied to the reinforced base coat to ensure a watertight seal. For large expanses of EIFS, expansion joints must be provided to allow for movement due to expansion and contraction. Expansion joints are designed for a minimum of four times the anticipated movement and are a minimum of ¾″ wide. Where sleeves or other penetrations are required through EIFS construction, manufacturer specifications must be followed.

Plaster. Exterior surfaces may be coated with various types of plaster finishes. The plaster surface is troweled smooth, swirled, or left with a rough finish as indicated in the specifications or on exterior elevations.

Portland cement plaster is secured in place with lath. Lath is available in sheets of expanded metal or gypsum. Expanded metal lath is tied or fastened to metal furring channels using screws. Gypsum lath is installed using screws in a manner similar to sheets of gypsum board. Metal expansion joints and control joints are installed at locations shown on the prints or detailed in the specifications. A base coat of plaster is applied to the lath and left with a rough finish. The finish coat is applied over the surface of the base coat with the surface finish indicated on the elevation drawings, details, or specifications.

Doors and Windows. Door, window, and hardware schedules in the specifications and on the prints provide most of the information regarding the types and styles of exterior doors, windows, and hardware. Information regarding the location of doors and windows is shown on architectural plan views. Details provide information about joining door and window jambs to the various

structural and wall finishes. **See Figure 7-10.** Details are also provided regarding proper placement of hardware and security devices.

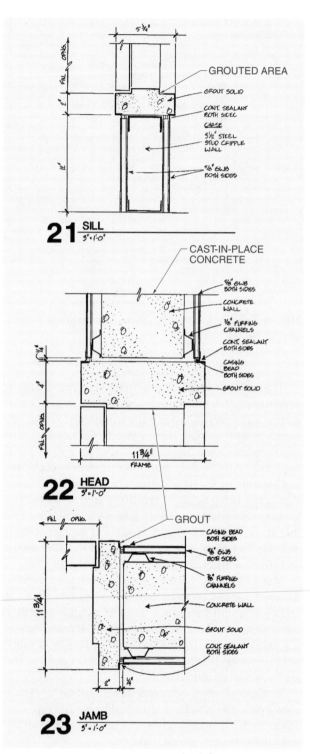

Figure 7-10. Voids behind hollow metal window jambs and doorjambs may be filled with grout to provide greater stability.

The International Building Code (IBC) requires that the minimum width of commercial egress doors be adequate for the potential occupancy load and be a minimum of 32″ wide and 80″ high. Variances to these dimensions for hotels or resident sleeping units are allowed as detailed in the IBC. Egress doors must be easily distinguished from adjacent construction. The IBC provides additional information related to revolving doors, power-operated doors, horizontal sliding doors, and access-controlled egress doors. The architect must ensure that door specifications comply with the building code in effect in the particular jurisdiction.

Door hardware commonly includes exit devices designed for both security and safety such as panic bars and door closers. **See Figure 7-11.** A *panic bar* is a door hardware device with a horizontal bar that releases a latch or bolt when pushed. Panic bars are mounted to the inside face of commercial doors a minimum of 34″ and a maximum of 48″ above the finished floor to provide for a safe emergency exit. As the horizontal bar is pushed, the latch or bolt retracts and allows the door to swing outward.

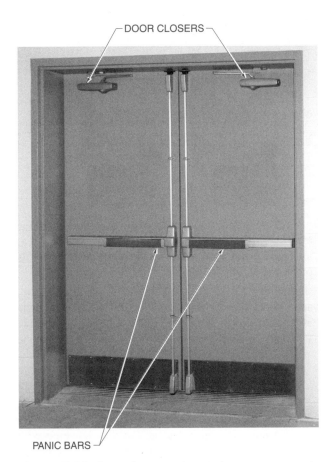

DOOR CLOSERS

PANIC BARS

Figure 7-11. Door closers and panic bars are commonly installed on commercial egress doors to provide for occupant safety.

A *door closer* is a hardware device that closes a door and controls the speed and closing action of the door. Door closers are commonly face-mounted to commercial doors and frames to allow for passage and to return doors to a closed position. Electrically activated door closers may be specified for installations that normally need to remain open, such as stairwells. Electrically activated door closers hold fire doors open during normal conditions through the use of an electromagnetic plate. When activated, the magnetic force of the plate is released, allowing the door to automatically close.

For other security concerns, metal detectors, sensors, and access control systems are installed near door egress points. Metal detectors use a magnetic field to find metal objects on individuals who pass through the detector. Based on operator settings, the amount of metal that activates the detector varies. An audible and/or visible alarm is activated when the amount of metal surpasses predetermined limits.

Sensors include motion, vibration, acoustic, and thermal detection units. The appropriate detection unit is selected based on environmental conditions, location, and application.

Access control systems include video surveillance and card readers. A video surveillance system consists of a camera, lens, and mounting hardware; a lighting system; transmission equipment; and a video monitor and recorder. The camera is mounted in a well-lighted area to ensure the images being transmitted are of adequate quality. Transmission equipment transmits the images to a video monitor wirelessly or via coaxial, optical, or two-wire cable. The images may be recorded and stored as necessary.

A variety of card technologies are available for different applications. Bar code cards have a series of parallel stripes. The spaces between the stripes are read optically by a photodetector cell. Magnetic stripe cards have stripes of magnetic material embedded between layers of the card or on the card surface. The card is read by a magnetic sensing device. Proximity cards incorporate radio frequency (RF) circuits that are read by a receiver that activates the system.

Roofs

Roofs for large commercial buildings are finished with bituminous coatings, such as tar and gravel, elastomeric coatings of various plastics, galvanized, treated, or decorative metal panels, or various sustainable roof systems. Architectural plan views indicate the roof covering, slope, and finish materials. Access doors, skylights, walk pads, and locations for vent pipes may also appear on roof level plan views. **See Figure 7-12.**

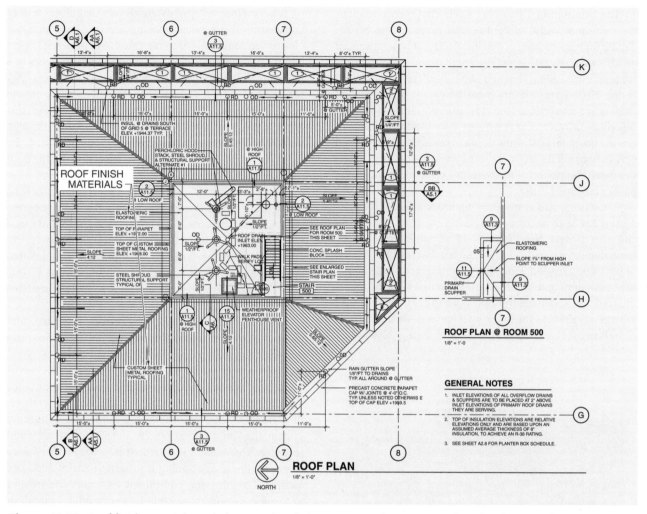

Figure 7-12. Roof finish materials, including metal and elastomeric roofing, are noted on the plan view for the roof.

A system of roof drains and pipes is installed to collect rainwater from large roof surfaces and channel it to the stormwater drainage system or gray water system. Roof drain and gutter locations are shown on plan views. Details show the methods for drain installation and joining roof finish materials to the drains to ensure a watertight seal. Pipe locations are shown on mechanical prints.

Roof details show methods for joining roof finishes to decking, walls, and parapets. **See Figure 7-13.** Flashing, gutters, roofing materials, and sealing materials are shown on detail sections.

Bituminous Coating. Roof decking is covered with a vapor barrier and rigid insulation prior to the application of a bituminous built-up roof. Several plies of felt paper saturated with asphalt or coal tar pitch are rolled out across the surface of the roof. Each layer is coated with hot asphalt or coal tar pitch. The final layer is coated with gravel while the asphalt or pitch is in a liquid state. Special types of light colored gravel may be specified to reduce heat absorption by the roof system. Details and specifications provide information about the bituminous materials to be used and the number of plies to be applied.

Elastomeric Coating. For an elastomeric roofing system, large sheets of chlorinated polyethylene (CPE), ethylene propylene diene monomer (EPDM), or polyvinyl chloride (PVC) are laid in place and sealed at the joints. PVC is the most commonly used elastomeric roofing material.

A vapor barrier and rigid roof insulation are installed over the roof decking. Elastomeric sheets are then rolled out across the entire surface of the roof and the joints are chemically sealed with a solvent to join them into a single unit. The entire surface may be covered with gravel after all joints are sealed. Precast concrete pavers may also be installed on the roofing surface as a walkway for maintenance personnel to prevent puncturing the elastomeric sheets.

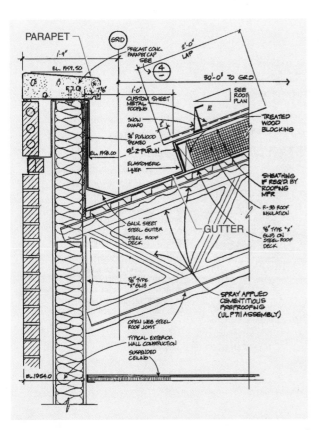

Figure 7-13. Parapet caps, parapet construction, roof decking, gutter construction, and dimensions for overlapping of materials are shown on roof details.

TECH TIP

Thermoplastic polyolefin (TPO) is a single-ply roofing material made from a flexible propylene/ethylene rubber and is resistant to tears, punctures, and UV degradation. Light-colored TPO materials have high solar reflectance, which lowers the roof surface temperature, reduces energy costs, and may contribute to LEED® certification.

Metal Roofing. Sloped roofs may be covered with metal roofing materials. Prefinished or decorative metal roofing materials, such as copper, are used for roof and ornamental coverings. Rigid insulation is applied to the top of the roof decking and covered with building paper. Sheets of prefinished or decorative metal are set in place, with the longer dimension placed parallel to rafters.

Joints along the sides of the metal sheets are fastened using a variety of methods, including flat, ribbed, or standing seams. **See Figure 7-14.** Metal roof sheets are fastened to the roof decking with clips or self-tapping screws. Overlapping joints are coated with various joint sealants.

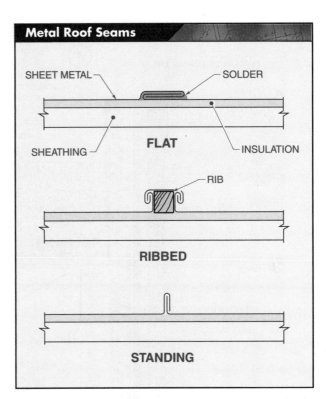

Figure 7-14. Flat, ribbed, or standing seams are commonly used to join metal roofing sheets.

Sustainable Roofing. Various types of materials may be specified as a roof covering to provide a sustainable roof finish. After a waterproof membrane, such as an elastomeric system, is applied to the roof structure, additional materials such as gravel, drain piping, geotextile fabric, and soil may be applied to the roof surface. **See Figure 7-15.** Plant materials are added to provide natural roof cooling and reduce the amount of stormwater run off from the roof area.

Solar panels may also be installed after watertight roof finish is applied for either solar thermal heating systems or photovoltaic systems. A *photovoltaic system* is an electrical system that uses crystalline silicon wafers that are sensitive to sunlight to directly convert solar radiation into electricity. Information on these systems is found on roof plans, detail drawings, and specifications.

Exposed Structural Elements

Structural members, such as columns and beams, may remain exposed in the finished structure. Elevations show the finish of these members. For road and bridge building, beams and columns are commonly exposed and finished. Stairways, landings, and handrails are shown on architectural plan views, elevations, and architectural and structural details.

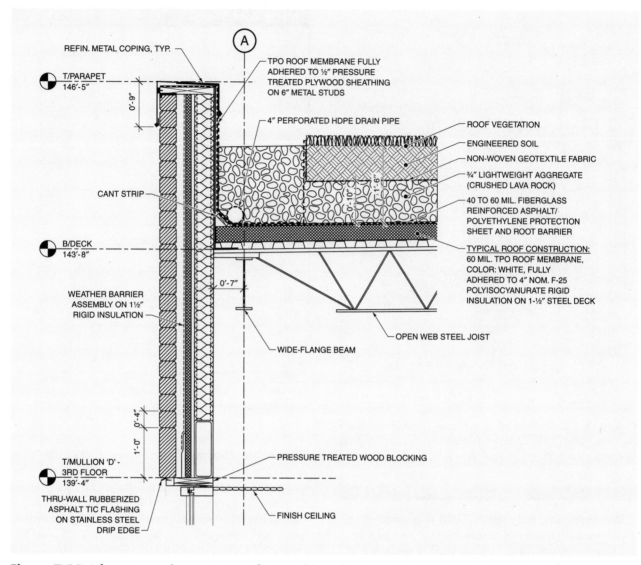

REFIN. METAL COPING, TYP.

T/PARAPET
146'-5"

0'-9"

TPO ROOF MEMBRANE FULLY
ADHERED TO ½" PRESSURE
TREATED PLYWOOD SHEATHING
ON 6" METAL STUDS

4" PERFORATED HDPE DRAIN PIPE

ROOF VEGETATION

ENGINEERED SOIL

NON-WOVEN GEOTEXTILE FABRIC

¾" LIGHTWEIGHT AGGREGATE
(CRUSHED LAVA ROCK)

CANT STRIP

40 TO 60 MIL. FIBERGLASS
REINFORCED ASPHALT/
POLYETHYLENE PROTECTION
SHEET AND ROOT BARRIER

B/DECK
143'-8"

TYPICAL ROOF CONSTRUCTION:
60 MIL. TPO ROOF MEMBRANE,
COLOR: WHITE, FULLY
ADHERED TO 4" NOM. F-25
POLYISOCYANURATE RIGID
INSULATION ON 1-½" STEEL DECK

0'-7"

WEATHER BARRIER
ASSEMBLY ON 1½"
RIGID INSULATION

OPEN WEB STEEL JOIST

WIDE-FLANGE BEAM

1'-0" 0'-4"

T/MULLION 'D' -
3RD FLOOR
139'-4"

PRESSURE TREATED WOOD BLOCKING

THRU-WALL RUBBERIZED
ASPHALT TIC FLASHING
ON STAINLESS STEEL
DRIP EDGE

FINISH CEILING

Figure 7-15. Information on the construction of sustainable roofing is commonly found in roof details and sections.

Columns and Beams. Reinforced concrete and structural steel columns and beams remain exposed in some commercial buildings and bridges. Concrete columns and beams are finished in a manner similar to exposed concrete walls. Chamfer strips may be set in the corners of concrete forms for square or rectangular beams and columns to create beveled edges. Concrete surfaces are finished as specified on elevation drawings or specifications. Structural steel columns and beams are coated with primer and paint to minimize rust. Structural steel columns, beams, and joists may be left exposed and finish painted or sprayed with fireproofing materials as noted on the details.

Exterior Stairways and Exit Ramps. Elevations show exterior stairway and exit ramp slopes and locations.

See Figure 7-16. Reinforced concrete is commonly used to construct exterior stairways and exit ramps constructed on-grade. Exterior concrete stairway and exit ramp dimensions are typically shown on site plans. Surface finish and reinforcing steel information is indicated in the specifications or on structural prints.

Stairways and exit ramps must be open on at least one side. The IBC indicates that the open side of an exterior stairway or exit ramp must have a minimum of 35 sq ft of total open area adjacent to each floor level and at each intermediate landing. This required open area must be less than 42" above the adjacent floor or landing level. The closed side of the stairway can be enclosed by an exterior wall. Uniform riser heights and tread depths must be maintained in exterior commercial stairways.

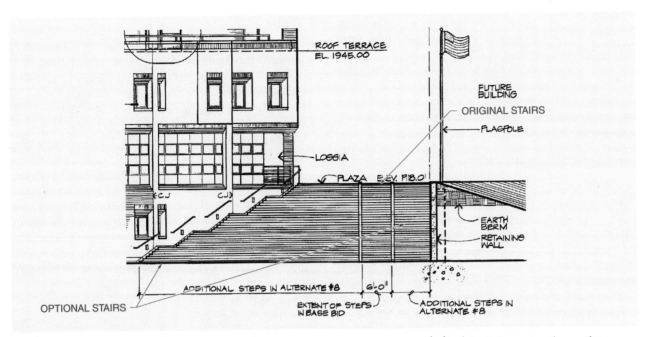

Figure 7-16. Exterior elevations show general exterior stairway appearance and alternate stair construction options.

Exit ramps must provide accessibility based on the provisions of the Americans with Disabilities Act (ADA). Exit ramps in commercial buildings must have a running slope of 8% or less, which is equal to 1 unit of vertical drop per 12 units of horizontal run. Ramps cannot be less than 36″ wide for a building with an occupancy capacity of 50 individuals or less. The side-to-side slope of ramps must be less than 2%, or 1 unit of vertical drop for 48 units of horizontal run. Landings for ramps must be a minimum of 60″ deep and must be slip resistant.

Structural steel is used for above-grade exterior stairs such as fire escapes. The stairway stringers are constructed of structural steel. The treads may be made of perforated or expanded steel mesh or steel pans filled with concrete. **See Figure 7-17.**

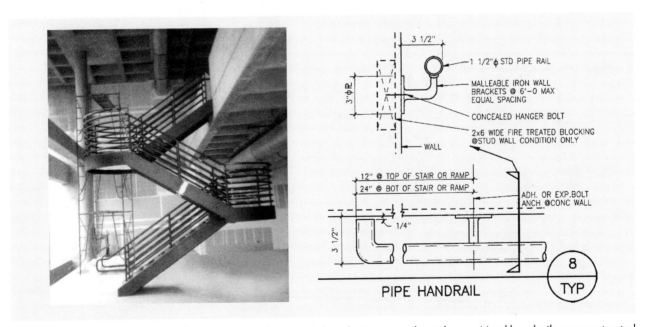

Figure 7-17. Concrete stair treads are supported by structural steel stringers and metal pans. Metal handrails are constructed of welded pipe and anchored to walls with support brackets.

Handrails for exterior stairs and ramps are shown on architectural elevations and details and structural print details. Exterior handrails are commonly made of steel or a corrosion-resistant metal. Pipe sections are welded together and finished to create handrails and posts.

Access ladders are shown on architectural plan views and architectural and structural details. Access ladders are constructed of welded structural steel members.

INTERIOR FINISH

Walls, floors, and ceilings inside a commercial building have a number of different finish requirements. Public areas are finished with appearance and durability in mind. Manufacturing and mechanical areas are finished with functionality as the primary concern. Office and classroom areas are finished to allow for flexibility of design, comfort, and functionality.

Plan views and interior elevations provide interior finish information. Wall locations and finishes, floor finishes, ceiling heights and finishes, and door and window locations are shown on plan views. Interior elevations present orthographic projections of various areas within a structure to show more complex finish treatments, cabinetry, and casework installations.

Walls

Interior and exterior walls between structural members in commercial buildings are commonly framed with metal studs. Light-gauge metal framing members, including tracks and studs, are fastened to structural members and floors. Batt or rigid insulation is placed between the studs as specified on architectural plan views.

Rough openings are framed-in for doors and windows. Architectural plan views and details indicate the on center spacing of the studs and rough opening dimensions. **See Figure 7-18.** Metal-framed walls are finished with gypsum board, lath and plaster, brick veneer, or other non-load-bearing finish materials. A schedule of wall framing and finishes is cross-referenced to the plan view. Interior elevations and a room finish schedule in the specifications indicate the types and locations of wall finishes.

Gypsum Board. *Gypsum board* is an interior surfacing material consisting of a fireproof gypsum core covered with heavy paper on both sides. Gypsum board is available in 4′ and 5′ widths and 8′ to 14′ lengths in 2′ increments. Gypsum board thicknesses range from ¼″ to 1″ in ⅛″ increments. Gypsum board is available in fireproof and waterproof grades.

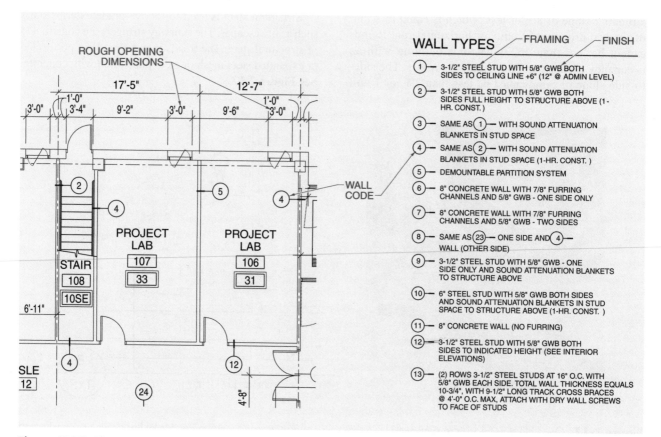

Figure 7-18. Plan views and architectural notes describe wall locations, framing and support members, and finish materials.

Gypsum board is attached to wood or metal framing members and furring channels with nails, screws, and/or construction adhesive. Joints between sheets of gypsum board and nail or screw holes are finished with a joint compound that is sanded smooth prior to application of surface finishes. **See Figure 7-19.** Surface finish materials include paint, vinyl, paper, fabric wallcoverings, and wood paneling. Gypsum board thickness and surface finish are shown on details and room finish schedules.

Figure 7-19. Gypsum board is attached to wood or metal framing members with nails or self-tapping screws to create a surface for a wall finish.

Lath and Plaster. Interior plaster wall finish materials include portland cement plaster on metal or gypsum lath and thin-coat finishes applied directly to gypsum board. Lath, control joints, and trim members are installed in a manner similar to exterior plaster applications.

Masonry. Brick, concrete masonry units, and manufactured and natural stone are used for interior fire break walls and for exposed interior walls. Finish information for interior masonry walls is similar to exterior exposed masonry information, in that they both include material, mortar joint color and finish, and bond pattern information.

Operable Walls. Large interior spaces may be divided into smaller areas using operable walls, such as movable and folding partitions or demountable partitions. Large movable wall sections are suspended from an overhead track. **See Figure 7-20.** The overhead ceiling track is attached to structural members to provide solid support for operable walls. The bottom edge of the ceiling track is flush with or recessed into the ceiling. Rollers that travel in the ceiling track are attached to the top of each wall section.

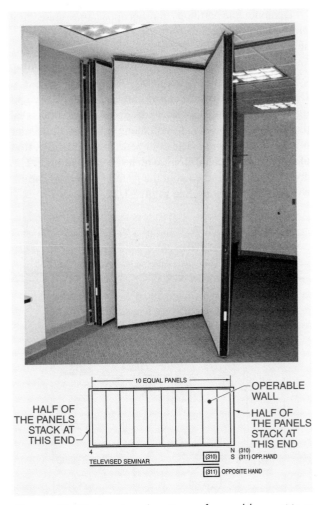

Figure 7-20. Interior elevations of movable partitions indicate the number of panels and the stacking plans.

Suspended operable wall panels may stack against one or several walls, depending on the design. Architectural plan views show the locations of tracks for operable walls. Interior elevations indicate the method for stacking operable wall panels. Recessed storage closets may be constructed in the wall at one or both ends of the wall panels to conceal the panels when not in use.

Demountable partitions are designed to be stationary for longer periods of time than operable walls. Floor and ceiling tracks are secured in place after ceiling, floor, and wall finishes are installed. The tracks can be installed and removed with minimal disturbance of finish materials. The demountable partitions are set into the tracks with intermediate struts or supports. Panels may be prefinished gypsum board or metal and finished with fabric, plastic laminate, prefinished metal cladding, or vinyl coatings. Electrical raceways within the partitions provide for installation of receptacles for work areas and lighting. Door frames may also be installed in demountable partitions.

Demountable partitions may not appear on architectural prints. Manufacturer drawings are provided for demountable wall installation. The drawings contain plan views and details indicating specific manufacturer identification codes for panels and framing members.

Doors and Borrowed Light Frames. Interior door and light (window) frame locations are shown on architectural plan views. Locations are commonly given to the center of each door opening. The door hand (swinging direction of a door) is indicated along with a schedule number. The door schedule is included in the general plan notes or the written specifications. **See Figure 7-21.**

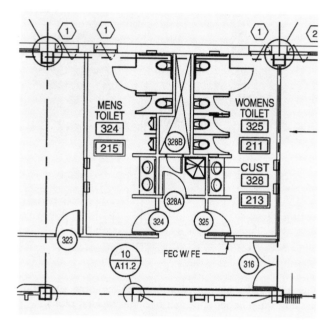

Figure 7-21. Door locations are shown on plan views along with schedule numbers, which are cross-referenced to a door schedule.

A *borrowed light frame* is a window opening in an interior partition between two interior areas. Borrowed light frames may be fixed or operable. Information regarding borrowed light frames is shown on architectural plan views, specifications, and interior elevations in a manner similar to interior doors.

Floors

Commercial floor finishes include concrete, carpet, resilient and ceramic tile, hardwood, bamboo, and raised floor systems. Each room and area on an architectural floor plan is numbered. The room number corresponds to a number on the specifications or general notes finish schedule that indicates the finish flooring to be installed. For large areas,

the architect may note the floor finish on the architectural plan views. Specific details may include a note about the floor finish material at a particular location.

Concrete. Floors in manufacturing and mechanical areas are commonly finished with exposed concrete. **See Figure 7-22.** Concrete slabs are troweled to a smooth finish as noted in the specifications. Super flat floors may be specified in warehouse applications where forklifts or automated material-handling equipment require close floor finish tolerances for optimal operation. Laser screed equipment is used in placing and finishing concrete for super flat floors to ensure minimal surface finish elevation variation.

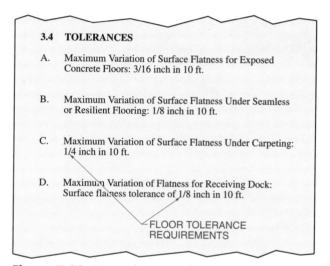

Figure 7-22. Exposed concrete finishes are noted in the project specifications.

Carpet. Commercial-grade carpet is glued to the supporting floor or stretched across padding and fastened around the perimeter of the floor. Seams are sewn or connected with heat-sensitive tape. Information regarding carpet type and placement is typically included in a room finish schedule or the specifications. Additional information in the specifications includes the manufacturer design, weight, backing, and installation instructions. For large commercial carpet or finish floor installations over a concrete floor, care must be taken to determine the moisture content of the concrete floor. Application of floor finish materials on a concrete floor with an excessively high moisture content exceeding manufacturer recommendations can result in failure of the finish floor materials to properly adhere to the floor.

Tile Products. Ceramic, resilient, or vinyl tile are commonly installed as floor finish materials in commercial buildings. Tile products are attached to the finished surface with mastic. For ceramic tile, grout is applied to fill the voids between the tile after the mastic has set. For resilient or vinyl tile, the primary installation concern is proper fitting of the seams between sheets or pieces of the tile. In a manner similar to carpet, room finish schedules in the general notes or specifications contain specific tile information. Bathroom elevations may show ceramic tile installation dimensions where tile is applied as a wall finish. **See Figure 7-23.**

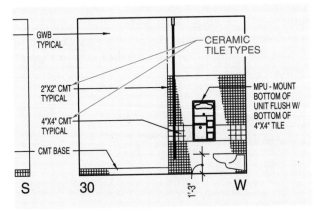

Figure 7-23. Ceramic tile sizes and locations are shown on interior elevations.

Wood. Oak, maple, and other decorative hardwood flooring are shown on architectural plan views. Individual interlocking pieces of hardwood are fastened to concrete with mastic or nailed to a wood subfloor. Wood may be prefinished or sanded, stained, and finished after installation. Enlarged floor plans are provided where specific wood finishes and patterns are required. In applications without special pattern requirements, hardwood floor information is contained in the room finish schedule.

Raised. A raised floor system is installed in areas where wiring, conduit, and ductwork are installed on top of the subfloor and below the finished floor. **See Figure 7-24.** A raised floor system, also referred to as an access floor or computer floor system, consists of base plates set in a regular grid layout equal to the dimensions of the raised floor panels. The bottoms of base plates are secured to the subfloor with mastic or mechanical fasteners.

Threaded pedestals are set into the base plates and adjusted to the proper height. A laser transit-level is used to verify the levelness of the floor. Floor panels are then set on pedestal heads at the top of each pedestal or fastened to stringers that span between pedestal heads.

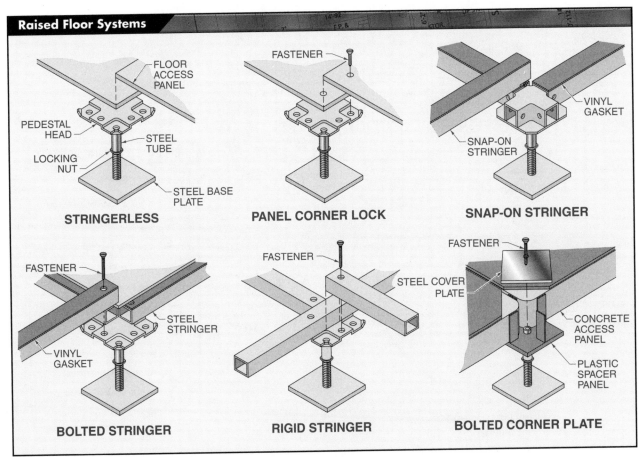

Figure 7-24. Raised floor systems facilitate the placement of and access to electrical wiring, conduit, and ductwork.

Raised floor panels may be prefinished with carpet or tile surfaces. The area below a raised floor system may serve as a plenum for air ventilation. When used for ventilation, some floor panels may contain vents to allow air to flow upward into occupied areas. Placement of these ventilated panels is provided on a floor plan view.

Ceilings

Ceilings in commercial buildings include suspended, furred, and exposed ceilings. Architectural prints include reflected ceiling plans. **See Figure 7-25.** A *reflected ceiling* *plan* is a plan view of a ceiling that indicates ceiling-mounted items such as air diffusers, exhaust fans, air intakes, and luminaires (lighting fixtures). Additional information concerning air diffusers, exhaust fans, and air intakes is shown on mechanical prints. Ceiling-mounted luminaires are further described on the electrical prints. Fire sprinkler locations may also be shown on reflected ceiling plans.

Suspended Ceilings. A suspended ceiling consists of a suspended light-gauge metal gridwork with ceiling tile placed between the grids. Hanger wires are fastened to structural members above the finished area to support the gridwork.

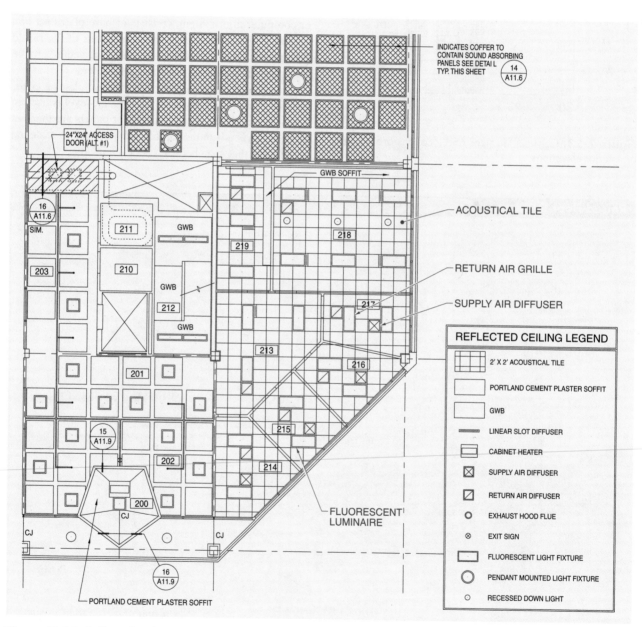

Figure 7-25. Reflected ceiling plans show ceiling finishes, including exposed ceilings, gypsum board ceilings, and suspended ceilings.

The gridwork supports lay-in tile, concealed-grid tile, metal runners, gypsum board, or lath and plaster. The height of the finished ceiling surface above the floor is indicated on reflected ceiling plan notes or the room finish schedule in the specifications.

The height of the metal gridwork is determined after hanger wires are installed. The levelness of the gridwork is verified using a laser transit-level that projects a level beam of light around the space for ceiling installation. **See Figure 7-26.** Wall channels are fastened to the interior walls around the perimeter of the area to support ceiling finish materials. Light fixtures are set into the gridwork and additional hanger wires are installed as necessary to support the weight of the fixtures. Insulation should not be installed within 3″ of luminaires (lighting fixtures) unless they are approved for use with insulation.

channels. Prefinished ceiling tile, gypsum board, or lath and plaster are attached to the runners.

Where soffits and coffered ceilings are indicated on the prints, framing is built to the required dimensions and attached to structural members. Details show the framing members to be installed, finishes to be applied, and width and height dimensions.

Exposed Ceilings. Structural steel or reinforced concrete beams, joists, and slabs may remain exposed to form the ceiling for the space below. Reflected ceiling plans indicate the locations of exposed ceiling areas. The exposed portions of the beams, joists, and floors above may be painted, sprayed with a decorative, acoustical, or fireproofing coating, or may remain unfinished. Luminaires (lighting fixtures), electrical and mechanical systems, sprinkler pipes, and ductwork remain visible to view.

Figure 7-26. Suspended ceilings use a metal gridwork to support lay-in tile or other finishes.

The location of lighting fixtures and fire protection sprinkler heads may be found on the reflected ceiling plan.

Lay-in tile are set into a gridwork that remains exposed. Concealed-grid tile are supported at each edge by splines that tie the tile together and conceal the gridwork. Metal runners designed to allow for prefinished metal channels may be clipped onto the underside of the grid. Gridwork may allow for the attachment of gypsum board with self-tapping screws. The gypsum board is then finished in a manner similar to interior walls. Metal, expanded wire, or gypsum lath may be attached to the suspended gridwork to support plaster ceilings.

Furred Ceilings. In furred ceilings, runners are attached directly to structural members, such as open web steel joists or concrete beams and slabs, to provide a base for fastening finish materials. Runners may be wood furring strips or metal members, such as tracks, studs, or furring

Interior Stairways

Interior stairways are shown on sections, which indicate landing elevations and the number of risers. **See Figure 7-27.** Grid lines that relate to the overall structural building grid may be shown on stairway sections. Details provide specific information about stairway construction, landing finishes, and finish materials for the top surface and underside of the stairs. Structural prints may provide additional stairway information.

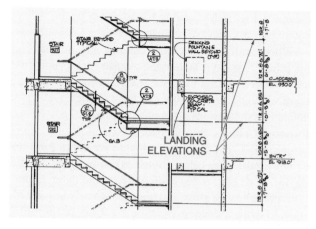

LANDING ELEVATIONS

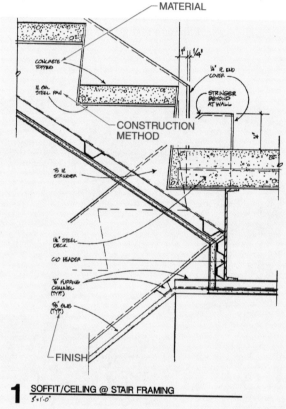

MATERIAL

CONSTRUCTION METHOD

FINISH

1 SOFFIT/CEILING @ STAIR FRAMING
3"=1'-0"

Figure 7-27. Stair elevations and sections indicate elevations, construction methods, materials, and finishes.

The IBC provides many specific dimensional requirements and finish requirements for interior commercial stairways. The walking surface of treads and landings must not slope more than 2% in any direction. A 2% slope is equal to 1 vertical unit for 48 horizontal units. Risers must be a minimum of 4″ high and a maximum of 7″ high. Treads must be at least 11″ deep. The vertical rise of stairways without a landing is a maximum of 12′ between floors. A landing must be provided where the distance between floors is greater than 12′. Handrail heights are a minimum of 34″ and a maximum of 38″ measured vertically from the stair nosing. Handrail height must be uniform throughout the run of the stair.

Specialized shop drawings are provided for spiral and circular metal or wood stairways. Based on the IBC, spiral stairways must have a minimum clear tread depth of 7½″ at a point 12″ from the narrowest edge. Maximum riser height is 9½″, and minimum headroom is 78″. For circular stairways, risers must be a minimum of 4″ high and a maximum of 7″ high. The smaller radius of the stairway must not be less than twice the stairway width. Treads must be a minimum of 11″ deep measured 12″ from the narrowest end, with the narrowest end not less than 10″ deep.

For spiral or circular wood stairways, assemblies may be prefabricated by a stairway manufacturer and delivered unassembled to the job site. The stringers, treads, risers, and handrails are installed according to the shop drawings. Structural steel stairways may be constructed at a fabricating shop and delivered to the job site as a single unit or in sections ready for assembly. The radius of the stairway curves, overall height, and number of risers are indicated on stairway details when spiral or circular stairways are to be constructed on the job site.

Cabinetry and Casework

Architectural symbols included on plan views provide orientation for interior elevations. Arrows or other symbols indicate the interior walls shown on elevations. For example, an arrow pointing toward an interior wall with the number "19" refers to interior elevation 19. **See Figure 7-28.**

Interior elevations and schedules provide location, dimension, and finish information for cabinetry, casework, and other interior specialties, such as white boards and projection screens. The size and manufacturer code for many of these fixtures are noted in the specifications and the room finish schedule. Architectural notes and codes on interior elevations relate to a schedule for cabinetry and casework. The schedule indicates the width, height, and depth of each cabinet and countertop treatment. Cabinetry may be metal, finished wood, or plastic laminate over medium density fiberboard (MDF) or particleboard. **See Figure 7-29.**

Cabinets are commonly prefabricated at a cabinet shop and delivered to the job site ready for installation. Cabinet manufacturers develop shop drawings from the architectural drawings to fabricate the cabinets and casework. These shop drawings may also provide additional installation information. Some custom-built cabinetry may be built in place. Countertops are constructed of MDF or particleboard and covered with plastic laminate or ceramic tile. Solid-surface materials or natural stone, such as granite, may also be used as finish countertop material.

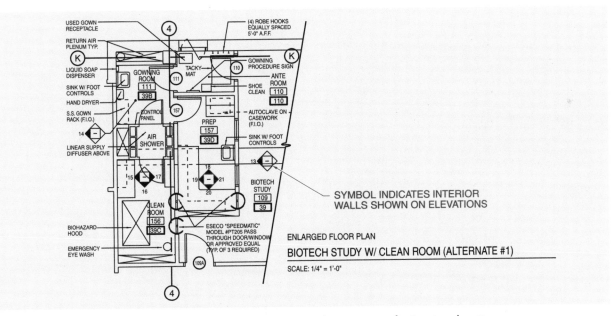

Figure 7-28. Architectural symbols included on plan views provide orientation for interior elevations.

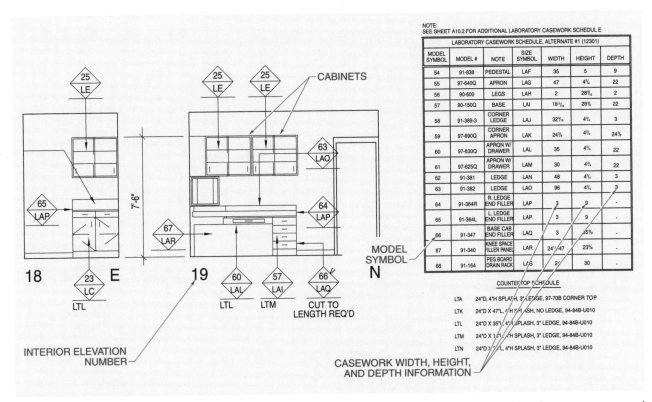

Figure 7-29. Elevations and schedules provide information about cabinets and casework, including placement, sizes, and finishes.

Refer to the CD-ROM in the back of the book for Chapter 7 Quick Quiz®and related printreading and reference material.

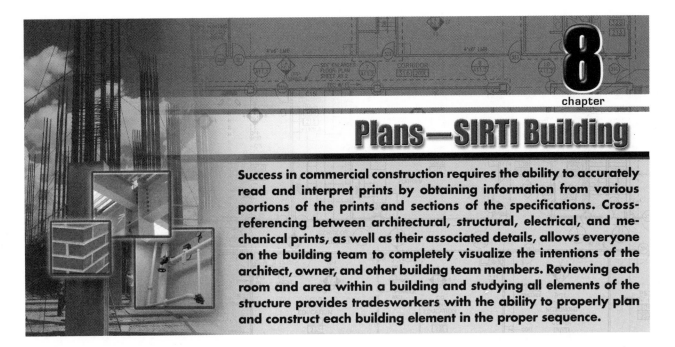

chapter 8

Plans—SIRTI Building

Success in commercial construction requires the ability to accurately read and interpret prints by obtaining information from various portions of the prints and sections of the specifications. Cross-referencing between architectural, structural, electrical, and mechanical prints, as well as their associated details, allows everyone on the building team to completely visualize the intentions of the architect, owner, and other building team members. Reviewing each room and area within a building and studying all elements of the structure provides tradesworkers with the ability to properly plan and construct each building element in the proper sequence.

BUILDING DESCRIPTION

The SIRTI (Spokane Intercollegiate Research and Technology Institute) building is a four-story educational building covering an area of 23,700 sq ft. **See Figure 8-1.** The lowest level, identified as the laboratory level, contains 23,518 sq ft and provides room for mechanical equipment and instructional space. The first-floor level, identified as the plaza/entry level, contains 8284 sq ft and provides room for administrative offices and additional instructional space. The second-floor level, identified as the classroom level, contains 16,583 sq ft and is primarily used for instructional space. The third-floor level, identified as the office/administration level, contains 7440 sq ft and provides additional classroom and administrative space. Based on the building size and usage, the maximum allowable occupancy of the SIRTI building is 1076 people.

The Index Sheet provides general information about the SIRTI building, including architectural abbreviations, reference and material symbols, project data, and a drawing index. **See Figure 8-2.** The building is constructed on a five-acre tract of land composed of three parcels. The building is constructed in accordance with the 1988 edition of the *Uniform Building Code (UBC)*. Twelve fire-rated assemblies are used in the structure based on the UBC, Underwriters Laboratories, Inc.® (UL), and *National Evaluation Report (NER)* guidelines.

The drawing index indicates that 151 sheets are included in the set of prints, which are divided into architectural, structural, civil, landscaping, mechanical, and electrical categories. The classroom level is the primary focus of study in this chapter. Extensive cross-referencing of the architectural, structural, mechanical, and electrical drawings related to this level provides specific information about each of the rooms.

Figure 8-1. The SIRTI (Spokane Intercollegiate Research and Technology Institute) building is a multistory reinforced concrete structure with a brick-veneer exterior finish.

The flat roof of the classroom level is finished with concrete pavers installed over rigid insulation and an elastomeric roofing membrane.

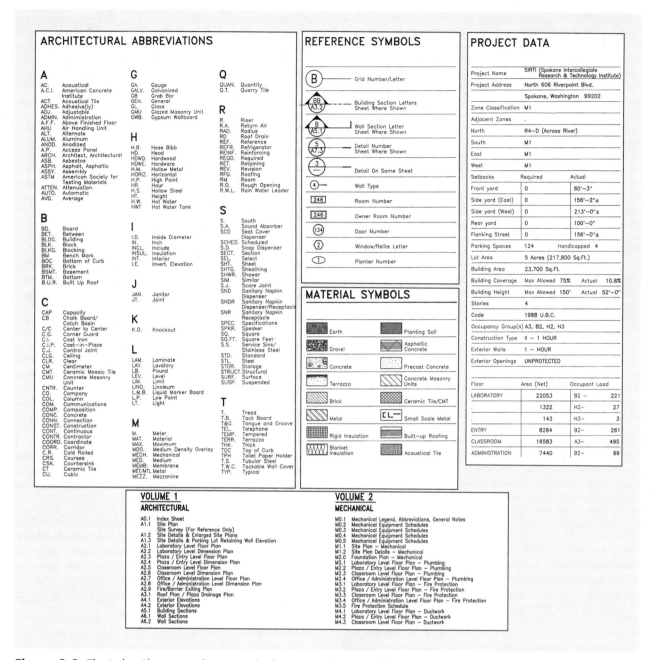

Figure 8-2. The Index Sheet provides general information about the SIRTI building, including architectural abbreviations, reference and material symbols, project data, and a drawing index.

TECH TIP

The Spokane Intercollegiate Research and Technology Institute (SIRTI) is a Washington state economic agency focused on the development and growth of innovative technology-based companies in the Northwest United States.

Two sets of room numbers are used to identify the classrooms and other spaces in the building, with one set used for construction purposes and the other set used by the owner. **See Figure 8-3.** For example, classrooms and other spaces on the classroom level (second floor) are identified with 300-series numbers for construction use and 200-series numbers for owner use. The 300-series numbers are used for the purpose of this chapter.

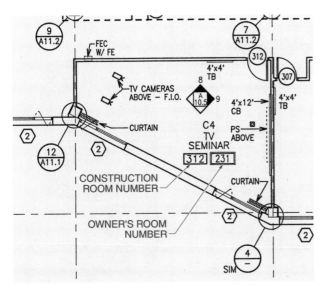

Figure 8-3. Two sets of room numbers identify classrooms and other spaces on the prints. The construction room numbers are used in this chapter.

Each of the numbered areas on the classroom level is described with architectural, structural, mechanical, and electrical drawings. Schedules applying to each classroom or other space provide information about building materials and finishes. Contractors, subcontractors, project managers, supervisory personnel, and tradesworkers must visualize and understand the structural and finish materials in each building space to ensure that construction processes conform to the building design. When reading and studying the chapter, refer to the prints identified in the text to gain a full understanding of the information presented on the prints.

Room 300—Elevator Lobby

Room 300 (Elevator Lobby) provides access from the elevator to the classrooms and other spaces on the classroom level. As indicated on Sheet A2.5, the elevator opening is located at the west end of the south wall of Room 300. Access to Rooms 301, 302, and 305 is also provided from the Elevator Lobby.

Architectural and Structural. As shown in General Notes 1 and 2 on Sheet A2.6, dimensions for metal stud walls are indicated to the centers of the studs (unless otherwise indicated) and dimensions for masonry walls are indicated to the face of the masonry. The General Notes are applicable to all classrooms and other spaces on the floor plan.

As indicated on Sheet A2.6, the dimensions of Room 300 are 31'-0" × 12'-3⅝" (4⅛" + 9'-4¼" + 4'-5¼" + 2'-5⅝" + 6'-10⅜" + 6'-10⅜" + 8" = 31'-0"; 11'-4¾" + 10⅞" = 12'-3⅝"). The north wall, which is a Type 4 wall, is framed with 3½" metal studs and covered with ⅝" gypsum wallboard (GWB) on both sides. The wallboard extends the full height of the

wall to the structure above. Sound-attenuation blankets are placed between the studs to minimize the amount of sound vibrations transmitted into Room 305. An interior elevation key on Sheet A2.5 indicates that Sheet A10.4, Elevation 15 provides additional information about the north wall of Room 300. Elevation 15 indicates a display case is to be installed on the north wall with recessed fluorescent luminaires (lighting fixtures) above. **See Figure 8-4.**

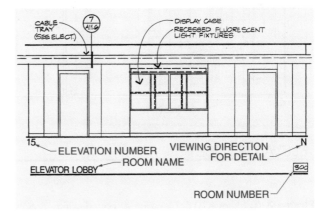

Figure 8-4. A display case with a recessed fluorescent luminaire is installed along the north wall of Room 300.

As indicated on Sheet A2.6, the south wall of Room 300 is constructed using four different combinations of metal studs and gypsum wallboard. The portion of the wall along the mechanical chase is a Type 13 wall, which is constructed of two rows of 3½" metal studs spaced 16" OC with ⅝" gypsum wallboard on both sides. The walls along Rooms 301 and 302 are Type 2 walls, which are constructed with 3½" metal studs with ⅝" gypsum wallboard on both sides that extend full height to the structure above. The Type 23 walls are elevator shaft walls consisting of 2½" metal shaft wall studs with 1" coreboard and one layer of ⅝" gypsum wallboard with a 1 hr fire rating. The Type 8 wall is constructed in the same manner as the Type 23 wall on the elevator side with 1" coreboard and ½" gypsum wallboard, and in the same manner as Type 4 walls on the Elevator Lobby side with 3½" metal studs, sound-attenuation blankets between the studs, and ⅝" gypsum wallboard to the full height of the structure above.

Ceramic tile measuring 8" × 8" is used as the wall finish on the south wall of Room 300 as shown on Sheet A10.4, Elevation 4. **See Figure 8-5.** A 20 min fire-rated door is to be installed in a recess along Room 300. An electromagnetic door holder is to be installed behind the door. The electromagnetic door holder is actuated by the fire alarm system, automatically closing the door in the event of a fire. Additional information regarding construction of the wall and frame at the elevator door is shown on Sheet A11.9 (not included with this set of prints).

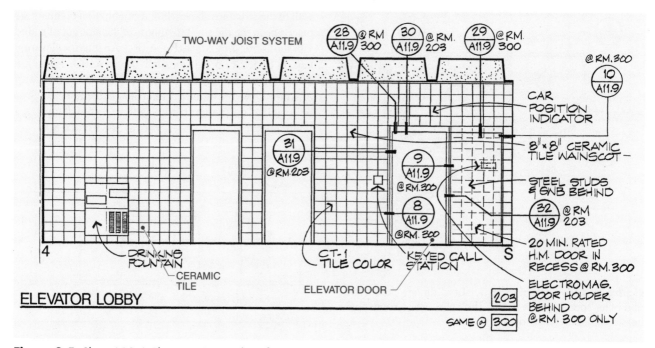

Figure 8-5. Sheet A10.4, Elevation 4, provides information regarding the wall finish materials for the south wall of Room 300.

Sheet A11.2, Detail 8 indicates that wood blocking and gypsum wallboard is to be installed over the concrete columns along the north wall of Room 300 at gridline intersections 6/J and 6/H. Tackable wallcovering (TWC) is to be applied over the wallboard. Sheet A11.2, Detail 10, indicates wood blocking and gypsum wallboard is to be installed at the corner of the concrete wall at the southeast side of Room 300 to tie with the metal stud wall and metal furring channels on either side. The studs are offset 1½″ from the face of the column to align the gypsum wallboard with the wood blocking at the column corner. The gypsum wall finish within the open mechanical shaft is adhered to the concrete wall with backer rod and sealant. The exterior corner is finished with a corner bead (CB).

As shown on Sheet A2.6, Tile & Carpet Floor Pattern Layout, the floor of Room 300 is to be finished with vinyl composition tile (VCT) of alternating designs, with 1′-0″ wide border tile of a different design and a rubber cove base. **See Figure 8-6.** As noted in the specifications, a 3′-6″ × 7′-0″ wood door, with the door width verified by the elevator manufacturer, is to be installed to provide access to the elevator.

Column positions are indicated on the Framing Plan on Sheet S2.4. The Column Schedule on Sheet S5.1 provides structural and reinforcing information about the columns. Columns J6 and H6 are identified in Room 300, and are 16″ square concrete columns that extend to an elevation of 1962.7′. Column reinforcement for the classroom level consists of four #8 rebar overlapping the eight #11 rebar from the plaza/entry level by 3′-9″. Sheet S5.1, Detail 17,

indicates the position of the rebar and notes that columns with four vertical rebar require 1½″ of clear space between the rebar and the face of the concrete columns. Stirrups are used to tie together the vertical rebar in the columns. A *stirrup* is reinforcement used to resist shear and diagonal tension stresses in a structural member. Stirrups at the classroom level for columns J6 and H6 are #3 rebar bent into a box shape and spaced 16″ OC. Additional information concerning the finish at the tops of the columns is provided on Sheet S3.3, Detail 4 (not included with this set of prints).

Figure 8-6. The floor of Room 300 is to be finished with 1′-0″ square vinyl composition tile (VCT) of alternating designs, with 1′-0″ wide border tile of a different design and a rubber cove base.

The floor for the classroom level is a cast-in-place concrete two-way joist system (waffle beam and slab). **See Figure 8-7.** The slab is typically 6″ thick throughout the classroom level, with #4 rebar spaced 15″ OC in each direction. As shown on Sheet S2.4, Detail 12, the overall typical floor thickness from the bottom of the two-way joists to the top of the floor slab is 20½″.

Portland Cement Association

Figure 8-7. Domes are used as concrete forms to create a two-way joist system.

Joists in the floor of Room 300 and throughout the classroom level are formed with domes, which are described on Sheet S2.3A, Detail 18. Type 4 domes are used in this area as noted on Sheet S2.4. Most of the domes used to form the two-way joists are typical domes measuring 16″ high by 52″ square. Other domes are indicated on the Framing Plan on Sheet S2.4 and are described on Sheet S2.3A, Detail 18. **See Figure 8-8.**

Joist marks on the Framing Plan represent the joist type and direction. The Joist Schedule on Sheet S2.4 describes the joist size and reinforcement requirements. Type C6 joists are 8″ wide, with #8 rebar at the bottom and #6 rebar at the top. A series of three diagonal lines is shown on Sheet S2.4 at the corners of the elevator shaft and mechanical chase floor openings. The diagonal lines represent short pieces of rebar to be installed at the corners of the opening to provide additional reinforcement in these areas. Additional information concerning the installation of this diagonal floor reinforcing is shown on Sheet S1.2, Detail 14 (not included with this set of prints).

Sheet S2.3, Detail 16, indicates the typical web reinforcing for the two-way joist system. Joists are spaced 5′-0″ OC, with solid slabs at the columns. Web mesh is to be installed in the joists to provide additional shear stress protection and reinforcement around each of the columns. Details 28 and 29 specify that additional W7 wire reinforcement be installed for shear stress reinforcement in the cast-in-place joists in these areas. Dimensions for horizontal and vertical spacing of the shear stress reinforcement are shown. Concrete coverage should extend 2″ above the top distributed rebar. The upper surface of the concrete slab may be recessed for tile where necessary.

TECH TIP

Two-way concrete joist systems are commonly used in commercial and institutional buildings. Column spacing is typically multiples of dome form spacing to ensure uniformity of the drop panels at each column.

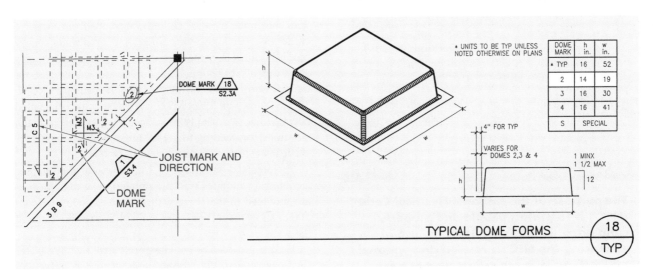

Figure 8-8. Typical domes measuring 16″ high by 52″ square are used for the majority of the forming work for the classroom level two-way joist system. Other dome marks are indicated on the Framing Plan, and the sizes are noted on Detail 18.

Mechanical and Electrical. As indicated on Sheet E2.3, four Type B luminaires provide illumination in Room 300. As indicated on the Lighting Fixture Schedule (not included with this set of prints), Type B luminaires are recessed 2′ × 4′ fluorescent luminaires fitted with a regressed acrylic lens and are provided with two energy-saving lamps. **See Figure 8-9.** As indicated with the crosshatching on Sheet E2.3, the center two Type B luminaires are connected to emergency circuit E2 through two separate junction boxes. The luminaires in the corridor adjacent to Room 300 are connected to circuit 2 of panelboard 4B, which is located in Room 301.

As noted on Detail 15, a recessed fluorescent luminaire is also to be installed in the display cabinet and a cable tray is to be installed along the north wall. Additional information concerning the construction of the cable tray is found on Sheet A11.6, Detail 7 (not included with this set of prints). An elevator car position indicator is to be installed above the elevator on the south wall. **See Figure 8-10.** A junction box for powering the water cooler is installed on the south wall and powered by circuit 4 on panelboard 2D. Sheet E4.3 shows that a power supply for the key-activated access system is installed in Room 300 on the south wall. Sheet E4.3 also indicates that a smoke detector is installed in the ceiling outside the elevator door.

TYPE 'A'	RECESSED MOUNTING 2'X4' FLUORESCENT FIXTURE WITH A 3" DEEP, 18 CELL PARABOLIC LOUVER. PROVIDE THREE F40T12 ENERGY SAVING LAMPS PER FIXTURE. LITHONIA 2PM3-GH-340-18-S-277-ES-GLR, DAYBRITE, COLUMBIA, METALUX OR APPROVED SUBSTITUTION.
TYPE 'B'	RECESSED MOUNTING 2'X4' FLUORESCENT FIXTURE WITH REGRESSED FRAME AND ACRYLIC LENS. PROVIDE TWO F40T12 ENERGY SAVING LAMPS PER FIXTURE. LITHONIA 2SPGH-240-RN-A12.125-277-ES-GLR, DAYBRITE, COLUMBIA, METALUX OR APPROVED SUBSTITUTION.
TYPE 'C'	RECESS MOUNTING 1'X4' FLUORESCENT FIXTURE WITH ACRYLIC LENS AND PLASTER FRAME. PROVIDE TWO F40T12 ENERGY SAVING LAMPS PER FIXTURE. LITHONIA SP-FH-240-RN-A12.125, DAYBRITE, COLUMBIA, METALUX OR APPROVED SUBSTITUTION.
TYPE 'D'	SUSPENDED INDIRECT FLUORESCENT LIGHTING FIXTURE. FIXTURE TO BE 10" WIDE BY 3 5/8" HIGH AND 24' LONG FIXTURE TO BE SUSPENDED 18" BELOW CEILING LINE. SUSPEND USING AIRCRAFT CABLE. PROVIDE TWELVE F40T12 ENERGY SAVER LAMPS. PEERLESS LD3-020459-24-277V OR APPROVED SUBSTITUTION.
TYPE 'D1'	SAME AS TYPE 'D', EXCEPT 20' LENGTH. PROVIDE TEN F40T12 ENERGY SAVER LAMPS. PEERLESS LD3-020459-20-277 OR APPROVED SUBSTITUTION.
TYPE 'D2'	SAME AS TYPE 'D', EXCEPT 8' LENGTH. PROVIDE FOUR F40T12 ENERGY SAVER LAMPS. PEERLESS LD3-020459-8-277 OR APPROVED SUBSTITUTION.
TYPE 'E'	RECESSED CEILING MOUNTED SINGLE FACE EXIT FIXTURE. PANEL TO BE CLEAR 1/4" PLEXIGLASS WITH ROUTED-IN AND SCREENED LETTERING. GREEN LETTERS ON WHITE. PROVIDE ONE 8 WATT T-5 FLUORESCENT LAMP PER FIXTURE. ALKCO RPC-110E, SILTRON, EMERGI-LITE, LITHONIA OR APPROVED SUBSTITUTION.
TYPE 'E1'	SAME AS TYPE 'E', EXCEPT WITH ARROWS AS SHOWN ON DRAWINGS.
TYPE 'E2'	SAME AS TYPE 'E', EXCEPT DOUBLE FACE WITH ARROWS AS SHOWN ON DRAWINGS.
TYPE 'G'	CHAIN HUNG SURFACE MOUNTING TWO LAMP OPEN STRIP FIXTURE WITH WIRE GUARD. PROVIDE TWO F40T12 SUPER SAVER LAMPS PER FIXTURE. LITHONIA C240, DAYBRITE, KEYSTONE, METALUX OR APPROVED SUBSTITUTION.
TYPE 'L'	SURFACE MOUNTING 8' FLUORESCENT LIGHTING FIXTURE WITH FLAT BOTTOM ACRYLIC DIFFUSER, TANDEM WIRED 8' LENGTH. PROVIDE FOUR F40T12 ENERGY SAVER LAMPS PER FIXTURE. LITHONIA 8TAW240 OR APPROVED SUBSTITUTION.
TYPE 'S'	4' SURFACE MOUNTING MARCO SINGLE CIRCUIT LIGHTING TRACK. PROVIDE TWO MARCO T1734 LIGHTING HEADS PER 4' TRACK COMPLETE WITH 120W/ER40 LAMPS. MARCO, CAPRI, LITHONIA OR APPROVED SUBSTITUTION.
TYPE 'S1'	SAME AS TYPE 'S', EXCEPT 8' TRACK WITH THREE LIGHTING HEADS.
TYPE 'Y'	9" WIDE RECESSED LINEAR LIGHTING FIXTURE WITH PARABOLIC LOUVER. PROVIDE TWO F40T12 ENERGY SAVING LAMPS PER FIXTURE. LITHONIA 9RL FH1 RSS-277-ES-PBL-8 OR APPROVED SUBSTITUTION.

Figure 8-9. The Lighting Fixture Schedule provides information regarding the luminaires to be installed.

Figure 8-10. The elevator car position indicator is connected to the low-voltage lighting circuit.

Drinking fountains are to be installed on the south wall of Room 300 east of the Room 302 door. As shown on Sheet M7.2, Detail 1, the drinking fountains are supplied with cold water by a ½″ chilled water pipe from a remote chiller installed above the ceiling. A 1″ cold water supply pipe is installed from the west, extending towards the men's and women's restrooms. The 1″ hot water supply pipe for the restrooms is also installed above the ceiling in Room 300. As shown on Sheet M2.3, a 4″ waste pipe is installed between the two supply pipes, with a cleanout near the elevator door, feeding to a 1½″ horizontal pipe and 1¼″ vertical waste pipe up to the next floor level.

As indicated on Sheet M3.3, the elevator shaft is protected by a wet pipe (WP) fire protection system. Sleeves are provided in the shear wall at the east side of Room 300 to allow the wet pipe fire main piping to pass through the wall.

Heat is provided to the classroom level by a hot water (hydronic) system with terminal units, ductwork, and ceiling diffusers. Sheet M8.3 provides an overview of the hot water heating system for the entire building. **See Figure 8-11.**

TECH TIP

The purpose of the National Electrical Code® (NEC®) is to protect people and property from hazards that arise from the use of electricity. The NEC® is revised and updated every three years to reflect current trends and developments in the electrical industry.

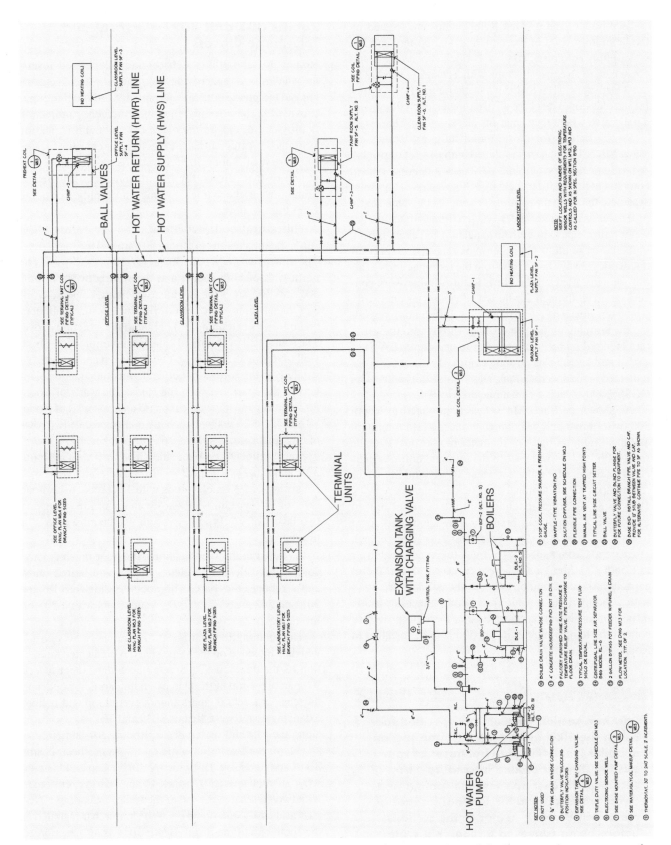

Figure 8-11. The Heating Piping Schematic on Sheet M8.3 provides an overview of the hot water heating system for the building.

Two boilers and two hot water pumps are located on the laboratory level of the building. The boilers and hot water pumps are installed on a 4″ thick concrete pad and connected to an expansion tank with a charging valve. The hot water supply (HWS) and hot water return (HWR) pipes for the classroom level are 4″ in diameter, each fitted with a ball valve where they enter the classroom level.

Piping details for the terminal units are shown on Sheet M8.5, Detail 4 (not included with this set of prints). A *terminal unit,* or terminal device, is a heating system component that transfers heat from hot water in a hot water system to the air in the building system. A hot water supply pipe extends from the boiler to each terminal unit and a hot water return pipe exits each terminal unit and extends back to the boiler to allow the water to be reheated. Supply air ductwork conveys heated air from the terminal units to air diffusers installed in the ceiling. Return air ductwork conveys return air through a grille and into return air ductwork, which terminates at the terminal unit. Information regarding the terminal units to be installed throughout the structure is based on an entering water temperature (EWT) of 180°F and an entering air temperature (EAT) of 55°F. As indicated in the specifications, all terminal units are Model DESV-3000, which are manufactured by Titus.

As shown on Sheet M4.3, heat is provided to Room 300 by terminal unit #33, which has a 330 cfm minimum and maximum rating. A 6″ diameter blower duct extends from a 36″ × 16″ oval duct to the terminal unit. As indicated on Sheet M5.3, the hot water supply and hot water return pipes for the terminal unit are ¾″ diameter. A 14″ × 10″ supply air trunk extends from the terminal unit to one 12″ × 8″ supply air branch to provide heated air to Rooms 301 and 302. An 8″ diameter supply air branch also extends from the 14″ × 10″ trunk to provide heated air to Room 300 through a 24″ long diffuser with a 150 cfm capacity. All supply air branches in the building are fitted with a damper and flexible duct before attaching to the diffusers. As shown on Sheet E4.3, a thermostat used to control the temperature in Room 300 is to be installed in Room 301.

TECH TIP

Cast iron sectional boilers are specified for use in the SIRTI building. A cast iron sectional boiler is a boiler with hot gases of combustion flowing around internal cast iron sections. The cast iron sections contain and direct water around the internal furnace of the boiler. Water is heated in the sections before being conveyed through hot water supply pipes to the terminal units.

Rooms 301 and 302—Electrical and Telephone Closets

Rooms 301 and 302 (Electrical and Telephone Closets) are small rooms that provide secure areas to access electrical and telephone control equipment. Minimal lighting is provided in these rooms. Fire and emergency protection is provided to ensure the electrical and communication systems are safeguarded. A mechanical chase is located to the east of Room 302. A *chase* is an enclosure in a structure that allows for the placement of electrical, plumbing, or mechanical wiring and piping extending from floor to floor.

Architectural and Structural. As indicated on Sheet A2.6, the dimensions of Room 301 are 6′-10⅞″ × 8′-6⅝″. The dimensions of Room 302 are 6′-10⅜″ × 8′-6⅝″. The north wall of both rooms and the wall between Rooms 301 and 302 are Type 2 walls, which are constructed of 3½″ metal studs and covered with ⅝″ gypsum wallboard on both sides extending full height to the structure above. The south wall of Room 301 and west portion of the south wall of Room 302 are Type 4 walls, which are constructed similarly to the north wall but with sound-attenuation blankets in the space between the studs. The walls of Room 302 abutting the mechanical chase and east wall of Room 302 are Type 13 walls, which are constructed of two rows of 3½″ metal studs spaced 16″ OC and covered with ⅝″ gypsum wallboard on both sides. The west wall in Room 301, identified as a Type 23 wall, is an elevator shaft wall consisting of 3½″ metal studs with 1″ coreboard and one layer of ⅝″ gypsum wallboard with a 1 hr fire rating and two layers of ½″ Type X gypsum wallboard with a 2 hr fire rating.

As noted in the specifications, the floor finish is vinyl composition tile with a clean, waxed, and buffed finish and a rubber cove base. The doors leading into Rooms 301 and 302 are 3′-0″ × 7′-0″ wood doors with a 20 min fire rating.

The Framing Plan on Sheet S2.4 refers to Section 12, which is included on the same sheet. As indicated on Section 12, the floor is a 6″ concrete slab with #4 rebar spaced 9″ apart at the top of the slab extending north to south and #4 rebar spaced 18″ apart and staggered top and bottom extending east to west. Beams 3B14 and 3B15 are indicated along the east and west sides of Rooms 301 and 302. As shown on the Beam Schedule on the same sheet, beams 3B14 and 3B15 are 12″ wide by 20½″ deep and have #4 rebar stirrups spaced 9″ apart. Beam 3B14 is reinforced with two #6 rebar at the top of the beams extending 10′-6″ past grid line 7 and one #7 × 8′-0″ long rebar installed at the south end of the beam. Beam 3B14 is also reinforced with two #6 rebar at the bottom of the beam projecting 3″ past grid line 7 and beam 3B12. Beam 3B15 is reinforced with two #5 rebar at the top that are hooked on the south end 4″ past grid line 7. Two #5 rebar are installed at the bottom of beam 3B15, projecting 3″ past adjoining beams 3B12 and 4B13.

Mechanical and Electrical. As shown on Sheet E2.3, illumination in Room 301 is provided by a Type G luminaire, which is a chain-hung, surface-mounting, two-lamp luminaire with a lamp guard. As indicated in Note 9 on Sheet E2.3, the luminaire is controlled by a red-handled switch finished with a cover plate. **See Figure 8-12.** The luminaire is connected to emergency circuits 2 and 4 on panelboard 4X1.

Figure 8-12. Red-handled switches and cover plates clearly identify emergency circuits in the SIRTI building.

As shown on Sheet E3.3, four panelboards (2B, 2D, 2X1, and 4B), one low-voltage relay control cabinet, one service drop panelboard (SD-5), and two transformers (TX1 and TB) are to be installed in Room 301. Sheet E2.3, Detail 3 provides information regarding switching and relay arrangements for the low-voltage circuits. Each relay mounted in the low-voltage relay cabinet and the related equipment and classroom are shown on Detail 3. For example, the outboard lamps in Room 311 are controlled by relay 10 and powered by circuit 20 on panelboard 4B. Transformer TX1 is to be installed above panelboards 2B and 2X1, while transformer TB is to be floor mounted. As indicated on Sheet E4.3, Room 301 also contains a smoke detector and thermostat.

As shown on Sheet E2.3, illumination in Room 302 is provided by a Type G luminaire, which is a chain-hung, surface-mounting, two-lamp luminaire with a lamp guard. The luminaire is controlled by a switch inside the room. As shown on Sheet E4.3, conduit risers are to be installed in the northeast corner of Room 302. As indicated by Note 7 on Sheet E4.3, a ceiling-mounted TV monitor is to be installed in Room 302 with ¾″ conduit routed from

the monitor to the cable tray. As shown on Sheet E3.3, six duplex receptacles are to be installed in Room 302 and connected to various circuits on panelboard 2X1. As indicated on Sheet E4.3, a rate-of-rise heat detector is also to be installed in Room 302. A rate-of-rise function detects heat quickly by responding to a rapid temperature increase. The pneumatic element in the heat detector responds to a rapid rise in temperature (approximately 15°F per minute) when the air expands faster within the sealed chamber than it can escape through the calibrated vent. The increase in pressure depresses the diaphragm, causing the electrical contacts to close the circuit and activate the device.

As shown on Sheet E2.3, two keyless porcelain luminaires with a switch are to be installed in the mechanical chase. Locations of the luminaires and switch are to be coordinated with Division 15 of the specifications.

As indicated on Sheet M3.3, the main header for the preaction fire protection system is to be installed in the northwest corner of the mechanical chase. A *preaction fire protection system* is a type of dry pipe system in which two separate events—smoke detection and heat development—must occur for the sprinkler system to activate. Preaction fire protection systems are commonly used in locations where there is a concern of accidental discharge of water, such as in computer rooms or control rooms. A preaction valve is installed in the water supply piping. Smoke detection allows water to flow into the sprinkler piping, but the sprinkler heads do not open until they are activated by heat from a fire. The mechanical chase is protected with a wet pipe fire protection system. **See Figure 8-13.**

Figure 8-13. Wet pipe fire protection systems are installed in the mechanical chase.

Sheet M7.2, Detail 1, provides information regarding pipe to be installed in the mechanical chase, including a ¾" cold water pipe and a ¾" drain pipe for a perchloric hood washdown unit, a 4" vacuum (VAC) vent, a 4" acid-resistant vent (ARV) stack, an 8" roof drain and overflow drain stacks, a 4" waste pipe extending down, a 6" vent stack, a 2½" cold water riser, a 1¼" hot water riser, and a ½" hot water circulating (HWC) riser. Various water supply and vent pipes to the men's and women's restrooms are installed through the north wall of the mechanical chase, extending eastward toward the restrooms. Supply and vent pipes for the water fountain in Room 300 are also to be installed in the mechanical chase.

Heat is provided to Rooms 301 and 302 by a 12" × 8" branch duct extending from the 14" × 10" trunk in Room 300. Six-inch square diffusers, each with a 100 cfm capacity, are supplied with air by a 6" flexible duct extending from the 12" × 8" duct. Return air is provided by a 12" × 8" duct with a 6" square grille.

A roof drain receives stormwater collecting on a roof surface and discharges it into a drain line.

Room 303—Information Networking Resource Center

Room 303 (Information Networking Resource Center) is a small room located near the southwest corner of the SIRTI building. The room is not designed for use as a classroom; rather, the room is utilized by students for research and is equipped with a computer printer and several tables.

Architectural and Structural. The dimensions of Room 303 are 25'-0" × 23'-2⅝" based on dimensions provided and gridline spacing. The 25'-0" dimension is determined by adding the dimensions along the west wall (1'-0" + 3'-0" + 2'-0" + 8'-0" + 2'-0" + 8'-0" + 1'-0" [one-half column width] = 25'-0"). The 23'-2⅝" dimension is calculated by adding the 30'-0" column grid spacing with 14" for the floor projection, and subtracting 7'-4¾" for the corridor width and 6⅝" for the metal-framed wall at the exterior wall (30'-0" + 14" − [7'-4¾" + 6⅝"] = 23'-2⅝"). The walls of Room 303 are Type 4 walls, which are constructed with 3½" metal studs and covered with ⅝" gypsum wallboard on both sides extending full height to the structure above, with sound-attenuation blankets between the studs.

Sheet A10.4, Elevation 16, shows the casework and shelving units to be installed along the north wall of Room 303. **See Figure 8-14.** The Plastic Laminate Casework Schedule provides detailed information regarding each cabinet. The coat closet is constructed of two Type K cabinets, each measuring 30" wide by 84" high by 25" deep. Three Type B cabinets with five shelves and one Type A cabinet with two shelves are also installed along the north wall. The north wall is finished with five equal-width panels of tackable wallcovering and one section that is 1'-6" wide.

Sheet A10.4, Elevation 17, provides an elevation of the east wall of Room 303. **See Figure 8-15.** A 4'-0" × 8'-0" liquid marker board is to be installed 3'-0" above the floor on the east wall. Five equal-width panels of tackable wallcovering and two panels cut to fit the remaining space are also installed on the east wall.

The floor plan on Sheet A2.5 refers to Detail 1, which is also on the same sheet. Detail 1 shows the exterior wall and column construction at grid line intersection 6/G. Wood blocking is fastened to the interior corners of the concrete column and covered with gypsum wallboard. The finished surface of the gypsum wallboard is to be 9½" from the center of the concrete column. The gypsum wallboard is then finished with tackable wallcovering. Additional construction information for a typical column finish is provided on Sheet A2.5, Detail 2.

TECH TIP

Diffusers and grilles are selected and sized to provide proper airflow into each building space to offset the heat loss or gain. Diffusers and grilles must be properly located to efficiently distribute the air to the building spaces. Diffuser size is determined by the airflow rate and required air velocity in a building space. Grille size is based on the air velocity at the grille face.

As indicated in the specifications, the floor finish is to be conductive vinyl composition tile (CVCT), which is cleaned, waxed, and buffed, with a rubber cove base. One 3'-0" × 7'-0" wood door is to be installed at the entrance to Room 303. One Type 1 and two Type 2 windows are shown for Room 303 on Sheet A2.5. As indicated in the specifications, Type 2 windows are 8'-0" × 5'-0" windows consisting of a 5'-0" square fixed light and a 3'-0" wide operable window that pivots vertically. The Type 1 window is a 3'-0" × 5'-0" vertically pivoting window. **See Figure 8-16.**

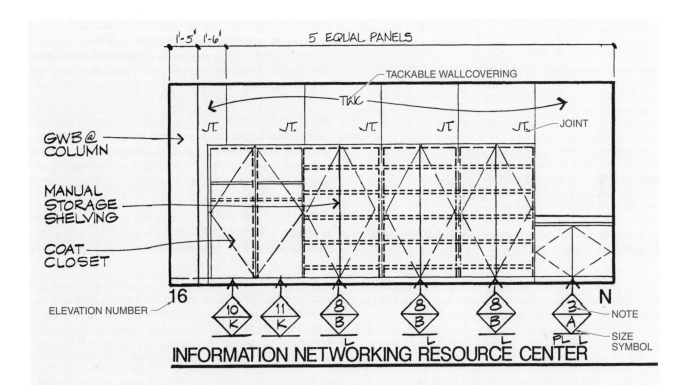

PLASTIC LAMINATE CASEWORK SCHEDULE (12302)						
Model Symbol	Model #	Notes	Size Symbol	Width	Height	Depth
1	10011	1	A	47	36	25
2	10013	2	B	47	84	25
10	33011	10	J	47	30	30
11	33012	10	K	30	84	25
12	55012	11	L	36	30	13

Notes

1. 1 shelf-base cabinet
2. 1 shelf
3. 2 shelves
4. sink base
5. 4 equal drawers - single bank
6. 4 equal drawers - double bank
7. 2 file drawers
8. 5 shelves
9. Vertical partition & five shelves
10. Coat rod with shelf
11. 1 shelf - wall cabinet
12. Apron width 60"

Figure 8-14. Interior elevations show cabinet and casework configurations in orthographic projection. Schedules and notes provide details of items identified on the elevations.

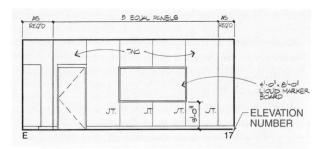

Figure 8-15. Elevation 17 provides an elevation of the east wall of Room 303 and identifies finish materials such as tackable wallcovering and liquid marker boards.

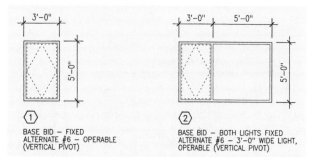

Figure 8-16. Type 1 and 2 windows are the predominant window types to be installed on the classroom level.

The windows in Room 303 are also shown on the West Elevation on Sheet A4.2. Construction details for the areas around the windows are referred to on the West Elevation and shown on Sheet A11.3, Details 2, 7, and 8 (not included with this set of prints). A brick-veneer control joint (CJ) is indicated on the West Elevation at grid line 6.

As shown on Sheet A2.4, a structural column is indicated at grid line intersection 6/G. The Column Schedule on Sheet S5.1 indicates that column G6 is a 16″ square concrete column that extends to an elevation of 1944′. Reinforcing steel for this column at the classroom level consists of eight #7 rebar, which overlap four #11 rebar extending up from the plaza/entry level by 3′-6″. Detail 23 on Sheet S5.1 specifies that columns with eight vertical rebar require 1½″ of clear space between the rebar and the face of the concrete columns. Stirrups are fabricated from #4 rebar and are spaced 8″ OC along the entire height of the classroom level for this column. Sheet S5.1, Detail 16, shows the connection between the column and beam above. The top surface of the column is to be roughened to provide a good joint between the column and beam. Detail 16 also indicates the reinforcing steel arrangement where the column joins the concrete beam and deck above.

Type C5, C7, M1, and M4 joists are indicated on Sheet S2.4 for Room 303. The Joist Schedule provides information regarding the width, depth, and reinforcing steel requirements of the joists. All joists for the classroom level floor are 8″ wide by 20½″ deep. C5 joists are reinforced with 16 #6 rebar beneath the top exterior surface, one #9 rebar and one #10 rebar along the bottom, and 36 #6 rebar along the top interior surface. C7 joists are reinforced with eight #5 rebar along the top exterior surface, two #8 rebar along the bottom, and eight #6 rebar along the bottom interior surface. M1 joists are reinforced with two #7 rebar along the bottom and 16 #5 rebar along the top interior surface. M4 joists are reinforced with two #8 rebar along the bottom and 16 #5 rebar along the top interior surface.

As shown on Sheet S2.4, perimeter beam 3B4 is specified for Room 303. A 3B4 beam is 18″ wide by 20½″ deep and is reinforced with #6, #8, and #10 rebar as indicated on the Beam Schedule. Stirrups are fabricated from #4 rebar spaced 9″ apart, similar to the size and spacing of stirrups for the other beams in the building. A ⅜″ camber is built into the 3B4 beam during the forming process. *Camber* is a slight upward curvature in a horizontal structural member that is designed to compensate for deflection of the member under load. Sheet S2.3A, Detail 22, provides typical information for perimeter concrete beams. The outer edge of the beam extends 1′-2″ beyond the grid line, or 8″ if indicated otherwise. Two inches of concrete coverage is required over the rebar at the top of the beam and 1½″ of coverage is required on the bottom, inner, and outer edges of the beam. Reinforcing steel information is provided in the Beam Schedule on Sheet S2.3A.

Mechanical and Electrical. In addition to the natural light provided through the windows, illumination in Room 303 is provided by nine Type A and two Type K luminaires as indicated on Sheet E2.3, Classroom Level Lighting Plan. Type A luminaires are recessed 2′ × 4′ fluorescent luminaires with 3″ deep parabolic louvers and three energy-saving lamps per luminaire. Type K luminaires are 8″ diameter recessed incandescent luminaires with clear reflectors and one 150 W lamp per luminaire. Type K luminaires are connected to circuit 16 on panelboard 2D and controlled by a dimmer switch.

As shown on Sheet E3.3, seven duplex electrical receptacles are to be installed in Room 303 and connected to various circuits on panelboard 2D. Two of the receptacles are isolated-ground receptacles as indicated by the "IG" next to the receptacle symbol. An *isolated-ground receptacle* is a receptacle in which the grounding terminal is isolated from the device yoke or strap. The isolation ensures a clean equipment ground for the electrical equipment that may be adversely affected by noise in the equipment grounding path. Isolated-ground receptacles are identified by an orange triangle on the receptacle face.

Three data/power poles are to be installed in Room 303, one at the north end of each of the sets of workbenches. Data/power poles are vertical raceways that extend from the ceiling to the floor. The data/power poles enclose data and power conductors and allow data and/or power to be delivered to the workbenches. Details of the data/power poles are shown on Sheet E5.2, Detail 2 (not included with this set of prints). Each set of workbenches is prewired with isolated-ground receptacles.

As indicated on Sheet A4.3, a clock outlet, mini-horn, and thermostat are to be installed on the east wall in Room 303. A door intrusion sensor is to be installed near the door. Two telecommunications outlets are to be installed on the south and east walls of Room 303.

As shown on Sheet M4.3, heat is provided to Room 303 by terminal unit #26, which has a 680 cfm maximum and 350 cfm minimum rating. **See Figure 8-17.** A 9″ diameter blower duct extends from a 36″ × 16″ oval duct to the terminal unit. The 36″ × 16″ duct is to be secured tight to the structure above the ceiling. As indicated on Sheet M5.3, the hot water supply and hot water return pipes for the terminal unit are ¾″ diameter. A 14″ × 12″ supply air trunk extends from the terminal unit to four 8″ diameter supply air branches and four 48″ long diffusers, each with a 170 cfm capacity. Return air is routed through a 22″ square grille and 22″ × 8″ duct and into a 45″ × 16″ return duct. The 45″ × 16″ return duct is also to be secured tight to the structure above the ceiling at the west side of the room.

As shown on Sheet M3.3, a wet pipe fire protection system is specified for Room 303. The fire protection system is installed along the north wall of the room.

Room 304—Student Lounge/Vending Area

Room 304 (Student Lounge/Vending Area) is a triangular-shaped informal area for student and faculty access. Vending machines are to be located along the east wall of Room 304. Four windows along the southwest side of the room provide natural light for the room.

Terminal units and ductwork must be properly supported.

TECH TIP

Per the SIRTI specifications, a minimum of 3′ of straight rigid steel duct is to be provided immediately upstream of the terminal unit to provide proper airflow.

TERMINAL UNIT BOX SCHEDULE
MFR: TITUS

#	INLET DIA. (IN)	MAX CFM	MIN CFM	MIN SP (IN)	NOISE CRI AT ROOM(2)	NOISE(1) CRITERIA RADIATED	EAT = 55° LAT	LWT	GPM	HOT WATER COIL WATER PD (IN)	EWT = 180° AIR PD (IN)	# OF ROWS	MBH	MODEL NO.	NOTES
TU-1	5	180	180	0.13	37	27	99	160	0.07	0.03	–	1	–	CESV-3000	5°-4 (6)
2	9	760	760	0.19	39	27	105	158	3.8	1.05	–	2			(6)
3	6	290	110	0.17	33	25	102	159	5.4	20		1			
24	10	850	270	0.20	39	26	123	153	1.46	.19		2			
25	10	850	270	0.20	39	26	123	153	1.46	.19		1			
26	9	680	350	0.15	39	26	117	153	1.7	25		1			
27	8	510	250	0.18	39	26	117	152	1.22	34		2			
28	6	300	120	0.18	34	26	100	160	55	22		1			(6)
29	14	1400	420	0.23	32	26	105	160	2.27	33		1			
30	14	1400	420	0.23	32	26	105	160	2.27	33		1			

HOT WATER COIL CABINET

AIR INLET

Titus

Figure 8-17. Terminal units transfer heat from hot water supply piping to ductwork installed above the ceilings.

Architectural and Structural. As shown on Sheet A2.6, the dimensions of Room 304 are irregular due to its location at an angled portion of the building. The north wall is located 5′-0″ north of grid line 7. The east wall of the room is centered on the grid line H. The southwest window wall is 43′-10½″ long and is at a 45° angle to grid line G. Interior walls are Type 4 walls, which are constructed of 3½″ metal studs and covered with ⅝″ gypsum wallboard on both sides extending full height to the structure above, with sound-attenuation blankets between the studs.

Sheet A2.5 indicates that two elevations of the room are included in the prints. **See Figure 8-18.** Tackable wall-covering is to be installed along the north wall of the room. A 3′-4″ high borrowed light frame with Type W glass is to be installed 4′-0″ above the finished floor along the north and east walls. The edge of the borrowed light frame on the north wall is installed 3′-6″ from the adjoining vending area east wall. The borrowed light frame on the east wall near the door is spaced equally between the adjoining north wall and the door frame. Vending machines are to be furnished and installed by the owner (FIO) along the east wall.

As referenced on Sheet A2.5, Detail 4 on the same sheet describes the columns at the ends of the angular wall. The diagonal distance from the center of the column to the finished surface of the interior wall is 1′-0″. The finished surface of the exterior brick-veneer wall structure is 1′-9″ from the center of the column. Special brick shapes "C" and "D" are to be installed at the 45° angle corner. The exterior wall is constructed with metal studs and covered with gypsum wallboard on both sides. Batt insulation is to be installed between the studs. As indicated on Sheet A2.5, Type 1 and 2 windows, similar to the windows installed in Room 303, are to be installed in the southwest wall of Room 304. The windows appear foreshortened on the West Elevation on Sheet A4.2 because the wall is constructed at an angle. Solid wood blocking is to be fastened to the interior column corners to support the gypsum wallboard.

As noted in the specifications, the floor finish is direct glue-down carpet with a rubber cove base. Door 304 is a 3′-0″ × 7′-0″ wood door with a 20 min fire rating.

Structural columns G7 and H8 are shown in this area on Sheet S2.4. As indicated on the Column Schedule on Sheet S5.1, columns G7 and H8 are 16″ square concrete columns that extend to an elevation of 1944′. Reinforcing steel for these columns at the classroom level consists of eight #7 rebar overlapping four #11 rebar extending up from the lower level by 3′-6″. Stirrups are fabricated from #4 rebar and are spaced 8″ OC along the entire height of the classroom level. Sheet S5.1, Detail 16, provides information about roughening the top of the columns and the arrangement of reinforcing steel where the columns join the concrete beam and deck above.

As indicated on Sheet S2.4, type C5 and M3 joists are specified in the area of Room 304. Typical domes measuring 16″ high by 52″ square are used to form most of the concrete joists for the floor of Room 304. Type 2 domes are used to form the perimeter joists. Type 2 domes are 14″ high by 19″ square. Note that Type 2 domes are 2″ shorter than other dome forms used to form the two-way joist system, allowing the slab to be 2″ thicker in these areas. Perimeter beam 3B9 is 18″ wide. The 1′-2″ dimension noted along beam 3B9 indicates that the face and edge of the beam and the floor slab project 1′-2″ beyond the center of the column grid lines. As noted in the Beam Schedule on Sheet S2.4, perimeter beam 3B9 is reinforced with #6, #8, and #10 rebar and has a ½″ camber.

TECH TIP

Per the SIRTI specifications, unfaced R19 batt insulation is to be friction fit between the wall studs.

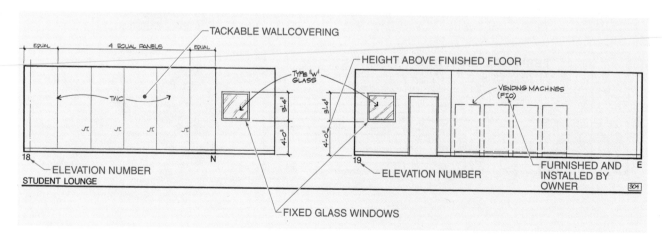

Figure 8-18. Two interior elevations are included in the prints to describe the wall finish materials for Room 304.

The South Elevation on Sheet A4.2 shows the exterior wall finish for Room 304. A masonry control joint is to be placed in the exterior brickwork between grid lines G and H. Sheet A4.2, Detail 2, shows the construction of a typical control joint. Brick ties fastened through the gypsum sheathing and into the metal studs are secured to the brick on each side of the control joint. The joint is filled with a 3″ wide by ½″ thick piece of neoprene and finished with a sealant at the face of the joint. The West Elevation also shows the exterior wall of Room 304 at the far right.

Mechanical and Electrical. In addition to the natural light provided through the windows, illumination in Room 304 is provided by eight Type B luminaires. Type B luminaires are recessed 2′ × 4′ fluorescent luminaires fitted with a regressed acrylic lens and are provided with two energy-saving lamps. As shown on Sheet E2.3, four of the luminaires are connected to switch "a" and the other four luminaires are connected to switch "b." Circuit 2 on panelboard 4B provides power to the luminaires in Room 304. As shown on Sheet E3.3, ten double duplex electrical receptacles are to be installed in Room 304 and are provided with power by various circuits on panelboard 2B. As indicated in Note 4 on Sheet E3.3, a separate neutral and insulated-ground conductor is provided for each information outlet receptacle circuit.

As indicated on Sheet E4.3, a clock outlet and mini-horn are to be installed on the east wall above the vending machines. A door-intrusion sensor and a thermostat are to be installed near the door and a telecommunications outlet is to be installed on the north wall.

As shown on Sheet M2.3, waste and vent piping comprise the majority of the piping associated with Room 304. A 4″ waste pipe extends east to west along grid line 7 and terminates with a 4″ cleanout. A 2″ waste pipe extends up to the level above from the 4″ waste line. A 4″ horizontal waste branch pipe extends to the south from the 4″ waste pipe, with two 2″ waste pipe and one 4″ waste pipe extending up to the level above. Three 2″ vent pipes are installed above Room 304. Another 2″ waste pipe is installed along the east wall in Room 304 near the vending machines. A 2″ floor drain is installed to the south of the vending machines.

As shown on Sheet M2.3, a 1½″ cold water supply pipe is to be installed near the floor drain, with a ½″ chrome stop valve to be installed 12″ above the finished floor. The 1½″ cold water supply pipe continues to the north above Room 304 to provide a water supply for plumbing fixtures to be installed in other rooms. As shown on Sheet M3.3, a wet pipe for the fire protection system is installed along the east wall of Room 304.

As shown on Sheet M4.3, heat is provided to Room 304 by terminal unit #27, which has a 510 cfm maximum and 250 cfm minimum rating. An 8″ diameter blower duct extends from a 36″ × 16″ oval duct to the terminal unit. As indicated on Sheet M5.3, the hot water supply (HWS) and hot water return (HWR) pipes for the terminal unit are ¾″ diameter. A 14″ × 12″ supply air trunk extends from the terminal unit to three 8″ diameter supply air branches and three 48″ long diffusers, each with a 170 cfm capacity. Return air is routed through a 22″ square grille and into a 22″ × 18″ return duct.

TECH TIP

Ductwork for the SIRTI building is to be fabricated from galvanized steel or aluminum HVAC distribution ductwork.

Room 305—Studio Classroom

The design and audio/video equipment in Room 305 allow instruction to be provided at a remote location. An instructor can teach a class while video cameras around the room capture the information for broadcast to other remote learning locations or can allow for recording of the lessons. **See Figure 8-19.** Cameras, monitors, and the room layout are designed to facilitate distance learning.

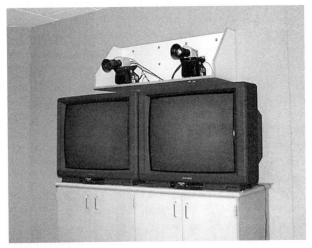

Figure 8-19. Cameras and monitors are installed in Room 305 to facilitate distance learning.

Architectural and Structural. The dimensions of Room 305 are 35′-4⅛″ × 30′-0″ based on a 30′-0″ distance between grid lines. The 35′-4⅛″ dimension is determined by subtracting the 10⅞″ dimension along the south end of the room from 30′-0″ and adding the 6′-3″ dimension at the north end of the room (30′-0″ − 10⅞″ + 6′-3″ = 35′-4⅛″).

Two 45° 1'-6" deep offsets are to be constructed along the south wall of Room 305 beginning 6'-9⅜," from the east and west walls. The offsets allow the egress doors to open outward without impeding pedestrian traffic flow in Room 300 (Elevator Lobby).

As indicated on Sheet A2.6, Type 4 walls are specified for Room 305. Type 4 walls are constructed with 3½" metal studs and covered with ⅝" gypsum wallboard on both sides extending full height to the structure above. Sound-attenuation blankets are to be installed between the studs.

Sheet A2.5 references Elevations 21, 22, and 23 on Sheet A10.4 for interior elevations of Room 305. **See Figure 8-20.** Detail 21 shows the north wall construction. A 12'-0" long chalkboard is installed 3'-0" above the raised platform in the center of the north wall. A 7" high by 8'-0" wide raised platform with two ramps, a landing, and a handrail is located at the north end of the room. Two monitors that are furnished and installed by the owner are placed on two custom monitor cabinets that are detailed on Sheet A9.2 (not included with this set of prints). Doors on the monitor cabinets are hinged to swing toward the outside of the cabinets, thus providing adequate storage space. The teaching console on the platform is also detailed on Sheet A9.2. Elevation 22 represents the east and west walls of Room 305. The west interior elevation of Room 305 is a reverse view (opposite hand) of the east interior elevation. The tops of the monitor cabinets at the north end of Room 305 are 4'-6" high above the teaching platform to facilitate viewing. Elevation 23 shows another monitor cabinet installed along the south wall, two egress doors, and tackable wallcovering.

Sheet A11.2, Detail 7, describes the two columns along the east and west walls in Room 305. The columns are finished in a similar manner to other columns with solid wood blocking at the column corners to support gypsum wallboard. A corner bead is to be installed at each corner to provide protection for the gypsum wallboard corners. The metal stud walls are finished with gypsum wallboard and tackable wallcovering.

Wall Section C on Sheet A6.1 is referenced on the floor plan on Sheet A2.5. Wall Section C is a structural detail taken through the west wall of Room 305 and extending vertically up and down through other levels of the building. Wall Section C provides general information regarding the overall building structure. Starting from the ground level, a footing supports exposed concrete columns, which have a light sandblast finish. A concrete slab-on-grade is placed over a vapor barrier and compacted fill.

The entry-level floor system is a cast-in-place concrete two-way joist system and slab. The wall on the entry level below Room 305 consists of a series of 2'-0" high porcelain enamel-on-steel panels and glass in hollow metal frames. The floor-to-floor height of the entry level is 12'-0".

A cast-in-place concrete two-way joist system and slab are constructed over the entry-level columns, providing a 14'-0" floor-to-floor height. The finished ceiling for the classroom level is an acoustical tile, which is suspended from the concrete joists and slab above.

Above Room 305, the overall roof height extends 28'-0" (15'-6" + 12'-6" = 28'-0") above the office/administration level. A structural steel beam is supported by the concrete column at grid line H. Cementitious fireproofing material is sprayed onto the steel beam and underside of the roof and open web steel joists that support the roof. The flat roof to the east of the wall is 1½" thick metal roof decking with a concrete topping that supports R-38 rigid roof insulation and elastomeric roofing. The rigid roof insulation is tapered to provide proper drainage for the flat roof.

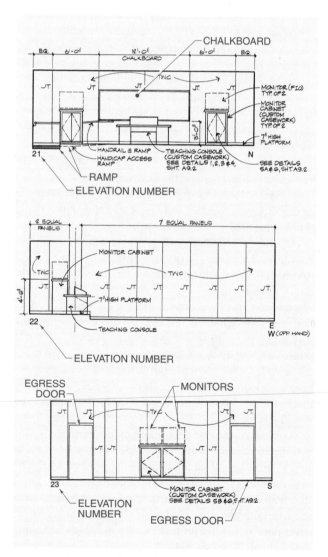

Figure 8-20. On Sheet A10.4, Elevations 21, 22, and 23 provide information regarding wall finishes and other equipment to be installed in Room 305.

The parapet projects 9'-0" above the concrete roof deck and 4'-0" above the sloping portion of the west roof finish. The parapet is framed with metal studs and is finished with sheet metal vertical covering and prefinished metal coping. Additional details regarding framing of the parapet are included on Sheet A11.5, Details 1 and 2 (not included with this set of prints).

The sloped portion of the roof is supported by 1½" steel roof decking fastened to open web steel joists. A layer of ⅝" Type X gypsum board is applied over the roof decking, and rigid roof insulation with a minimum R-38 rating is applied to the top of the gypsum board. Two layers of #15 felt building paper are applied to the top of the rigid roof insulation, and the roof is finished with custom sheet metal roofing.

As indicated on Sheet A2.6, Tile & Carpet Floor Pattern Layout, the floor finish in Room 305 is carpet with a 1'-0" wide carpet border. The floor plan on Sheet A2.5 indicates two doors for Room 305—300A and 300B—which are 3'-0" × 7'-0" wood doors.

As shown on Sheet A2.5, columns J5 and H5 intersect the east and west walls of Room 305. As noted in the Column Schedule on Sheet S5.1, columns J5 and H5 are identical at the classroom level. Columns J5 and H5 are 16" square concrete columns that extend to an elevation of 1944'. The reinforcing steel at the classroom level consists of four #8 rebar overlapping the eight #11 rebar extending up from the floor below by 3'-9". Stirrups are fabricated from #3 rebar and are spaced 16" OC along the entire height of the classroom level.

As indicated on the Framing Plan on Sheet S2.4, type C5 joists are to be formed near column J5, and type C6 and M1 joists are to be formed throughout the remainder of Room 305. A construction joint is shown on the Framing Plan in the concrete floor extending in an east-west direction. As shown on Sheet S2.3A, Detail 23, a 3½" × 6" keyway is to be formed at each joist along the construction joint. The construction joint is to be formed at the center of the dome form. Reinforcing steel along the bottom of the slab is to continue through the construction joint. Along the top of the slab, #3 × 2'-0" long dowels spaced 12" OC are used to tie the first and second concrete sections together at the construction joint. The dowels should overlap 2'-0".

Mechanical and Electrical. Illumination is provided in Room 305 by 16 Type A luminaires and six Type K luminaires (similar to the luminaires in Room 303). The 12 Type A luminaires at the south end of the room are controlled by switches "a" and "b," and the other four Type A luminaires are controlled by switches "c" and "d." This switching arrangement allows the four luminaires at the north end of the room to be turned off when lighted presentations are made from the raised platform. The Type A luminaires are connected to circuits 30 through 40 on panelboard 2B.

Type K luminaires are controlled by an incandescent wall box dimmer switch and are connected to circuits 22 and 24 on panelboard 2D. Low-voltage luminaires at the north end of Room 305 are indicated on Sheet E2.3, Detail 3—Low-Voltage Control on relays 1 through 4, and are connected to circuits 38 and 40 on panelboard 2B.

As indicated on Sheet E3.3, 11 duplex electrical receptacles are to be installed around the perimeter of Room 305 with one of the receptacles being an isolated-ground receptacle. The receptacles are connected to circuits 7, 9, or 11 on panelboard 2D1. Four data/power poles are to be installed at the west end of the workbenches and connected to circuits 13 through 20 on panelboard 2D1. Sheet E5.2, Detail 2 (not included with this set of prints), provides information regarding the typical connection of data/power poles to a cable tray above the ceiling. A *cable tray* is an assembly of sections and associated fittings that form a rigid structural system used to support cables and raceways. **See Figure 8-21.** As noted on Sheet E4.3, the cable tray is 12" wide by 4" deep with an open trough and is installed with the top of the tray 11'-0" above the floor.

As shown on Sheet E4.3, a clock outlet is installed on the west wall of Room 305, with three #16 conductors enclosed in ½" conduit extending to the cable tray. As indicated in Note 6, this clock installation is typical of all clocks on this floor and the clocks are to be routed on one three-conductor #16 cable to a master clock. A mini-horn is also installed on the west wall. A thermostat used to control the temperature in Room 305 is located near the southeast door. Two telecommunications outlets are installed at the north and south walls. Two video cameras are to be installed at the south end of Room 305, and one video camera is to be installed at the north end of the room.

Figure 8-21. Cable trays allow for quick installation of electrical cable and provide easy access for electricians.

A symbol on Sheet A2.5 indicating a down-pointing camera to be furnished by the owner and installed in a well in the ceiling is shown at the north of the room. A specific note in Room 320 identifies the symbol as a down-pointing camera.

As shown on Sheet M4.3, heat is provided to Room 305 by terminal unit #31, which has a 1120 cfm maximum and 330 cfm minimum rating. A 12″ diameter blower duct extends from a 35″ × 10″ oval duct to the terminal unit. As indicated on Sheet M5.3, the hot water supply and hot water return pipes for the terminal unit are ¾″ diameter. A 16″ × 15″ supply air trunk extends from the terminal unit to seven 8″ diameter supply air branches and seven 48″ long diffusers, each with a 160 cfm capacity. Return air is routed through a 22″ square grille and into a 22″ × 16″ return duct. As indicated on Sheet E4.3, a thermostat located near the northeast door in Room 305 controls the temperature in the room.

As indicated on Sheet M3.3, the preaction main for the fire protection system is installed at the northeast corner of Room 305. The preaction main provides fire protection for Room 308.

Rooms 306 and 307—Continuing Education

Rooms 306 and 307 are designed as traditional classrooms. Operable wall panels between the two rooms can be stacked along the east side of Rooms 306 and 307 to combine the rooms and provide adequate space for larger groups.

Architectural and Structural. As shown on Sheet A2.6, the dimensions of each room are 30′-0″ × 23′-2⅝″ based on standard grid line spacing of 30′-0″ and dimensions indicated. Two 5′-5″ × 3′-0″ offsets are to be constructed along the east wall so the doors do not project into the corridor when opened. A six-panel operable wall that stacks against the east wall separates Rooms 306 and 307. **See Figure 8-22.**

As indicated on Sheet A2.6, the south and east interior walls are Type 4 walls, which are constructed with 3½″ metal studs and covered with ⅝″ gypsum wallboard on both sides extending full height to the structure above. Sound-attenuation blankets are to be installed between the studs. The north wall of Room 307 is a demountable partition system consisting of eight 2′-0″ wide panels.

The interior elevations for Room 307, which are referenced on Sheet A2.5, are similar to the interior elevations for Room 306, with the opposite hand indicated for interior elevations 2 and 3. On Sheet A10.5, Elevations 1, 2, and 3 show interior elevations of Rooms 306 and 307. Elevation 1 shows the operable wall that extends between Rooms 306 and 307. The wall panels stack against the interior partition. As shown on Elevation 2, a

4′-0″ × 12′-0″ chalkboard is to be installed 3′-0″ above the finished floor on the east walls of Rooms 306 and 307. A projection screen above the chalkboard is to be installed in an 18″ deep soffit. Details for construction of the soffit are provided on Sheet A11.6, Detail 8 (not included with this set of prints). A map rail and a recessed chalkboard luminaire are to be installed on the east wall of each room. As shown on Elevation 3, 4′-0″ square tackboards are to be installed 3′-0″ above the finished floor on the north wall of Room 307 and the south wall of Room 306.

Sheet A2.5, Detail 3, is referenced on the floor plan on Sheet A2.5. Detail 3 indicates the finish to be applied to column G5 between Rooms 306 and 307. Solid wood blocking is to be installed at the corners of the concrete column and along the east face of the column for attachment of gypsum wallboard and the operable wall receiver.

Sheet A2.5, Detail 4, is also referenced on the floor plan on Sheet A2.5. Detail 4 describes the construction at the angular northwest corner of Room 307. The construction is similar to the construction at the southeast corner of Room 304, which was previously discussed.

Figure 8-22. Operable walls allow one large room to be separated into two small rooms.

Sheet A5.1, Section AA, is referenced on the floor plan on Sheet A2.5. Section AA is a section taken from east to west extending through the entire SIRTI building. The cutting planes for Section AA are shown along the exterior walls for Rooms 306 and 323 on Sheet A2.5. Section AA shows the orientation of rooms on all levels of the building. The elevation of the classroom level floor is 1930.0' with a 14'-0" floor-to-floor height to the level above. The typical suspended ceiling height on the classroom level is 10'-0".

Sheet A6.1, Section B is also referenced on the floor plan on Sheet A2.5. Section B is a section of the west wall of the SIRTI building at Room 306. Beginning with the ground level, a concrete footing and column provide support for the exterior wall. At the entry level, the column is finished with brick veneer that is fastened to the column. A recessed entry area on the entry level consists of an aluminum storefront system, concrete pavers over rigid insulation and an elastomeric roofing membrane, and a portland cement plaster soffit. The area above the soffit is insulated with a minimum of R-38 fiberglass batt insulation.

Exterior walls on the classroom and office/administration levels are constructed in a similar manner. The exterior wall at the classroom level is constructed with 6" metal studs and covered with ½" gypsum sheathing on the exterior face and ⅝" gypsum wallboard on the interior face. The walls are insulated with 6" fiberglass batt insulation with an R-19 rating. Building paper is to be applied to the exterior surface of the exterior sheathing, and a 3" air space is provided between the metal studs and the brick veneer. Windows are thermally broken anodized aluminum frames. A thermally broken frame is a two-piece metal frame that has a vinyl or rubber separation member to prevent transmission of heat and cold from the exterior frame components to the interior frame components. The windowsill is 3'-0" above the classroom level floor, and the windows are 5'-0" high.

On Sheet A11.3, Details 7 and 8 provide additional information regarding the masonry at the exterior window headers and windowsills. **See Figure 8-23.** Window headers are supported by steel angles with 2½" flat bars welded to the horizontal legs of the angles. Exterior masonry is corbelled (stepped back) in succeeding courses from 3" to 2" to 1". Flashing is installed above the final course of stepped-back masonry and below the soldier course of brick. For the interior header, a corner bead is to be installed at the edge of the gypsum wallboard and an aluminum deflection channel installed at the top of the window frame. For the exterior sill, a brick rowlock sill is to be installed with a neoprene gasket and through-wall flashing below the sill. A ¼" × ¼" drip cut is sawn in the bottom edge of the rowlock sill, which projects 1" from the face of the masonry wall. A drip cut diverts rainwater away from the

structure. The rowlock sill brick are attached to the brick veneer with corrugated masonry anchors. A 3" air space is provided between the back of the brick-veneer wall and the face of the gypsum sheathing on the framed wall.

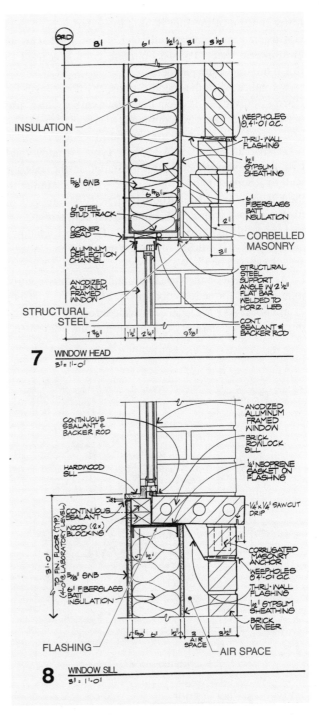

Figure 8-23. Sections provide information regarding special masonry work, flashing, and structural steel supports at windows.

The sloped roof shown along the top of Section B on Sheet A6.1 is constructed with open web steel joists, 1½″ thick corrugated steel decking, ⅝″ thick Type X gypsum wallboard, and rigid roof insulation with an R-38 (minimum) rating. The roof assembly is waterproofed with two layers of 15 lb felt building paper and custom sheet metal roofing. The rainwater gutter system at the lower portion of the roof behind the exterior parapet is constructed with galvanized steel and is lined with an elastomeric sheet. Precast concrete coping is used to finish the top of the parapet.

As indicated in the specifications, the floor finish for Rooms 306 and 307 is direct glue-down carpet with a rubber cove base. The egress doors for the rooms are 3′-0″ × 7′-0″ wood doors with 20 min fire ratings.

As indicated on Sheet A2.5, four Type 2 windows and two Type 1 windows are to be installed in the exterior wall of Rooms 306 and 307. Type 2 windows, similar to the windows installed in Rooms 303 and 304, are 8′-0″ × 5′-0″ windows with a 5′-0″ square fixed light and a 3′-0″ wide operable light that pivots vertically. Type 1 windows are 3′-0″ × 5′-0″ vertically pivoting windows. The windows are also shown on the West Elevation on Sheet A2.5.

As indicated on the West Elevation, a masonry control joint is to be formed at column G5. Sheet A11.3, Detail 2, and Sheet A11.5, Detail 5, provide information regarding the exterior masonry construction at the bottom and top of the wall in Room 307. **See Figure 8-24.** Detail 2 indicates that the exterior brick-veneer wall is supported by structural steel angles fastened to the cast-in-place concrete slab. Similar to window headers, the exterior brick are corbelled from 3″ to 2″ to 1″ to a soldier brick course with through-wall flashing. The air space along the concrete floor slab is insulated with 2″ rigid phenolic board insulation to 6″ above floor level. The soffit finish below the floor is supported by metal clips installed 12″ OC, supporting a framework of ¾″ cold rolled channel and metal lath. A ¾″ layer of portland cement plaster is applied to the metal lath frame.

Detail 5 shows that the top of the parapet is finished with a precast concrete cap. One course of brick is set back from the face of the wall, just below a soldier course. The parapet is constructed to an overall height of 3′-6″ from the top of the floor slab. A 1½″ diameter steel pipe guardrail is installed at the exterior wall above Room 307 on this wall. The guardrail projects 1′-0″ above the parapet and is set back 6″ from the interior face of the parapet. The pipe guardrail is supported by 2 × 6 wood blocking installed between the studs of the parapet framing and fastened with three ⅝″ lag screws at each connection in a 6″ diameter plate. Vent holes are drilled along the bottom edge of the pipe guardrail to allow for air flow to remove moisture and prevent corrosion. The parapet framing is reinforced by ¾″ treated exterior-grade plywood.

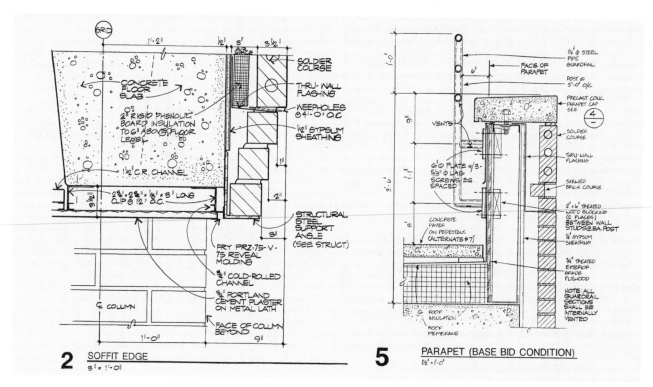

Figure 8-24. Details 2 and 5 provide information regarding the exterior masonry construction at the bottom and top of the wall in Room 307.

As shown on Sheet S2.4, column G5 is located along the exterior wall between Rooms 306 and 307. As indicated on the Column Schedule on Sheet S5.1, column G5 is identical to column G6 in Room 303, which was previously discussed. The structural construction of the floor includes type C5, C7, M1, and M4 joists, which are identical to the concrete joists in Room 303. The perimeter beam for Rooms 306 and 307 are Types 3B4 and 3B8, respectively. As indicated in the Column Schedule on Sheet S5.1, the primary difference between the two beams is the reinforcing steel configuration at the bottom of the beams. As indicated on Sheet S2.4, Classroom Level Framing Plan, additional information regarding types 3B4 and 3B8 is included on Sheets S3.4 and S1.3, respectively (not included with this set of prints).

Mechanical and Electrical. In addition to the natural light provided through the windows, illumination in Rooms 306 and 307 is provided by nine Type A and four Type K luminaires. As shown on Sheet E2.3, the lighting pattern is identical in each room. Type A luminaires are fluorescent 2′ × 4′ luminaires and Type K luminaires are incandescent luminaires (previously discussed). The Type A fluorescent luminaires have an "ab" and "cb" switching arrangement. Type K incandescent luminaires are connected to a dimmer switch. The fluorescent luminaires are connected to circuits 4 and 6 on panelboard 4B, and the incandescent lights are connected to circuit 2 on panelboard 4B.

Rooms 306 and 307 are also equipped with Type Y luminaires in the soffit above the chalkboard. **See Figure 8-25.** Type Y luminaires are 9″ wide recessed linear luminaires with parabolic louvers and require two energy-saving lamps per luminaire. A switch for each Type Y luminaire is provided near the chalkboard.

isolated-ground receptacles in each room. A connection for an electric projection screen is also installed in each room according to manufacturer specifications.

As shown on Sheet E4.3, clock outlets and mini-horns are to be installed on the east walls of Rooms 306 and 307. Two telecommunications outlets are to be installed in each room. Conductors for the clock outlets, mini-horns, and telecommunications outlets are to be routed through the cable tray. Thermostats are located near the door in each room. A thermostat is also to be installed outside the door leading into Room 306. Two sections of ¾″ conduit are to be installed in Room 306 with the junction boxes along the exterior wall positioned 36″ above the finished floor.

As indicated on Sheet M2.3, a 3″ roof drain line (RDL) is to be installed up to a 3″ roof drain above the suspended ceiling in Room 307. A 3″ overflow drain line (ODL) is also to be installed up to a 3″ roof drain in this area. As cited in the General Notes, all drain lines are installed at a 1%, or 1″ in 100″, slope throughout the classroom floor level to provide proper drainage.

As indicated on Sheet M4.3, heat is provided to Rooms 306 and 307 by terminal units #24 and #25, which are located above the ceiling. Terminal units #24 and #25 are identical units with an 850 cfm maximum and 270 cfm minimum rating. A 10″ diameter blower duct extends from a 25″ × 16″ oval supply duct to the terminal units. As indicated on Sheet M5.3, the hot water supply and hot water return pipes for the terminal unit are ¾″ in diameter. In Room 306, a 16″ × 15″ supply air trunk extends from the terminal unit to five 8″ diameter supply air branches and five 48″ long diffusers, each with a 170 cfm capacity. In Room 307, a 16″ × 15″ supply air trunk extends from the terminal unit to three 8″ diameter and one 10″ diameter supply air branches.

Figure 8-25. Recessed fluorescent luminaires are installed above the chalkboards and marker boards in several classrooms of the SIRTI building.

As indicated on Sheet E3.3, five duplex electrical receptacles are to be installed in each room, including two

A clock outlet and mini-horn are to be installed on the east wall of Room 307.

Return air is routed through a 22″ square grille and into a 22″ × 8″ return duct. The height of the return air duct changes from 16″ to 12″ near the wall between Rooms 306 and 307. A thermostat located near the door into each room controls the temperature in the room.

Room 308—Control Room

Room 308 (Control Room) is located to the north of Room 305 and provides adequate space for a video technician to broadcast or record presentations being made in Rooms 305, 310, 311, 312, and 320. Room 308 is designed to accommodate the additional wiring and air conditioning requirements for the electrical equipment to be installed in the room.

Architectural and Structural. As indicated on Sheet A2.6, the dimensions of Room 308 are 14′-0″ × 30′-0″ based on the dimensions shown and the grid line spacing. The walls of Room 308 are Type 4 walls, which are constructed with 3½″ metal studs and covered with ⅝″ gypsum wallboard on both sides extending full height to the structure above, with sound-attenuation blankets between the studs. A 3′-7¼″ wide by 12′-0″ long ramp is to be constructed to the elevation of the raised floor, which is 1931.0′ (1′-0″ above the elevation of the finished concrete floor). Handrails are to be installed on both sides of the ramp.

As indicated on Sheet A2.5 and further described in the specifications, a raised access floor is to be installed throughout most of the room and finished with carpet tiles and a rubber cove base. A 3′-0″ × 7′-0″ wood door is to be installed at the west end of the room leading to the corridor.

The structural floor includes type C6 and M1 joists. C6 joists are 8″ wide by 20½″ deep and are reinforced with two #8 rebar along the bottom and 36 #6 rebar along the top. M1 joists are 8″ wide by 20½″ deep and are reinforced with two #7 rebar along the bottom and 16 #5 rebar along the top. Sheet S2.3A, Detail 30B, indicates the typical position of reinforcing steel for the middle strips of joists. Dimensions between grid lines 7 and 8 on the Framing Plan on Sheet S2.4 indicate that the middle strips of domes are 15′-0″ wide. One short and one long rebar per floor rib are shown on Sheet S2.3A, Detail 30B, with 6″ overlap of the top rebar at the grid lines and the bottom rebar positioned 3′-0″ away from the grid lines.

Mechanical and Electrical. As shown on Sheet E2.3, Room 308 is illuminated with six Type A fluorescent and two Type K incandescent luminaires. The Type A luminaires, which are connected to circuit 18 on panelboard 4B, are also connected to the low-voltage system through relays 5 and 6 and switches "e" and "f." The master light control switch shown at the lower left of Sheet E2.3,

Detail 3, is located on the west wall of this room as indicated by Note 10. Three junction boxes are to be installed above the ceiling in Room 308.

Five duplex and seven double duplex electrical receptacles are to be installed in Room 308 as shown on Sheet E3.3. As indicated in Note 10, the double duplex receptacles are to be mounted under the 12″ raised access floor. The receptacles are to be connected to circuits 4 and 6 on panelboard 2D1. Panelboard 2D1 is installed on the west wall of Room 308. A motor connection is to be installed on the north wall of the room to provide a connection for an air conditioner.

As indicated on Sheet E4.3, a thermostat is to be installed along the west wall of Room 308. A telecommunications outlet is to be installed along the north wall. Two open-trough horizontal cable trays measuring 12″ wide by 4″ deep are to be installed above the ceiling along the south wall and perpendicular to it across the middle of the room. Two vertical cable tray tees are to be installed at the southeast and southwest corners of the room to provide power to the area beneath the raised floor. **See Figure 8-26.** A 4′ square piece of ¾″ plywood is to be installed at the east end of the north wall for microwave head end equipment.

Figure 8-26. Vertical drops deliver electrical power from overhead cable to the floor level.

As shown on Sheet M2.3, a 3″ roof drain line is installed above the ceiling in Room 308 extending up to a 3″ roof drain in the roof above. A 2″ floor drain is installed near the southwest corner of the room under the raised floor, and a 2″ vent is installed in the northwest corner of the room extending downward.

Sheet M3.3 provides the fire protection plan for Room 308. The room is protected by a preaction fire protection system. As shown on Sheet M3.3, preaction mains are to be installed under the floor in Room 308. A pipe extending from Room 305 provides a connection for the preaction main. The preaction system is connected to the alarm valve header with an air pressure sensor, a 4″ air check valve, and a 4″ preaction valve.

As shown on Sheet M4.3, heat is provided to Room 308 by terminal unit #34, which has a 160 cfm maximum and minimum rating. A 5″ diameter blower duct extends from a 25″ × 16″ oval duct to the terminal unit. As indicated on Sheet M5.3, the hot water supply (HWS) and hot water return (HWR) pipes for the terminal unit are ¾″ diameter. A 14″ × 10″ supply air trunk extends from the terminal unit and transitions to an 8″ diameter duct and a 6″ diameter duct. Flexible ductwork extends from the 8″ diameter duct to a 48″ long air diffuser with a 100 cfm capacity, and from the 6″ diameter duct to a 24″ long diffuser with a 60 cfm capacity. A thermostat to control heat to the room is to be installed on the west wall of this room.

Cooled air is supplied to Room 308 by an air conditioning unit installed above the ceiling in Room 314 (corridor). A 16″ square supply duct along the north wall of the room is connected to three 12″ diameter ducts. The ducts are connected to two 10″ round diffusers, each with a 250 cfm capacity and one 12″ round diffuser with a 300 cfm capacity. Two 22″ × 8″ return air ducts with 22″ square grilles feed into the 16″ × 16″ return air duct along the south side of the room.

Rooms 310 and 311—Televised Seminar Rooms

Rooms 310 and 311 are designed for presentation of distance learning sessions similar to Room 305. Instruction can be captured by the cameras in this area and broadcast or recorded as necessary. An operable wall between Rooms 310 and 311 allows the area to be utilized as two small spaces or one large space.

Architectural and Structural. As shown on Sheet A2.6, the combined dimensions of Rooms 310 and 311 are 29′-8¼″ × 30′-0″, based on dimensions indicated and grid line spacing. An operable wall of 10 equal panels separates the combined space of Rooms 310 and 311 into two equal-size areas. Five panels stack against the east wall and five panels stack against the west wall.

The north, west, and south walls and a portion of the east wall of Room 310 are Type 4 walls, which are constructed with 3½″ metal studs and covered with ⅝″ gypsum wallboard on both sides extending full height to the structure above. Sound-attenuation blankets are installed between the wall studs. An 8″ thick concrete wall with ⅞″ furring channels for gypsum wallboard on both sides is shown north of column J4. A shaft wall system is to be constructed at the mechanical chase consisting of 2½″ shaft wall studs with 1″ coreboard and one layer of ⅝″ gypsum wallboard.

Interior elevations for Rooms 310 and 311 are referenced on Sheet A2.5. Elevations 4 to 7 on Sheet A10.5 provide information regarding wall finishes and specialty installations. Elevation 4 is an elevation of the operable wall separating the rooms. The gypsum wallboard on column H4 is exposed. Elevation 5 indicates three 4′-0″ wide tackable wall panels with one section of exposed gypsum wallboard at the column. A 4′-0″ square tackboard is installed 3′-0″ above the floor on the west wall of Room 310.

Elevation 6 shows the chalkboard placement and projection screen soffit construction along the south wall of Room 310 and the north wall of Room 311. The soffit, which houses a projection screen and a recessed chalkboard luminaire, is 18″ deep and finished with gypsum wallboard. A map rail is installed on the wall below the soffit. A note indicates that the chalkboards in each room are 4′-0″ high, with a 20′-0″ long chalkboard in Room 310 and a 16′-0″ long chalkboard in Room 311. The hidden lines on Elevation 6 show the differences in chalkboard and soffit lengths in Room 311 to allow for proper door placement. A 4′-0″ square tackboard is installed 3′-0″ above the finished floor on the west wall of each room.

Sheet A11.1, Detail 1, is referenced on the floor plan on Sheet A2.5. Detail 1 shows the construction at column J4, located on the east side of Room 310. Column J4 is a transition from the 3½″ metal stud wall to an 8″ concrete wall. Solid wood blocking is fastened to column J4 to bring the wall framing flush with the ⅞″ furring channels on both sides of the cast-in-place concrete wall, which provides a true surface when attaching the ⅝″ gypsum wallboard. One-half inch reveals are to be provided at three identified corners of the column where tackable wallcovering is to be applied.

Metal studs covered with gypsum wallboard are used to construct the interior walls of the SIRTI building.

Sheet A11.2, Detail 7, and Sheet A6.2, Section F, are also referenced on Sheet A2.5. Detail 7 provides information regarding the finish at column H4, which is similar to column H5 in Room 305. Section F (not included with this set of prints) shows a wall section through the north wall of Room 311.

As indicated in the specifications and on the Tile & Carpet Floor Pattern Layout, the floor finish is carpet with a 1'-0" wide carpet border. A 2'-0" wide border is to be installed along the line of travel of the operable wall. Five doors are to be installed in Rooms 310 and 311. Doors 310A, 310B, 311A, and 311B are 3'-0" × 7'-0" wood doors with a 20 min fire rating. Door 311C is a 2'-6" × 7'-0" door that is not fire rated.

As indicated on Sheet S2.4, columns J4 and H4 are indicated along the east and west walls of Room 310. As shown on Sheet S5.1, columns J4 and H4 are 16" square concrete columns that may extend to an elevation of 1957' if Alternate #7 is approved by the owner and architect. Reinforcing steel for the classroom level of these columns consists of four #8 rebar overlapping the eight #11 rebar extending up from the floor below by 3'-9". Stirrups are fabricated from #3 rebar and are spaced 16" OC along the entire height of the classroom level. Column H4 is also shown on Sheet A4.2, West Elevation, covered with brick veneer and supporting a cast-in-place concrete trellis.

The Framing Plan on Sheet S2.4 indicates type C6 and M1 joists are part of the two-way joist system. Most of the floor joists are formed with typical domes measuring 16" high by 52" square. Type 4 domes, measuring 16" × 41", are used to form the joists along the north end of Room 311. In addition to the typical reinforcing steel used for the joists and floor, angled reinforcing steel is to be installed near the west end of the stairway landing, which is adjacent to Room 311. Reinforcing steel requirements for the joists in Rooms 310 and 311 are identical to the requirements for Room 305. Sheet S2.3A, Detail 30 indicates the position of the reinforcing steel in the joists and floor. Three layers of reinforcing steel are placed at column grids. Rebar are hooked at the ends to tie into perimeter beams. Extension of reinforcing beyond grid lines is based on a proportion of the end and interior spans.

Mechanical and Electrical. As shown on Sheet E2.3, illumination is provided in each of the rooms by eight Type A luminaires and two Type K luminaires, similar to luminaires installed in other rooms. The arrangement of the luminaires in each room is a mirror image of the other room. The Type A luminaires are controlled by switches "g" and "h" or "i" and "j" to allow for the control of different lighting patterns in each room. Circuit 20 on panelboard 4B provides power for the Type A fluorescent luminaires. The Type K incandescent luminaires in each room are controlled by dimmer switches near the east doors and connected to circuits 22 and 24 on panelboard 2D.

As indicated on Sheet E3.3, Rooms 310 and 311 each have five duplex electrical receptacles, including two isolated-ground receptacles, and one data/power pole. Sheet E2.3 shows recessed lighting with a Type Y1 16' long luminaire installed above the chalkboards in each room.

As shown on Sheet E4.3, four video cameras and two ceiling-mounted video monitors are to be installed in Rooms 310 and 311. Conduit (¾") is to be routed from the equipment to the cable tray. As noted on Sheet A2.5, symbols indicating two additional down-pointing cameras in a ceiling well, to be furnished and installed by the owner, are shown at the north and south ends of the rooms. **See Figure 8-27.**

As shown on Sheet E4.3, clock outlets and mini-horns are to be installed on the east wall of Rooms 310 and 311. A thermostat is located in each room by the west door. Two telecommunications outlets are installed in each room at the east and west walls. A 12" wide by 4" deep open-trough cable tray above the ceiling in Rooms 310 and 311 is to be installed with the top of the tray 11'-0" above the floor.

As indicated on Sheet M2.3, one 3" overflow drain line and 3" roof drain is to be installed above Room 310. Stormwater entering the overflow drain line from this room and the roof drain from Room 308 is fed into two 6" drain pipes extending above Room 311 and directed toward the mechanical chase near Room 313 (stairway). Sheet M7.2, Detail 2, provides additional information regarding the supply and waste piping for Room 311, including the mechanical chase.

As specified on Sheet M3.3, a wet pipe fire protection system is installed in an east-west direction above the ceiling of Room 310. A sleeve for passage of fire protection piping through the concrete shear wall is provided when the concrete forms for the wall are set in place.

As indicated on Sheet M4.3, heat is provided to Room 310 by terminal unit #35, which is located above the ceiling. A 7" diameter blower duct extends from a 19" × 16" supply duct to the terminal unit. As indicated on Sheet M5.3, the hot water supply and hot water return pipes are ¾" diameter. A 14" × 12" supply air trunk extends from the terminal unit to three 7" diameter supply air branches and three 48" long diffusers, each with a 170 cfm capacity. Heat is provided to Room 311 by terminal unit #32, which is located above the ceiling in the corridor to the east of Room 311. An 8" diameter blower duct extends from a 19" × 12" supply duct to the terminal unit. The hot water supply and hot water return pipes are ¾" diameter. A 14" × 12" supply air trunk extends from the terminal unit to three 7" diameter supply air branches and three 48" long diffusers, each with a 170 cfm capacity. Return air is routed through 22" square grilles and into a 22" × 10" duct for Room 310 and a 22" × 6" duct for Room 311. A thermostat is located near the entry door to each room to control the temperature in each room.

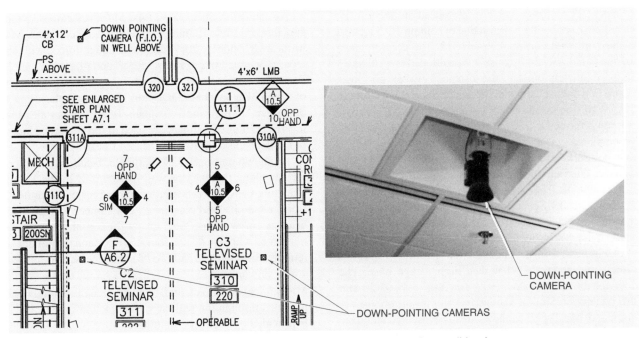

Figure 8-27. Down-pointing cameras are to be furnished and installed in a ceiling well by the owner.

Room 312—TV Seminar

Room 312 is designed for presentation of distance learning sessions, similar to Room 305. Instruction can be captured by the cameras in this area and broadcast or recorded as necessary. Room 312 is the smallest of the televised seminar rooms.

Architectural and Structural. As shown on Sheet A2.6, Room 312 is an irregularly shaped room. The dimensions of the room are 30'-0" along the interior east wall and 23'-2⅝" along the south wall. The exterior wall of Room 312 extends at a 26.57° angle from the lettered column line and is 33'-2½" long (1'-9½" + 8'-0" + 15'-0" + 8'-5" = 33'-2½").

As indicated on Sheet A2.6, the north and east walls are Type 4 walls, which are constructed with 3½" metal studs and covered with ⅝" gypsum wallboard on both sides extending full height to the structure above. Spaces between studs are filled with sound-attenuation blankets.

Sheet A2.5 references Sheet A10.5, Elevations 8 and 9, for information regarding wall finish in Room 312. Elevation 8 shows a 4'-0" square tackboard installed 3'-0" above the floor on the east wall. The wall is to be finished with tackable wallcovering. As shown in Elevation 9, the south wall of Room 312 is a demountable partition, consisting of ten 2'-0" wide panels. A 4'-0" × 12'-0" sectional chalkboard is to be installed 3'-0" above the floor on the demountable partition. A projection screen is to be suspended from ceiling brackets along the demountable partition.

TECH TIP

Per the SIRTI specifications, Type L drawn copper tube is to be used for hot water supply and hot water return pipes less than 2″ in diameter.

The Control Room (Room 308) provides adequate space for a video technician to broadcast or record presentations being made in Rooms 305, 310, 311, 312, and 320.

Detail 12 on Sheet A11.1 is referenced on Sheet A2.5. Detail 12 describes the construction of column 3G.5 located at the northwest corner of Room 312. The column has a unique name—3G.5—because it is positioned on grid line 3 at approximately one-half the distance between grid lines G and H. Per the Column Schedule on Sheet S5.1, Column 3G.5 is 16″ square and extends to an elevation of 1944′. The column is formed at a 26.57° angle to the lettered grid lines. Wood blocking is installed at all interior column corners to provide a surface for fastening gypsum wallboard. Metal stud walls are fitted and framed around the column to support wall finish materials. The exterior face of the stud wall is finished with brick veneer. Special-shape 3″ × 4″ brick is used to form the interior wall corner, with the inside corner of the brick set on a line 8¼″ from the centerline of the column.

As indicated in the specifications, the floor finish is direct glue-down carpet with a rubber cove base. The door for this room—Door 312—is a 3′-0″ × 7′-0″ 20 min fire-rated door. Similar to other exterior rooms previously discussed, two Type 2 windows are to be installed along the exterior wall. Type 2 windows are 8′-0″ × 5′-0″ windows consisting of 5′-0″ square fixed lights and 3′-0″ wide operable panels that pivot vertically. As shown on Sheet A4.2, West Elevation, a control joint is to be formed in the brick veneer at column 3G.5.

As specified in the Column Schedule on Sheet S5.1, all columns for the SIRTI building are to be 16″ square reinforced concrete.

Columns 3G.5 and G4 are located at the northwest and southwest corners of Room 312, respectively. Columns 3G.5 and G4 are identical to column G6, discussed as part of Room 303. Sheet S5.1, Column Schedule, provides additional information regarding columns 3G.5 and G4.

As shown on Sheet S2.4, type M4 joists comprise the structural concrete framing of the floor for Room 312. M4 joists are 8″ wide by 20½″ deep and are reinforced with #8 rebar along the bottom and #5 rebar along the top. Type 3 domes are used to form the joists along the exterior edge of Room 312. Sheet S2.3A, Detail 18, indicates that Type 3 domes are 16″ high and 30″ square. Typical domes measuring 16″ high by 52″ square are used to form the remainder of the floor joists for Room 312.

As shown on Sheet S2.4, perimeter beam 3B7 is to be formed along the exterior wall of Room 312. The Beam Schedule provides information regarding beam dimensions, rebar size and placement, and camber. A type 3B7 beam is 18″ wide by 20½″ deep. The top of the beam is reinforced with two #8 and four #10 rebar and the bottom is reinforced with two #8 and two #6 rebar. The beam is formed with a ⅜″ camber.

Mechanical and Electrical. In addition to the natural light provided through the windows, illumination in Room 312 is provided by six Type A and three Type K luminaires as shown on Sheet E2.3. Type A luminaires are fluorescent luminaires, which are controlled by switches "a," "b," and "c." Note 8 describes the switching of legs "a" and "c" to control the center lamps of these luminaires, and leg "b" to control the outboard lamps in each luminaire. Type A luminaires are connected to circuits 4 and 6 on panelboard 4B. Type K luminaires are incandescent luminaires that are controlled by a dimmer switch and connected to circuits 2 and 6 on panelboard 2B.

As shown on Sheet E3.3, two duplex and three double duplex electrical receptacles are to be installed in Room 312. Two double duplex receptacles are isolated-ground receptacles. A connection for an electric projection screen is to be installed on the south wall per manufacturer specifications. As indicated on Sheet A2.5, two video cameras installed at the north end of Room 312 and one down-pointing camera at the south end of the room are to be furnished and installed by the owner.

As indicated on Sheet E4.3, a clock outlet and mini-horn are to be installed on the east wall of Room 312. A thermostat is located near the door. Two telecommunications outlets are to be installed, one on the north wall and one on the south wall. An open-trough cable tray is to be installed above the ceiling in this area. Sheet E5.2, Detail 4 (not included with this set of prints), provides information regarding the cable tray along the east end of the room.

As shown on Sheet M2.3, a 3″ roof drain line and a 3″ overflow drain line extend up to 3″ roof drains installed in the roof above Room 312. Sheet M3.3 indicates that a wet pipe fire protection system is to be installed near the south wall.

As indicated on Sheet M4.3, heat is provided to Room 312 by terminal unit #23, which is located above the ceiling in Room 307. A 9″ diameter blower duct extends from a

19″ × 16″ supply duct to the terminal unit. The hot water supply and hot water return pipes are ¾″ in diameter. A 14″ × 12″ supply air trunk extends from the terminal unit to four 8″ diameter supply air branches and four 48″ long diffusers, each with a 175 cfm capacity. Return air is routed through a 22″ square grille and into a 22″ × 8″ duct that transitions into a 22″ × 16″ return air duct. The 22″ × 16″ duct then transitions to a 45″ × 12″ return air duct that is to be held tight to the ceiling along the west side of the room.

Rooms 313, 322, and 327—Stairways

Three U-shaped stairways with landings, identified as Rooms 313, 322, and 327, provide access to the classroom level. The stairways and landings are equipped with handrails that conform to Americans with Disabilities Act (ADA) requirements. **See Figure 8-28.** Two stairways are accessed through a set of double doors, and one stairway is accessed through a single swinging door.

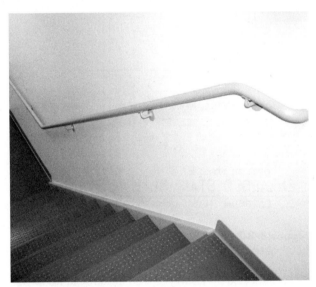

Figure 8-28. Stairway handrails conform to ADA requirements.

Architectural and Structural. As indicated on Sheet A2.6, the dimensions of Stairway 313 are 9′-5⅛″ × 21′-6⅜″ (4′-9″ + 4′-8⅛″ = 9′-5⅛″). The north wall of Room 313 is a Type 7 wall, which is an 8″ thick concrete wall with ⅞″ furring channels for fastening gypsum wallboard on both sides. The south and west walls are Type 4 walls, which are constructed with 3½″ metal studs and covered with ⅝″ gypsum wallboard on both sides extending full height to the structure above with sound-attenuation blankets between the studs. The east wall and center wall in Room 313 are Type 2 walls, which are constructed with 3½″ metal studs and covered with ⅝″ gypsum wallboard on both sides extending full height to the structure above.

As indicated in the specifications, the ceiling finish is gypsum wallboard and the floor finish is exposed concrete. Stair treads and risers are finished with rubber floor covering, with nonslip rubber tile at the landing. A set of wood double doors measuring 5′-4″ × 7′-0″ with a 1 hr fire rating leads to the stairway. The door includes a 4″ wide by 2′-0″ high fixed light to provide a view into the stairway before opening the door.

The Column Schedule on Sheet S5.1 provides information about the dimensions and reinforcing steel for column H3, which is located at the northwest corner of Room 313. Column H3 is a 16″ square concrete column that extends to an elevation of 1944′. Reinforcing steel for the classroom level of column H3 consists of four #8 rebar overlapping eight #11 rebar extending up from the floor below by 3′-9″. Stirrups are fabricated from #3 rebar and are spaced 16″ OC along the entire height of the classroom level. Sheet S3.2, Detail 29 (not included with this set of prints), provides additional information about column H3.

Sheet A11.2, Detail 9 is referenced on Sheet A2.5. Detail 9 provides information regarding the wall finish and door frame construction at column H3. A 3″ long Type 4 metal stud wall supports the door frame at the stairway. Wood blocking is installed on the outside column corners, with corner beads installed over the gypsum wallboard. A tackable wallcovering is installed over the wallboard as noted on the detail.

TECH TIP

The American with Disabilities Act states the minimum width of a stair tread for an accessible stairway must be at least 11″ measured from riser to riser.

Enlarged stairway plans provide additional information regarding stairway dimensions and construction. **See Figure 8-29.** The landing at the top of Room 313 is 5′-4″ deep from the door to the top step. The landing at the upper turn in the stairway is 5′-0″ deep. The stairway leading up to the next floor is comprised of 12 treads with an 11″ unit run equaling a total run of 11′-0″ (12 × 11″ = 132″ = 11′-0″). The stairway leading down to the floor below is comprised of 11 treads with an 11″ unit run equaling a total run of 10′-1″ (11 × 11″ = 121″ = 10′-1″). The landing at the lower level turn is 5′-11″ deep. The handrail, which extends along the stairs and landing, projects 2′-0″ beyond the risers along the north wall and 1′-0″ beyond the top riser along the south wall. Additional information regarding overall stairway construction for Room 313 is included on Sheet A7.2, Sections A and B (not included with this set of prints).

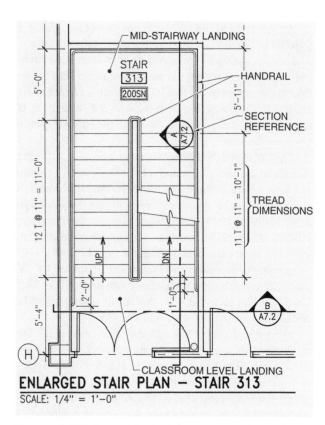

ENLARGED STAIR PLAN – STAIR 313
SCALE: 1/4" = 1'-0"

Figure 8-29. Room 313 is the stairway at the north end of the classroom level. Enlarged stair plans provide stairway dimensions, including landings and handrails.

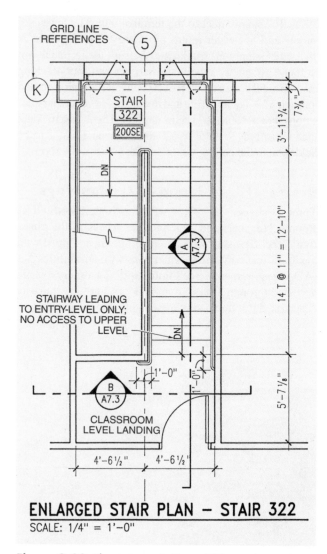

ENLARGED STAIR PLAN – STAIR 322
SCALE: 1/4" = 1'-0"

Figure 8-30. The stairway in Room 322 travels in a downward direction from the classroom level only.

As indicated on the Enlarged Stair Plan, the dimensions of Stair 322 are 9'-1" × 23'-0¼" (4'-6½" + 4'-6½" = 9'-1"; 5'-7⅛" + 12'-10" + 3'-11¾" + 7⅛" = 23'-0¼"). **See Figure 8-30.** The north, south, east, and center wall in the stairway are Type 4 walls, which are constructed with 3½" metal studs covered with ⅝" gypsum wallboard on both sides extending full height to the structure above, with sound-attenuation blankets between the studs. As shown on Sheet A2.6, a 5'-7" deep landing is provided at the top of this stairway.

As noted in the specifications, the ceiling finish in Room 322 is suspended acoustical tile and the floor finish is exposed concrete. Stair treads and risers are finished with rubber floor covering, with nonslip rubber tile at the landing. The stairway door is a single wood door measuring 3'-0" × 7'-0" and has a 1 hr fire rating. Two Type 1 windows, which are 3'-0" × 5'-0" vertically pivoting windows, are to be installed in the east wall of the stairway.

Sheet 11.2, Section 12, is referenced on Sheet A2.5. Section 12 provides information regarding typical column treatment. "SIM" is noted on Sheet A2.5 next to the section reference to indicate the column treatment at stairway columns is similar to other exterior wall columns.

As indicated by a heavy dashed line around Room 322 on Sheet A2.5, an enlarged stairway detail is provided on Sheet A7.1. The stairs of Room 322 extend downward from the classroom level only; the office/administration level above cannot be accessed from Room 322. The landing area at the top of the stair from the entry door to the top riser is 5'-7⅛" deep. The stair is comprised of 14 treads with an 11" unit run equaling a total run of 12'-10". The landing at the turn in the stairway is 4'-7⅛" deep. Handrails project 1'-0" beyond the top riser along the south wall and continue around the entire perimeter of the stair at the lower landing. Additional information concerning construction of Room 322 is provided on Sheet A7.3, Details A and B (not included with this set of prints).

The Framing Plan on Sheet S2.4 identifies the columns along the west side of Room 322 as columns K4.9 and K5.1. Columns K4.9 and K5.1 are 16" square concrete columns that extend to an elevation of 1944′. The reinforcing steel for the columns at the classroom level consists of eight #7 rebar overlapping four #11 rebar extending up from the floor below by 3′-6″. Stirrups are fabricated from #4 rebar and are spaced 8″ OC along the entire height of the classroom level.

Detail 6 on Sheet S2.4 is also referenced on the Framing Plan on Sheet S2.4. The landing at the top and middle of the stairway floor slab is 6″ thick. A #8 × 20′-0″ rebar is to be installed in the bottom of the joist at the top stairway landing. A 4″ × 3″ × ¼″ × 9′-4″ long steel angle is to be installed along the edge of the landing with ½″ diameter welded studs spaced 12″ OC extending into the bottom of the concrete floor slab for anchorage. Perimeter beam 3B17 forms the exterior of Room 322. Perimeter beam 3B17 is 18″ wide by 20½″ deep with no camber. As indicated in the Beam Schedule on Sheet A2.4, the beam is reinforced with #6, #8, and #10 rebar.

The dimensions of Room Stairway 327 are 9′-4⅝″ × 21′-6⅜″. A portion of the north wall and the west wall are noted as Type 4 walls, which are constructed with 3½″ metal studs and covered with ⅝″ gypsum wallboard on both sides extending full height to the structure above, with sound-attenuation blankets between the studs. The other portion of the north wall is a Type 8 wall, which is constructed with 3½″ metal studs and ⅝″ gypsum wallboard on one side the full height of the wall, and a shaft wall system of metal studs, 1″ coreboard, and one layer of ⅝″ gypsum wallboard for an increased fire rating at the elevator shaft. The south wall of Room 327 is a Type 7 wall, which is an 8″ thick concrete wall with ⅞″ furring channels on both sides and is covered with gypsum wallboard on both sides. The east wall is a Type 13 wall, which is constructed of two rows of 3½″ metal studs spaced 16″ OC with ⅝″ gypsum wallboard on both sides. The total wall thickness is 10¾″ with 9½″ metal track cross-bracing spaced 4′-0″ OC and fastened with drywall screws to the face of the studs. The center wall of the stairway is constructed with 3½″ metal studs with ⅝″ gypsum wallboard on both sides extending full height to the structure above.

As indicated in the specifications, the ceiling finish for Room 327 is gypsum wallboard and the floor finish is exposed concrete. Stair treads and risers are finished with rubber floor covering, with nonslip rubber tile at the landing. The set of double doors for entry into Room 327 is identical to the doors to be installed at Room 313 and is a wood 5′-4″ × 7′-0″ double door with a 1 hr fire rating.

Sheet S5.1 provides information regarding column H7, which is located at the southwest corner of Room 327. Column H7 is a 16″ square concrete column that extends to an elevation of 1962′-7″. The reinforcing steel for column H7 at the classroom floor level consists of four #8 rebar overlapping eight #11 rebar extending up from the floor below by 3′-9″. Stirrups are fabricated from #3 rebar and are spaced 16″ OC along the entire height of the column at this level.

Sheet A11.1, Detail 11, is referenced on Sheet A2.5. Section 11 provides information about the finish for column H7 at the southwest corner of Room 327. Wood blocking is installed at the column corners and along the door frame to support gypsum wallboard. Tackable wallcovering is applied over the gypsum wallboard in Room 326. A 3″ metal-framed wall is built to support the door frame 5″ from the finished wall surface.

Sheet A11.2, Detail 1, is referenced on Sheet A2.5 at the northwest corner of Room 327. Detail 1 describes the construction of the shaft wall and Type 8 wall at grid line 6.7/H. Type 8 walls are constructed with 3½″ metal studs and covered with ⅝″ gypsum wallboard on one side with 1″ coreboard and one layer of ⅝″ gypsum wallboard over the coreboard. A shaft wall with a 2 hr fire rating is constructed directly over the 12″ wide concrete beam below. The shaft wall is constructed with 2½″ metal studs with the wall flush with the beam below. One layer of 1″ coreboard and one layer of ⅝″ gypsum wallboard are applied over the coreboard and to the opposite side of the wall. Additional information regarding stairway construction is included on Sheet A7.1, which is indicated by the heavy dashed line shown on Sheet A2.5.

TECH TIP

Rubber tile measuring ³⁄₁₆″ × 24″ × 24″ are used to finish the stairway landing.

A 2½″ hose valve is joined to the 6″ standpipe in Rooms 313 and 327 for connection of a fire hose.

Additional information concerning Room 327 is shown on an enlarged stair plan. **See Figure 8-31.** The stairway landing is 5′-4″ deep at the entry door. The top landing in the stairway is 5′-11″ deep. The stairway leading up to the next floor is comprised of 11 treads with an 11″ unit run equaling a total run of 10′-1″. The stairway leading down is comprised of 12 treads with an 11″ unit run equaling a total run of 11′-0″. The landing at the lower turn is 5′-0″ deep. The handrail projects 2′-0″ beyond the risers along the north wall and 1′-0″ beyond the top riser along the south wall. Additional information regarding overall stairway construction for Room 327 is included on Sheet A7.2, Sections C and D (not included with this set of prints).

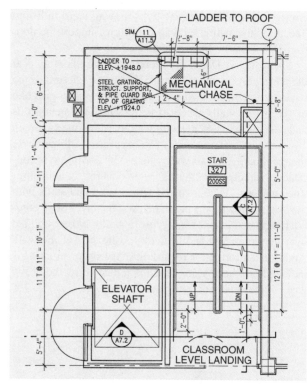

Figure 8-31. The Enlarged Stair Plan for Room 327 provides additional stairway information and details regarding the mechanical chase to the east, including information on a ladder to the roof with a steel grate landing.

Mechanical and Electrical. As indicated on Sheet E2.3, two Type L surface-mounted fluorescent luminaires with flat diffusers and four lamps provide light in Rooms 313 and 327. The luminaires at the east ends of the stairways are connected to emergency circuits. Light in Room 322 is provided by four Type B recessed fluorescent luminaires, with the luminaire at the east end connected to an emergency lighting circuit. There are no electrical receptacles in the stairways. As indicated on Sheet E4.3, a sprinkler flow switch and sprinkler tamper switch are to be installed near the door in Rooms 313 and 327.

As indicated on Sheet M3.3, Detail 2, a 6″ standpipe riser and 2″ drain pipe for the fire protection system are to be installed in the corner of the upper stairway landings in Rooms 313 and 327. The control valve, sectional test, and drain are to be installed as high as possible above the classroom floor level. **See Figure 8-32.** A 2½″ hose valve is installed 36″ above the finished floor in Rooms 313 and 327.

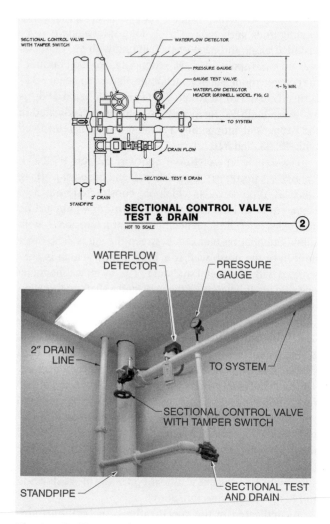

Figure 8-32. Detail 2 provides information regarding the control valve, sectional test, and drain to be installed in Rooms 313 and 327.

TECH TIP

Per the SIRTI specifications, brass or engraved laminated plastic identification tags must be provided and attached with chain to all control, drain, and test valves of the fire protection piping.

Rooms 314, 315, and 316—Corridors

The three corridors and Room 300 (Elevator Lobby) are connected to create a rectangular traffic flow on the classroom level. All classrooms (except Room 329) and stairways are accessible from the corridors and Room 300. Electrical cable trays are installed above corridor ceilings throughout the classroom level.

Architectural and Structural. As indicated on Sheet A2.6, Rooms 314 and 316 are 7'-4¾" wide and Corridor 315 is 13'-6" wide. Type 4 walls, constructed with 3½" metal studs and covered with ⅝" gypsum wallboard on both sides extending the full height to the structure above with sound-attenuation blankets between the studs, are specified for most corridor walls throughout the classroom level. Type 6 and 7 walls, constructed of 8" thick concrete with ⅞" furring channels for gypsum wallboard, are installed along the corridors at the stairways and mechanical chases.

Elevation 10 on Sheet A10.5 is referenced on Sheet A2.5. Elevation 10 is an interior elevation that depicts the east wall of Room 314 and the west wall of Room 316. The primary wall finish material is tackable wall-covering, with four areas of exposed gypsum wallboard at the columns. Six fire extinguisher cabinets with fire extinguishers are shown in the corridors, with three in Room 314 and three in Room 316. As shown on Elevation 10, fire extinguisher cabinets are to be installed with the top of the cabinets 5'-0" above the floor.

See Figure 8-33. Elevation 11 on Sheet A10.5 shows cable tray soffits installed near the ceiling in Rooms 314 and 316 along the classroom walls.

As shown on Sheet A2.6, one Type 1 and one Type 2 windows are to be installed on the west end of Room 315. As indicated in the specifications, Type 1 windows are 3'-0" × 5'-0" vertically pivoting windows and Type 2 windows are 8'-0" × 5'-0" windows consisting of a 5'-0" square fixed pane and a 3'-0" wide operable panel that pivots vertically. The floor finish is alternating pattern vinyl composition tile, with a 1'-0" wide border tile of a different design.

As shown on Sheet S2.4, type C3, C5, and C6 joists are to be formed below the corridor floors. The Joist Schedule on Sheet S2.4 provides information regarding joist size and reinforcing steel requirements. Several dome types are used to form the joists for the floors in Room 316 near the restrooms. A 6" concrete slab is cast at the west end of Corridor 315 in the offset formed by perimeter beams 3B10 and 3B11. Sheet S2.4, Detail 12 provides additional information regarding slab and beam reinforcement.

A note to the right of the Framing Plan on Sheet S2.4 indicates that the small circles shown along Rooms 315 and 316 represent locations for ½" diameter by 6" long ferrule loop inserts. The ½" diameter ferrules are secured to #4 × 1'-0" long rebar using wire ties. Hangers are screwed into the ferrules to support fixtures hung from the ceiling below.

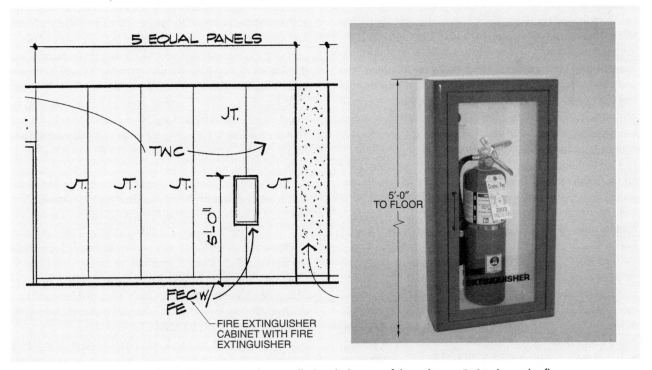

Figure 8-33. Fire extinguisher cabinets are to be installed with the top of the cabinets 5'-0" above the floor.

Mechanical and Electrical. As shown on Sheet E2.3, illumination is provided in Room 314 by 11 Type B fluorescent luminaires, with three luminaires connected to the emergency lighting circuit. Two Type E2 luminaires are to be installed at the doors leading to Rooms 313 and 327. Type E2 luminaires are recessed ceiling-mounted exit lights with ¼″ thick Plexiglas®, routed lettering and arrows, and an 8 W lamp. Arrows by the luminaire symbols indicate the direction of arrow installation in the luminaires to indicate the emergency escape path.

Illumination in Room 315 is provided by 12 Type B luminaires, with two luminaires connected to the emergency circuit, and one Type E2 exit light. Illumination in Room 316 is provided by 10 Type B luminaires, with three connected to the emergency circuit. One Type E2 exit light is to be installed at the intersection of Rooms 300 and 316. Power is provided by circuits 2 and 4 on panelboard 4X1.

As shown on Sheet E3.3, two duplex electrical receptacles are to be installed in Room 314. In addition, a motor connection for the air conditioner in Room 308 is to be installed above the ceiling in Room 314. Three duplex electrical receptacles and a junction box for the water cooler on the north wall are to be installed in Room 315. Transformer TC3 is also installed above the ceiling in Room 315. Three duplex electrical receptacles are to be installed in Room 316, with transformer TC1 to be installed above the ceiling at the south end of the corridor. Two of the receptacles are connected to circuits 4 and 10 on panelboard 2D, and one receptacle is connected to circuits 23, 25, 27, and 29 on panelboard 2D.

As indicated on Sheet E4.3, two audio and visual fire alarm signals and two manual fire alarm stations are to be installed in Room 314. **See Figure 8-34.** A thermostat is located near the middle of the corridor on the west wall and connected to terminal unit #28. A solid-bottom cable tray is to be installed above the ceiling along the west wall of Room 314. Sheet E5.2, Detail 3 (not included with this set of prints) provides information regarding the cable tray connections on the east wall of Room 314.

One audio and visual fire alarm signal is to be installed on the south wall of Room 315. A thermostat is to be installed near the middle of the corridor on the north wall and connected to terminal unit #22. Two key-activated door control units are installed above the ceiling at each end of Room 315.

Two audio and visual fire alarm signals are to be installed on the west wall and one manual fire alarm station is to be installed on the east wall of Room 316. A solid-bottom cable tray is installed above the ceiling along the entire east wall of Room 316 with the top of the tray 9′-0″ above the floor. Sheet E5.2, Detail 3 (not included with this set of prints), provides information regarding the cable tray connections on the west wall of Room 316.

Figure 8-34. Audio and visual fire alarm signals are to be installed in the corridors of the SIRTI building.

As indicated on Sheet M2.3, various types of water supply pipes are installed above the ceiling in Room 314, including a ½″ cold water supply pipe at the south end of the corridor and a 1″ cold water supply pipe near Room 313 extending north toward Room 315. One 2″ vent pipe starts near the entrance to Room 308 and continues north, and another 2″ vent pipe is installed at the south end of the corridor. As shown on Sheet M5.3, hot water supply and hot water return piping for the hydronic heating system is to be installed above the ceiling in Rooms 314, 315, and 316. Positions of the terminal units are also indicated on Sheet M5.3. As indicated by Note 9 on Sheet M4.3 and shown on Sheet M5.3, refrigerant lines and the suction and liquid pump for the air conditioner for Room 308 are also to be installed above Room 314.

As shown on Sheet M2.3 and Sheet A10.5, Detail 21, a drinking fountain is to be installed on the north wall of Room 315, with a remote chiller installed above the ceiling. A 2″ × 2″ × 1½″ tee on the 2″ cold water supply pipe provides a connection for a 1½″ supply pipe in Room 315 for drinking fountain P-5. Two 3″ overflow drain lines and one 3″ roof drain line are to be installed above the ceiling in Room 315, with additional drain lines from other overflow and roof drains extending toward the mechanical chase, including two 6″ drain pipes at the east end of the corridor.

As shown on Sheet M3.3, wet pipe fire protection piping is to be installed at the north and south end of Room 314. Wet pipe and preaction piping is to be installed in Room 315.

Preaction fire protection piping is also to be installed along the west side of Room 316, terminating at the rooms at the south end of the classroom level.

As indicated on Sheet M4.3, heat is provided to Room 314 by terminal unit #28, which has a 300 cfm maximum and 120 cfm minimum rating. An 8″ diameter blower duct extends from a 36″ × 16″ supply duct to the terminal unit. As indicated on Sheet M5.3, the hot water supply and hot water return pipes are ¾″ in diameter. A 14″ × 12″ supply air trunk extends from the terminal unit to two 7″ diameter supply air branches and two 24″ long diffusers, each with a 150 cfm capacity. A 22″ square return air grille is installed near the north end of Room 314 near the door leading into Room 311.

As indicated on Sheet M4.3, heat is provided to Rooms 315 and 316 by terminal unit #22. An 8″ diameter blower duct extends from a 19″ × 16″ oval supply duct to the terminal unit. As indicated on Sheet M5.3, the hot water supply and hot water return pipes are ¾″ diameter. A 14″ × 12″ supply air trunk extends from the terminal unit and branches to a 7″ supply air duct and a 48″ long diffuser with a 130 cfm capacity. A transition is made from the 14″ × 12″ branch to a 12″ × 10″ supply duct and branching to another 7″ supply duct with a 48″ long diffuser. Another transition is made from the 12″ × 10″ duct to a 9″ × 9″ supply duct and branching to two 7″ diameter supply ducts and two 48″ long diffusers. Return air is routed through two 22″ square grilles and two 22″ × 6″ ducts located in Rooms 314 and 316. A thermostat to be installed in Room 315 controls the temperature in Rooms 315 and 316.

Room 317—Computer Instruction Lab

Room 317 (Computer Instruction Lab) is designed as a traditional classroom, and also has an abundance of electrical receptacles around the perimeter of the room and at prewired workstations to be installed across the width of the room. The additional receptacles provide students with an adequate number of receptacles for computers to be used in the classroom. An electrical subpanel is installed in Room 317 to ensure sufficient electrical service for computer equipment.

Architectural and Structural. As shown on Sheet A2.6, the dimensions of this irregularly shaped room are 31′-6″ along the east wall, 37′-4¾″ along the south wall, 46′-10¼″ along the angled west wall (including a portion of the exterior wall that projects into Room 315), and 20′-10½″ along the angled north wall. A 3′-0″ × 4′-10¾″ offset is provided for the west egress door. A 3′-0″ deep offset is also provided for the east egress door. The north and west walls are exterior walls finished with brick veneer. The south wall of Room 317 is a Type 4 wall, which is constructed with 3½″ metal studs and covered with ⅝″ gypsum wallboard on both sides extending full height to the structure above

with sound-attenuation blankets between the studs. The east wall is a demountable partition with a door installed 3′-3″ from the exterior wall.

Sheet A11.2, Detail 11, which is referenced on Sheet A2.5, provides information regarding metal, wood, and masonry construction at column H2. The angle of the grid lines on the detail indicates that the column is to be constructed at an angle to the grid lines.

Sheet A6.2, Section A, also referenced on Sheet A2.5, provides information regarding the west wall of Room 317. **See Figure 8-35.** The exterior wall is framed with metal studs and finished with brick veneer. The studs are covered on the exterior side with ½″ gypsum sheathing and #15 building paper. The space between the studs is insulated with R-19 rated fiberglass insulation. Windows have anodized aluminum frames. The upper portion of the west wall has a 3′-0″ high parapet concealing a precast concrete planter. A steel pipe support is to be installed at the parapet to provide additional structural support. The pipe support is constructed with 2½″ diameter steel pipe spaced 4′-0″ OC along the length of the parapet.

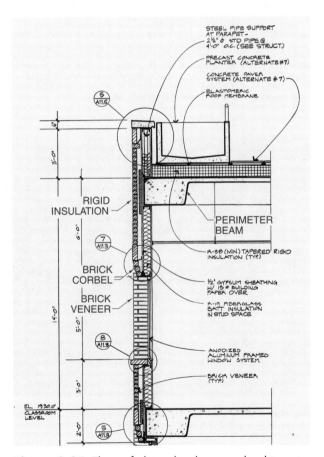

Figure 8-35. The roof above the classroom level is water-proofed with an elastomeric roof membrane and covered with concrete pavers. The precast concrete planter is an alternate bid item.

Interior elevations 12, 13, 14, and 15, shown on Sheet A10.5, are referenced on Sheet A2.5. As shown on Elevation 12, a 4'-0" × 6'-0" tackboard and 4'-0" × 10'-0" tackboard are to be installed 3'-0" above the floor on the demountable partition. A door is to be installed in the partition at the north end of the room. As shown on Elevation 13, a 4'-0" × 12'-0" liquid marker board is to be installed 3'-0" above the floor and centered on the wall between the doors. A projection screen and luminaire are to be installed above the marker board in an 18" deep gypsum wallboard soffit.

Sheet A5.1, Section BB is referenced on Sheet A2.5. Section BB shows a cutting plane extending through the building in a north-south direction, starting at the north wall of Room 317 and continuing through the south wall of Room 326. The Computer Instruction Lab (Room 317) is shown to the right side on the classroom level. Moving in an eastward (right to left) direction are the corridor (315), the stairway (313), televised seminar rooms (311 and 310), the control room with the raised access floor (308), the studio classroom with the ramp at the west side of the room (305), the Elevator Lobby (300), the mechanical chase, and the microcomputer classroom (326) at the far east wall. Numbered column grid lines are shown across the top of the building for proper orientation with other drawings in the set of prints. The building section also shows the structural members of the building, including concrete footings supporting concrete columns and a two-way joist system at each level. **See Figure 8-36.**

Figure 8-36. Concrete footings support the walls and columns of the SIRTI building.

The ceiling construction for the classroom level is shown on Sheet A5.1, Sections AA and BB. Inverted T-shaped joists are shown on both sections, indicating a two-way joist system. As shown on Section BB, the office/administration level is constructed above only a portion of the classroom level. The office/administration level extends northward to grid line 5 only. The remainder of the classroom level roof is a flat roof that is finished with concrete pavers. A cast-in-place concrete trellis may be installed over this area as alternate building work. Note that the classroom level has a suspended ceiling throughout all occupied areas. Elevations for all levels of the building are shown at the left (east) side of Section BB.

As indicated in the specifications, the floor is finished with conductive vinyl composition tile (CVCT) with a rubber cove base. Three doors are to be installed in Room 317. Doors 315A and 315B are identical wood doors measuring 3'-0" × 7'-0" with key-activated access systems. The door control units (DCU) are installed above the ceiling at each end of Room 315. Door 317 is also a 3'-0" × 7'-0" wood door, but is not part of the key-activated access system.

As shown on Sheet A2.5, two Type 2 windows and one Type 1 window are to be installed in the exterior walls of Room 317. Type 2 windows are 8'-0" × 5'-0" windows consisting of a 5'-0" square fixed light and a 3'-0" wide operable panel that pivots vertically. The Type 1 window is a 3'-0" × 5'-0" vertically pivoting window.

As shown on the West Elevation on Sheet A4.2, the guardrail along the planter over Room 317 is installed at a minimum height of 3'-6" above the roof deck level. A control joint is installed in the brick veneer at column H2.

As indicated on Sheet S2.4, columns H1.2, H2, J1.5, and J2 are located around the perimeter of Room 317. Columns H1.2, H2, and J1.5 are identical in size and reinforcing steel requirements, and extend to an elevation of 1944'. The 16" square columns are reinforced at the classroom level with eight #7 rebar overlapping four #11 rebar extending up from the floor below by 3'-6". Stirrups are fabricated from #4 rebar and are spaced 8" OC along the height of the classroom level. Column J2 terminates at the same elevation as the other columns, but the column is reinforced at the classroom level with four #8 rebar that overlap eight #11 rebar extending up from the floor below by 3'-9". Column J2 stirrups are fabricated from #3 rebar and are spaced 16" OC along the height of the classroom level.

Type C5, M1, and M4 joists are identified on the framing plan on Sheet S2.4. Dimensions and reinforcing steel requirements of the joists are shown on the Joist Schedule on Sheet S2.4. As shown on Sheet S2.3A, Detail 18, most of the joists in the area are formed with typical domes measuring 16" high by 52" square. Type 3 domes are used to form the joists along the outer edge of Room 317. Perimeter beams 3B1, 3B6, and 3B7 are specified on the Framing Plan and detailed in the Beam Schedule on Sheet S2.4. These perimeter beams are 18" wide and 20½" deep. Beam 3B1 is reinforced with two #6 × 18'-0" long rebar and two #6 × 20'-0" long rebar at the bottom, with four #5 rebar at the top hooked at the northwest end. Beam 3B6 is reinforced with two #8 × 24'-6" long rebar and two #6 × 28'-9" long rebar at the bottom, and four #5 rebar at the top with hooked ends. Beam 3B7 is reinforced with two #6 × 35'-3" long rebar and two #8 × 24'-6" long rebar at the bottom, and two #8 × 33'-3" long rebar and four #10 × 20'-0" long rebar at the top.

Mechanical and Electrical. In addition to the natural light provided through the windows in Room 317, illumination is provided by Type D, D1, D2, K, and Y luminaires. See Sheet E2.3. Type D, D1, and D2 luminaires are suspended indirect fluorescent luminaires, each 3⅝" high by 24" long, and suspended from the ceiling by aircraft cable. **See Figure 8-37.** Type D luminaires are 24' long, Type D1 luminaires are 20' long, and Type D2 luminaires are 8' long. Type K luminaires are recessed incandescent luminaires, which are controlled by a dimmer switch. A Type Y fluorescent luminaire is recessed in the soffit above the liquid marker board. Power is provided from circuits 4 and 6 on panelboard 4B.

Figure 8-37. Aircraft cable is used to suspend indirect fluorescent luminaires in the building.

As shown on Sheet E3.3, panelboard 2C is to be installed in the southwest corner of Room 317 due to the electrical requirements of the room. The 27 duplex electrical receptacles and five data/power poles at the west end of the prewired workbenches in Room 317 are connected to panelboard 2C.

As indicated on Sheet E4.3, a clock outlet and mini-horn are to be installed on the south wall of Room 317. A thermostat and telecommunications outlet are located midway across the width of the east wall. An open-trough horizontal cable tray is to be installed above the ceiling in

Room 317 with the top of the tray 8'-0" above the floor. A 60° bend is installed in the cable tray toward the west side of the room. Sheet E5.2, Detail 2 (not included with this set of prints), provides information regarding the typical power pole connections.

As shown on Sheet M2.3, two 3" roof drains and drain lines and one 3" overflow drain and drain line are to be installed in the roof above Room 317 to provide drainage for the flat roof. As indicated with a heavy dashed line on Sheet M3.3, a preaction fire protection system is used to provide fire protection in Room 317. The preaction main extends across the room in a north-south direction.

As indicated on Sheet M4.3, heat is provided to Room 317 by terminal unit #21, located above the ceiling in Room 315. A 9" diameter blower duct extends from a 14" diameter supply to the terminal unit. As shown on Sheet M5.3, hot water supply and hot water return pipes are ¾" in diameter. A 16" × 15" supply air trunk extends from the terminal unit to a 90° bend in the ductwork where the duct transitions to a 14" × 13" supply duct. Five 48" long diffusers with a 150 cfm capacity, each connected to 7" diameter supply air branches, provide air to the room. Return air is routed through a 22" square grille and 22" × 8" return duct and into a 22" × 12" return duct above the ceiling in Room 315. The thermostat installed midway across the width of the east wall controls the temperature in Room 317.

Galvanized steel or aluminum HVAC distribution ductwork comprises the majority of the ductwork for the SIRTI building.

Rooms 318 and 319—Classroom/Exhibit Rooms

Rooms 318 and 319 are multipurpose areas separated by an operable wall and can be used for classroom space or student or faculty exhibits. Track lighting is to be installed around the perimeter of the combined space to provide flexibility when lighting the exhibit space.

Architectural and Structural. As indicated on Sheet A2.6, Rooms 318 and 319 are irregularly shaped rooms consisting of two primary walls, a common wall, and a wall with an egress door. For Room 318, the exterior wall at the north end is 33'-7½" long, the wall adjoining Room 317 is 30'-6" long, and the short wall on the south side of Room 318 is 2'-6" long. For Room 319, the exterior wall at the east side of the room is 30'-0" long, the south wall is 23'-2⅝" long, and the east wall with the egress door is 13'-6" long with a 5'-4¾" × 3'-0" door offset.

Sheet A2.5 references Elevations 9, 14, and 15 on Sheet A10.5. Elevations 14 and 15 are related to Room 318. Elevation 14 shows that the common wall is an operable wall consisting of nine equal-width panels. As indicated on Sheet A2.5, the panels stack along the interior walls between Rooms 318 and 319. The egress door for Room 318 is also shown on the detail. Elevation 15 represents the west demountable partition of Room 318, which includes two 4'-0" × 6'-0" chalkboards and a 4'-0" × 4'-0" tackboards installed 3'-0" above the floor. As noted on Sheet A2.5, a projection screen is to be installed above one of the chalkboards. Elevation 9 represents the south wall of Room 319, including six 2'-0" × 4'-0" chalkboards with a projection screen above and a 3'-0" × 4'-0" tackboard to be installed 3'-0" above the floor.

The north and east walls of Rooms 318 and 319 are exterior walls with two Type 1 and two Type 2 windows as indicated on Sheet A2.5. Type 1 windows are 3'-0" × 5'-0" vertically pivoting windows and Type 2 windows are 8'-0" × 5'-0" windows consisting of a 5'-0" square fixed light and a 3'-0" wide operable panel that pivots vertically. Column K2 at the northeast corner of the building is finished in a manner similar to Detail 4 on Sheet A2.5. Special brick shapes C and D are used at the corners. **See Figure 8-38.**

Figure 8-38. Special brick shapes are used to form the corners of the building that are not 90° to one another.

As indicated in the specifications, the floor finish is direct glue-down carpet with a rubber cove base. Doors 315C and 315D are 3'-0" × 7'-0" wood doors with a 20 min fire rating. Door 317 is also a 3'-0" × 7'-0" wood door, but it is not fire rated.

Column K2 is referenced on the Framing Plan on Sheet S2.4. As shown on Sheet S5.1, Column Schedule, column K2 is a 16" square column that is reinforced with eight #7 rebar that overlap the #11 rebar extending up from the floor below by 3'-6". Stirrups are fabricated from #4 rebar and are spaced 8" OC at the classroom level.

Type C5, C6, C7, M1, and M4 joists are indicated on the Framing Plan on Sheet S2.4. All joists are 8" wide by 20½" deep with various rebar configurations. Typical domes are used to form the majority of the joists, with type 3 domes used to form the joists along the exterior edge of the slab.

Perimeter beams 3B2 and 3B3 are indicated on Sheet S2.4. Beam 3B2 is formed with ¼" camber and beam 3B3 is formed with ⅜" camber.

Mechanical and Electrical. In addition to the natural light provided through the windows, illumination in Rooms 318 and 319 is provided by fluorescent luminaires and track lighting. Ten Type A fluorescent luminaires, one Type Y fluorescent luminaire, and Type S and S1 track lights are indicated on Sheet E2.3 for Room 318. Ten Type A luminaires and Type S track lights on the east, west, and south walls are controlled by switches on the west wall. The Type A luminaires installed throughout Rooms 318 and 319 are dual switched, with one switch controlling the inside lamps and the other switch controlling the outboard lamps in each luminaire. Type A luminaires are connected to circuits 8 and 10 on panelboard 4B. Track lights in Room 318 are connected to circuits 22 and 24 on panelboard 2B. Track lights in Room 319 are connected to circuits 18 and 20 on panelboard 2B.

TECH TIP

Per the SIRTI specifications, a 50°F air temperature must be maintained during, and 48 hr after, completion of masonry work.

As indicated in the Lighting Fixture Schedule, Type S track lights are 4' long surface-mounted track lights with two lighting heads. **See Figure 8-39.** Type S1 track lights are 8' long surface-mounted track lights with three lighting heads. The track lights to be installed along the north and west walls in Room 318 are connected to circuits 22 and 24 on panelboard 2B. Sheet E2.3, Detail

2 shows the track lighting control switch arrangement in Room 318. The track lights installed in Room 319 are connected to circuits 18 and 20 on panelboard 2B. Sheet E2.3, Detail 1, shows the track lighting control switch arrangement in Room 319.

As shown on Sheet E3.3, each room includes seven duplex electrical receptacles, including one isolated-ground receptacle, and one junction box to provide a connection for the projection screen. The duplex receptacles are connected to various circuits on panelboard 2D. Circuit numbers are shown near several receptacles to note their circuit assignments. For example, the isolated-ground receptacle along the south wall of Room 319 is connected to circuit 23.

As indicated on Sheet E4.3, a clock outlet and mini-horn are to be installed on the south wall of Room 318, with a thermostat and telecommunications outlet installed in the middle of the west wall. An open-trough horizontal cable tray is to be installed above the ceiling in Room 318, with a 90° horizontal bend and a transition to a solid-bottom cable tray near Room 319. A clock outlet and mini-horn are to be installed on the west wall of Room 319, along with a thermostat near the door and a telecommunications outlet along the south wall.

As indicated on Sheet M2.3, two 3″ roof drain lines and roof drains and two 3″ overflow drain lines and roof drains are to be installed in the roof above Rooms 318 and 319. As shown on Sheet M3.3, a wet pipe fire protection system is to be installed in Room 318.

As indicated on Sheet M4.3, heat is provided to Rooms 318 and 319 by terminal units #20 and #19, respectively. For Room 318, an 8″ diameter blower duct extends from a 14″ diameter supply duct to terminal unit #20. A 14″ × 12″ supply air trunk extends from the terminal unit and transitions to a 12″ × 10″ duct. Four 48″ long diffusers with a 130 cfm capacity, each connected to 7″ diameter supply air branches, provide air to the room. For Room 319, an 8″ blower duct extends from a 14″ diameter

supply duct to terminal unit #19. A 14″ × 12″ supply air trunk extends from the terminal unit to two 7″ diameter supply air branches and 48″ long diffusers. A 9″ supply air duct also branches from the 14″ × 12″ supply air trunk and provides air to the room through two 7″ diameter ducts and two 48″ long diffusers, each with a 140 cfm capacity. Return air is routed through a 22″ × 12″ duct in Room 318 and a 22″ × 10″ duct in Room 319, each terminating in a 22″ square grille. A thermostat is located on the west wall of Room 318 to control the temperature in the rooms. As indicated on Sheet M5.3, hot water supply and hot water return pipes are ¾″ diameter.

Room 320—TV Seminar

Room 320 is equipped and configured similarly to Room 312 to provide for presentation of distance learning sessions. Video cameras focused on the instructor area allow for the projection of instructional sessions to remote sites.

Architectural and Structural. As indicated on Sheet A2.6, the dimensions of Room 320 are 25′-0″ × 23′-2⅝″. The north and south walls are Type 5 walls, which are demountable partitions. As represented on Sheet A10.5, Elevation 16, the north demountable partition consists of nine 2′-0″ wide panels with a smaller panel at each end of the wall. The south demountable partition consists of ten 2′-0″ wide panels with a smaller panel at each end of the wall as shown on Elevation 17. The west wall is a Type 4 wall, which is constructed with 3½″ metal studs and covered with ⅝″ gypsum wallboard on both sides extending full height to the structure above with sound-attenuation blankets between the studs. As shown on Sheet A10.5, Elevation 18, a 4′-0″ × 12′-0″ chalkboard is to be installed 3′-0″ above the floor along the west wall, with a projection screen above recessed in an 18″ deep gypsum wallboard soffit.

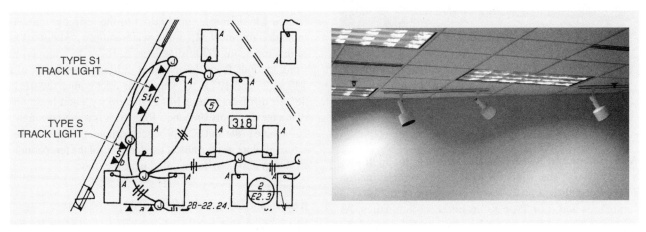

Figure 8-39. Track lights allow for flexible placement of fixtures and a variety of lighting configurations.

Sheet A2.5 references Detail 12 on Sheet A11.2. Detail 12 provides information regarding the finish of column K3, which is located at the northeast corner of Room 320. Solid wood blocking is to be installed along the interior column corners and also at the point where the demountable partition impacts the column.

As indicated in the specifications, the floor finish is direct glue-down carpet with a rubber cove base. Door 320, which is a 3'-0" × 7'-0" wood fire-rated door, is to be installed along the west wall of the room. One Type 2 window, which is an 8'-0" × 5'-0" window consisting of a 5'-0" square fixed light and a 3'-0" wide operable panel that pivots vertically, is to be installed in the east wall.

The Framing Plan on Sheet S2.4 identifies column K3 at the northeast corner of Room 320. Column K3 is a 16" square concrete column with a top elevation of 1944'. The reinforcement requirements of Column K3 are identical to column K2, which is in the northeast corner of Room 319. As indicated on the Framing Plan, type C5, C7, M1, and M4 joists are to be formed beneath the floor for Room 320. All joists are 8" wide by 20½" deep with varying reinforcing steel requirements as shown in the Joist Schedule on Sheet S2.4. Typical domes measuring 16" high by 52" square are used to form the floor joists. Perimeter beam 3B4 is similar in size and reinforcing steel requirements as beam 3B7 installed along the exterior wall of Room 317.

Type A luminaires are 2' × 4' fluorescent luminaires with 3" deep, 18 cell parabolic louvers.

Mechanical and Electrical. In addition to the light provided through the windows in Room 320, illumination is provided in Room 320 with nine Type A fluorescent luminaires and four Type K incandescent luminaires. As shown on Sheet E2.3, switch legs "a" and "c" control the center lamp of the Type A luminaires, while switch leg "b"

controls the outboard lamps. A dimmer switch controls the power to the Type K incandescent luminaires. Type A luminaires are connected to circuits 8 and 10 on panelboard 4B. Type K luminaires are connected to circuits 10 and 12 on panelboard 2B.

As shown on Sheet E3.3, a recessed fluorescent luminaire is to be installed in the soffit above the chalkboard in Room 320. Three duplex electrical receptacles and four double duplex electrical receptacles, including two isolated-ground receptacles, are to be installed in Room 320. The receptacles are connected to circuits on panelboard 2D as noted on Sheet E3.3. A symbol indicating a down-pointing camera to be furnished and installed by the owner in a well in the ceiling is shown on Sheet A2.5. Two ceiling-mounted video cameras are to be furnished and installed by the owner along the east side of the room.

As indicated on Sheet E4.3, a clock outlet, mini-horn, and thermostat are to be installed on the west wall of Room 320. Telecommunications outlets are to be installed on the north and south walls. An open-trough horizontal cable tray is to be installed above the ceiling in this area, with Sheet E5.2, Detail 4, providing information regarding the tray at the west wall of the room. The cable tray is to extend vertically downward on the east wall from the overhead tray to 12" above the finished floor.

TECH TIP

Conduit extending between HVAC units and thermostats is ¾" in diameter.

As indicated on Sheet M2.3, one 3" roof drain line and roof drain and one 3" overflow drain line and roof drain are to be installed in the roof above Room 320. The drain lines extend south to connect to a 4" drain pipe.

As shown on Sheet M4.3, heat is provided to Room 320 by terminal unit #18. A 10" diameter blower duct extends from a 14" diameter supply duct to the terminal unit. As indicated on Sheet M5.3, hot water supply and hot water return pipes are ¾" in diameter. A 16" × 15" supply air trunk extends from the terminal unit and transitions to a 14" × 12" and 12" × 10" supply air duct connected to six 48" air diffusers with a 140 cfm capacity, each fed by 8" diameter air supply branches. Return air is routed through a 22" square grille and 22" × 8" duct terminating in a 22" × 18" return air duct. A thermostat used to control the temperature in Room 320 is to be installed near the door.

Rooms 321 and 323—Project Labs

Specialized electrical equipment is installed in Rooms 321 and 323 to provide students with a hands-on learning environment. Special electrical requirements are indicated on the prints and in the specifications that address the power needs of the equipment.

Architectural and Structural. The dimensions of Rooms 321 and 323 are 30'-0" × 23'-2⅝". The north wall of Room 321 is a demountable partition, while the other interior walls in Rooms 321 and 323 are Type 4 walls. Type 4 walls are constructed with 3½" metal studs and covered with ⅝" gypsum wallboard on both sides extending full height to the structure above with sound-attenuation blankets between the studs.

Sheet A11.2, Detail 11, is referenced on Sheet A2.5. Detail 11 provides information regarding the finish at columns K4 and K6, which is similar to that of column H2 in Room 317. Since the wall is parallel to the grid lines, the lines marked as "GRID" on the print are used as a reference. Brick veneer is shown on both sides of the column.

Sheet A6.1, Section D, provides information regarding the construction of the east wall. Two-inch thick phenolic rigid insulation board is installed at the elevation of the two-way joist system around the perimeter of the building to minimize heat loss through the concrete floor. The insulation board must lap 6" above and below the floor slab at each level. Information is also provided regarding the precast concrete planter and concrete paver structure above the classroom level on the east side of the building. The planter is placed on a layer of elastomeric roofing material and R-38 tapered rigid roof insulation. A roof drain and roof drain line are installed beneath the concrete pavers. Steel pipe guardrail is installed on both sides of the planter. The parapet extends 3'-6" above the office/administration level elevation of 1944.00'. Sheet A11.3, Details 5, 7, 8, and 10 (not included with this set of prints), provide information regarding forming of the concrete floor slab and roof drain, and masonry windowsills and headers at window openings.

Elevations 17 and 19 on Sheet A10.5 are referenced on Sheet A2.5. As shown on Elevation 17, the north wall of Room 321 is a demountable partition consisting of ten 2'-0" panels and two equal-width filler panels at the ends. As noted on the detail, the south wall of Room 323 is a Type 4 wall constructed with 3½" metal studs and covered with ⅝" gypsum wallboard on both sides to the structure above, with sound-attenuation blankets between the studs. As shown in Elevation 19, a 4'-0" × 6'-0" liquid marker board is to be installed 3'-0" above the floor on the west wall of Rooms 321 and 323. Two wardrobe units are to be installed along the west wall near the stairway. Four coat hooks are to be installed on the doors of each room.

As noted in the specifications, the floor finish for Rooms 321 and 323 is sheet vinyl with a rubber cove base. One 3'-0" × 7'-0" wood door is to be installed in

each room as part of the key-activated access system. As indicated on Sheet A2.5, one Type 1 and one Type 2 window are to be installed in each room. Type 1 windows are 3'-0" × 5'-0" vertically pivoting windows and Type 2 windows are 8'-0" × 5'-0" windows consisting of a 5'-0" square fixed light and a 3'-0" wide operable panel that pivots vertically.

As indicated on Sheet S2.4, columns K4 and K6 are to be formed along the exterior walls of Rooms 321 and 323. Columns K4 and K6 are 16" square columns constructed to an elevation of 1944' in the same manner as column H2 in Room 317. Sheet S2.4 also identifies type C5, C7, M1, and M2 joists in both rooms. Typical domes are used to form the joists. Perimeter beams 3B16 and 3B18 are to be installed along the exterior edges of Rooms 321 and 323, respectively. As indicated on the Beam Schedule on Sheet S2.4, additional reinforcing steel is installed at the north end perimeter beam 3B16, including two #8 rebar and four #10 rebar.

Mechanical and Electrical. In addition to the natural light provided through the windows, illumination is provided in Room 321 by 12 Type A luminaires and four Type K luminaires. The arrangement of the luminaires in Room 323 is a mirror image of the luminaire arrangement in Room 321. For Type A fluorescent luminaires, switch legs "a" and "c" are connected to the center lamp of the luminaire while switch leg "b" is connected to the outboard lamps. The Type K incandescent luminaires are connected to dimmer switches. The luminaires are connected to circuits 10 and 12 on panelboard 2B, and circuits 8 and 10 on panelboard 4B.

As shown on Sheet E3.3, eight duplex electrical receptacles are to be installed in each room, including one insulated-ground receptacle. A 10' section of surface-mounted raceway is to be installed on the east wall of each room, including two gang device plates with blank covers for future outlets. A ¾" conduit-only installation is made from each of the raceways for future use.

Type 1 and 2 windows are specified for use on the classroom level.

Additional 10′ raceways with receptacles 12″ OC are installed in each room on the walls backing Room 322. The receptacles and raceways are connected to various circuits on panelboards L8 and L9. A 30 A, 3-phase enclosed thermal-magnetic circuit breaker is surface-mounted on the west wall of each room. Thermal-magnetic circuit breakers utilize both the thermal and magnetic effects associated with current flow to sense and operate the circuit breaker. Generally, the thermal element is used for overload protection while the magnetic element provides faster response to ground faults and short circuits.

A 4″ square junction box with a blank cover is to be provided 12″ above the finished floor. An 800 VA uninterruptible power supply is to be installed along the west wall of each room. An *uninterruptible power supply (UPS)* is an electrical power system that provides continuous current to designated circuits in the event of a power outage. The UPS in Room 321 is connected to circuits 1, 3, and 5 on panelboard 4B, and the UPS in Room 323 is connected to circuits 13, 15, and 17 on panelboard 4B.

As shown on Sheet E4.3, a clock outlet, mini-horn, and thermostat are to be installed on the west wall in each room. Two telecommunications outlets are to be installed on the north and south walls of Room 321. On Sheet E5.1, Details 3 and 4 (not included with this set of prints) provide typical wiring information for the telecommunications outlets. Room 323 has two telecommunications outlets on the north and west walls. A door control unit for key-activated access is located above the ceiling in each room.

As shown on Sheet M2.3, two 3″ roof drain lines and roof drains, and two 3″ overflow drain lines and roof drains are to be installed in the roof above Rooms 321 and 323. Sheet A6.1, Section D, shows the roof drain and drain line installed in the roof above Room 323 that drains the exposed concrete paver area.

The finish floor for Rooms 324 and 325 (Men's and Women's Toilets) is ceramic mosaic tile with a ceramic mosaic tile base.

As indicated on Sheet M3.3, a wet pipe fire protection system is to be installed above the finished ceiling in Rooms 321 and 323. The fire main extends from west to east.

As shown on Sheet M4.3, heat is provided to Rooms 321 and 323 by terminal units #17 and #16, respectively. A 12″ diameter blower duct extends from the supply air duct to the terminal units. A 16″ × 15″ supply air trunk extends from each of the terminal units and branches to 14″ × 12″ supply air ducts. Eight 7″ diameter ducts extend from the supply ducts to eight 48″ long diffusers, each with a 140 cfm capacity. Return air in Room 321 is routed through a 22″ square grille and 22″ × 10″ duct and a 22″ × 18″ duct. Return air in Room 323 is routed through a 22″ square grille and 22″ × 10″ duct and into a 45″ × 14″ duct. As indicated on Sheet M5.3, hot water supply and hot water return pipes for terminal units #17 and #18 are ¾″ in diameter. A thermostat is to be installed near the door leading into each room.

Room 324—Men's Toilet

Room 324 is a men's restroom on the classroom level that is entered through a small vestibule at the south end of Room 316. Most of the water supply piping for this restroom and Room 325 (Women's Toilet) is installed above the ceiling of the men's restroom.

Architectural and Structural. As shown on Sheet A2.6, the dimensions of Room 324 are 22′-7½″ × 10′-7¼″ at the widest point and narrowing to 9′-3⅝″ at the east end of the room. Two Type 1 walls are constructed along the south wall to support the lavatory. Type 1 walls are constructed with 3½″ metal studs and covered with ⅝″ gypsum wallboard on both sides to 6″ above the ceiling line. The walls project 2′-3⅞″ (4′-6″ − 2′-2⅛″ = 2′-3⅞″) from the south wall. One Type 1 wall is to be constructed 12′-4″ from the west exterior wall and the other wall is located 5′-2½″ from the first wall.

The north and west walls of Room 324 are Type 4 walls, which are constructed with 3½″ metal studs and covered with ⅝″ gypsum wallboard on both sides extending full height to the structure above, with sound-attenuation blankets between the studs. A portion of the south wall is a Type 9 wall, which is constructed with 3½″ metal studs and covered with ⅝″ gypsum wallboard with sound-attenuation blankets between the studs on one side only. The other portion of the south wall is a Type 4 wall, which is constructed with 3½″ metal studs and covered with ⅝″ gypsum wallboard on both sides extending full height to the structure above, with sound-attenuation blankets between the studs.

As noted in the specifications, the floor finish for Room 324 is ceramic mosaic tile (CMT) with a CMT base. The enlarged floor plan on Sheet A9.1 (not included with this set of prints), indicated by the dashed line around the area on Sheet A2.5, provides additional information about the floor finish and position of the restroom fixtures. As indicated in the specifications, Door 324 is a 3′-0″ × 7′-0″ wood door with a 20 min fire rating, and the Type 1 window is a 3′-0″ × 5′-0″ vertically pivoting window.

As indicated on Sheet S2.4, floor construction includes type C7, M1, and M3 joists, which are formed with typical domes measuring 16″ high by 52″ square. A symbol and notation on Sheet S2.4 indicates that the floor slab is to be recessed 1½″ below the surrounding slab to allow the ceramic mosaic tile to be flush with the surrounding finished floor. Plumbing blockouts through the floor slab are also shown on the Framing Plan. Perimeter beam 3B4 is specified along the exterior wall of Room 324. Beam dimensions and reinforcement requirements are shown on Sheet S2.4, Beam Schedule. The note "320 LB HUNG FROM BELOW" on the Framing Plan refers to equipment that is to be suspended from the ceiling on the entry level.

Mechanical and Electrical. Illumination is provided in Room 324 by three Type C luminaires and one Type H luminaire. Type C luminaires are recessed 1′-0″ × 4′-0″ fluorescent luminaires with acrylic lenses, each fitted with two lamps, and connected to circuit 2 on panelboard 4B. The luminaires are controlled by a key switch. The Type H luminaire is a round, recessed, compact fluorescent downlight luminaire with a clear reflector and is connected to emergency lighting circuits 2 and 4 on panelboard 4X1.

As shown on Sheet E4.3, one duplex electrical receptacle protected by a ground fault interrupter is to be installed in Room 324. A *ground fault interrupter,* or ground fault circuit interrupter, is an electrical device that automatically deenergizes a circuit or portion of a circuit when the grounded current exceeds a predetermined value that is less than the value required to operate the overcurrent protection device.

Two water closets, two urinals, and a lavatory with two wash basins are to be installed on the south wall of the men's restroom. The restroom is ADA-compliant, providing disabled individuals with adequate access to the fixtures and proper turning radii for wheelchair-bound individuals. The proper number of ADA-compliant fixtures to be installed in the restroom is based on the building occupancy and building type. The water closet and toilet stall partition on the east side of the restroom is ADA compliant. **See Figure 8-40.**

Information regarding the water supply and waste piping for Room 324 is shown on Sheet M7.2, Detail 1. A 2″ floor drain is installed between the water closets and urinals. A ¾″ hose bibb is installed 10″ above the finished floor beneath the lavatory and is supplied by a ¾″ cold water pipe. **See Figure 8-41.** The cold water supply for the men's and women's restroom is a 2½″ pipe that

extends from the mechanical chase, over the corridor ceiling, and to the restroom fixtures. A valve is installed above the ceiling in the corridor to provide a water shutoff for the restrooms. The 2½″ pipe is reduced to a 2″ pipe above the men's restroom ceiling and extends south into the chase between the men's and women's restrooms. A 2″ × 2″ × ¾″ tee is installed on the 2″ pipe to provide ¾″ water supply pipe to the lavatory fixtures. A 1¼″ hot water supply pipe extends from the mechanical chase, over the corridor ceiling, and to the restroom lavatories. A 1¼″ × 1¼″ × ¾″ tee and valve are installed in the ceiling over the corridor to reduce the fixture pipe size and provide a hot water shutoff to the restrooms.

Figure 8-40. The Americans with Disabilities Act (ADA) specifies requirements for handicapped-accessible restrooms.

Figure 8-41. A hose bibb is installed in each restroom under the lavatory.

The ¾″ hot water supply pipe is installed parallel to the cold water supply pipe. The lavatories are vented with 1½″ vents that are connected to a 2″ branch vent that extends to a 3″ vent running toward the plumbing chase. A 3″ vent is installed in an east-west direction through the plumbing chase between the restrooms with vents extending down to the water closets and urinals. A 3″ roof drain and roof drain line are installed in the roof over the men's restroom. A 3″ drain line extending to the north from the women's restroom passes over the ceiling in the men's restroom. Dashed lines in the chase between the restrooms show vent locations and diameters.

As shown on Sheet M4.3, heat is provided to Room 324 by terminal unit #15, which has an 840 cfm minimum and maximum rating. A 10″ diameter blower duct extends from a 36″ × 16″ duct to the terminal unit. As indicated on Sheet M5.3, hot water supply and hot water return pipes are ¾″ diameter. A 16″ × 15″ supply air trunk extends from the terminal unit to three 6″ diameter supply air branches and three 6″ diameter diffusers, each with a 90 cfm capacity. Return air is routed through a 12″ square grille and into an 18″ × 12″ return duct. An exhaust fan is to be installed in the restroom ceiling near the east wall.

Room 325—Women's Toilet

Room 325 is a women's restroom on the classroom level that is entered through a small vestibule at the south end of Room 316. Most of the water supply piping for the women's restroom is installed above the ceiling of the men's restroom. The thermostat that controls the temperature in the men's and women's restrooms is installed in the women's restroom.

Architectural and Structural. As indicated on Sheet A2.6, the dimensions of Room 325 are 23′-2⅝″ × 11′-4¼″. The room narrows to 9′-3⅝″ at the east end of the room. Two Type 1 walls are constructed along the south wall to support the lavatory. Type 1 walls are constructed with 3½″ metal studs and covered with ⅝″ gypsum wallboard on both sides to 6″ above the ceiling line. The walls project 5′-3″ from the north wall. One Type 1 wall is to be constructed 12′-4″ from the west exterior wall and the other wall is located 5′-2½″ from the first wall.

The south and west walls of Room 324 are Type 4 walls, which are constructed with 3½″ metal studs and covered with ⅝″ gypsum wallboard on both sides extending full height to the structure above, with sound-attenuation blankets between the studs. A portion of the north wall is a Type 9 wall, which is constructed with 3½″ metal studs and covered with ⅝″ gypsum wallboard and sound-attenuation blankets on one side only. The other portion of the north wall is a Type 4 wall, which is constructed with 3½″ metal studs and covered with ⅝″ gypsum wallboard

on both sides extending full height to the structure above, with sound-attenuation blankets between the studs.

As noted in the specifications, the floor finish is ceramic mosaic tile (CMT) with a CMT base. The enlarged floor plan on Sheet A9.1 (not included with this set of prints), indicated by the dashed line around the area on Sheet A2.5, provides additional information about the floor finish and position of the restroom fixtures such as lavatories, water closets, and urinals. As indicated in the specifications, Door 325 is a 3′-0″ × 7′-0″ wood door with a 20 min fire rating, and the Type 1 window is a 3′-0″ × 5′-0″ vertically pivoting window.

As indicated on Sheet S2.4, column K7 is to be formed at the southeast corner of the women's restroom. The 16″ square column is constructed to an elevation of 1944′ in the same manner as column H2 in Room 317.

As indicated on Sheet S2.4, floor slab construction includes type C7, M1, and M3 joists, which are formed with the typical domes measuring 16″ high by 52″ wide. A symbol and notation on Sheet S2.4 indicates that the floor slab is to be recessed 1½″ below the surrounding slab to allow the ceramic mosaic tile to be flush with the surrounding finished floor. Plumbing blockouts through the floor slab are also shown on the Framing Plan. Perimeter beam 3B4 is specified along the exterior wall of Room 324. Beam dimensions and reinforcement requirements are shown on Sheet S2.4, Beam Schedule.

Mechanical and Electrical. Illumination is provided in Room 325 by three Type C luminaires and one Type H luminaire. Type C luminaires are recessed 1′-0″ × 4′-0″ fluorescent luminaires with acrylic lenses, each fitted with two lamps, and connected to circuit 2 on panelboard 4B. The fluorescent luminaires are controlled by a key switch. Type H luminaires are round, recessed, compact fluorescent downlight luminaires with a clear reflector, and are connected to emergency lighting circuits 2 and 4 on panelboard 4X1.

As shown on Sheet E4.3, one duplex electrical receptacle protected by a ground fault interrupter is to be installed in Room 325. The thermostat is to be installed on the west wall of the room.

Four water closets and a lavatory with two wash basins are to be installed on the north wall of the women's restroom. The restroom is ADA-compliant, providing disabled individuals with adequate access to the fixtures and proper turning radii for wheelchair-bound individuals. The proper number of ADA-compliant fixtures to be installed in the restroom is based on the building occupancy and building type.

Information regarding the water supply and waste piping for Room 325 is shown on Sheet M7.2, Detail 1. A 2″ floor drain is installed between the middle two water closets. A ¾″ hose bibb is installed 10″ above the finished floor beneath the lavatory and is supplied by a ¾″ cold water pipe. Similar to the men's restroom, vent and water supply piping is routed through the chase between the restrooms. Lavatories are vented with 1½″ vents, which are connected to a 2″ vent that extends to a 3″ vent running toward the plumbing chase. A 3″ vent is installed in an east-west direction through the plumbing chase between the restrooms with vents extending down to the water closets.

As shown on Sheet M4.3, heat is provided to Room 325 by terminal unit #15, which has an 840 cfm minimum and maximum rating. A 10″ diameter blower duct extends from a 36″ × 16″ duct to the terminal unit. As indicated on Sheet M5.3, hot water supply and hot water return pipes are ¾″ diameter. A 16″ × 15″ supply air trunk extends from the terminal unit to a 10″ diameter air supply branch and a 12″ diameter diffuser with a 375 cfm capacity. Return air is routed through a 12″ square grille and into a 14″ × 8″ return duct. An exhaust fan is to be installed in the ceiling near the east wall.

Room 326—Microcomputer Classroom

Room 326 (Microcomputer Classroom) is the largest area on the classroom level and can be divided into two smaller areas by an operable wall. A large number of electrical receptacles are installed in this room to accommodate a large number of desktop computers. A subpanel in the room provides adequate power for the electrical requirements.

Architectural and Structural. As shown on Sheet A2.6, the dimensions of Room 326 are 60′-7⅜″ × 31′-0″, based on dimensions provided, grid line spacing, and exterior wall thicknesses. An operable wall consisting of nine equal panels stacks along the north wall.

The west wall and north wall of Room 326 along the women's restroom is a Type 4 wall, which is constructed with 3½″ metal studs and covered with ⅝″ gypsum wallboard on both sides extending full height to the structure above, with sound-attenuation blankets between the studs. The west portion of the north wall is a Type 6 wall along the mechanical shaft and a Type 7 wall along Room 327. A Type 6 wall is an 8″ thick concrete wall with ⅞″ furring channels and ⅝″ gypsum wallboard on one side. A Type 7 wall is an 8″ thick concrete wall with ⅞″ furring channels and ⅝″ gypsum wallboard on both sides.

Interior elevations for Room 326 are referenced on Sheet A2.5. Sheet A10.6, Elevations 1 and 2, and Sheet A10.5, Elevation 20, provide information regarding wall finishes and specialty installations. Elevation 1 shows the north wall of the room, which includes a double door. **See Figure 8-42.** The double door is Door 316, which is a 6′-0″ × 7′-0″ wood double door. A 3′-0″ × 4′-0″ tackboard is to be installed 3′-0″ above the floor toward the west end of the wall. Elevation 2 is an elevation of the operable wall panels and indicates that the panels stack at the north end of the wall.

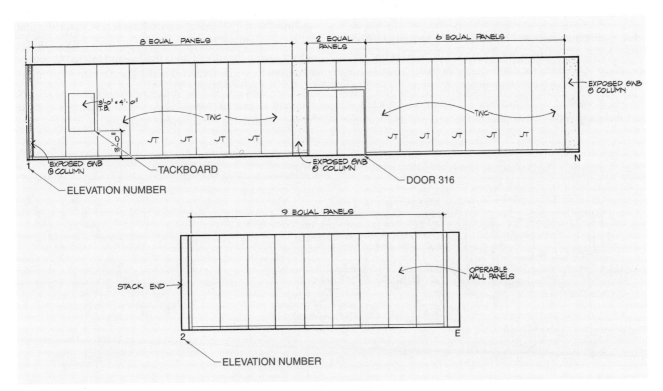

Figure 8-42. Elevations 1 and 2 are interior elevations of Room 326.

Sheet A10.5, Elevation 20, provides information regarding the west wall of Room 326. A 4'-0" × 12'-0" chalkboard is to be installed 3'-0" above the floor with a projection screen above installed in an 18" deep gypsum wallboard soffit. The door shown on the elevation is Door 314, which is a 3'-0" × 7'-0" 20 min fire-rated wood door. The wall is finished with tackable wallcovering.

Detail 2 on Sheet A2.5 shows the detail at column K7 at the northeast corner of Room 326. The construction of the exterior wall is similar to other columns, with metal framing, batt insulation, and a 3" air space along the outside edge of the column. The interior wall shown in the detail indicates sound-attenuation blankets installed in the wall between Rooms 325 and 326. As shown on the detail, dimensions for masonry and interior finishes are indicated from the column centerlines. Column J8 is finished in a manner similar to column G5 shown on Sheet A2.5, Detail 3.

Six Type 2 windows are indicated on Sheet A2.5. Type 2 windows, similar to other exterior windows at the classroom level, are 8'-0" × 5'-0" windows consisting of a 5'-0" square fixed light and a 3'-0" wide operable panel that pivots vertically. As indicated in the specifications, the floor finish is conductive vinyl composition tile with a rubber cove base.

The columns referenced in Room 326 not previously discussed are columns J7, J8, and K8. As shown on the Column Schedule on Sheet S2.4, column J7 is a 16" square concrete column that extends to an elevation of 1962.7'. The reinforcing steel at the classroom level consists of four #8 rebar overlapping eight #11 rebar extending up from the floor below by 3'-9". Stirrups are fabricated from #3 rebar and spaced 16" OC along the entire height of the column above the classroom floor level. Additional information regarding the top of column J7 is included on Sheet S3.3, Detail 4 (not included in this set of prints). Columns J8 and K8 are

constructed in a similar manner to column H2 in Room 317. Columns J8 and K8 extend to an elevation of 1944'.

As indicated on Sheet S2.4, type C3, C5, C7, M1, M3, and M4 joists are to be installed as part of the floor beneath Room 326. The Joist Schedule notes that these columns are 8" wide by 20½" deep, and have varying reinforcement requirements. Dimensions and notes between grid lines 7 and 8 at the top of the Framing Plan indicate the width of column and middle strips, which relate to Details 30A and 30B on Sheet S2.3A. Reinforcing steel must be overlapped as specified in these details to ensure proper strength characteristics and to support intended loads. These connections are the critical structural points for transferring and distributing horizontal and vertical loads between the columns and the beams and floor structure. When forming joists and installing structural steel prior to concrete placement, spacing and length of the reinforcing steel must be verified to ensure proper concrete coverage.

Perimeter beams 3B5 and 3B8 are noted on the Framing Plan on Sheet S2.4. The Beam Schedule indicates the beams are 18" wide by 20½" deep with a ⅜" camber. As noted in the Beam Schedule, the two beams have varying reinforcement requirements.

As indicated by the note to the right of the Framing Plan on Sheet S2.4, four ⅝" diameter ferrule loop inserts are to be installed at the bottom of the classroom level floor slab in a specific area to support a 2100 lb fan below. **See Figure 8-43.** No. 4 × 8'-0" long rebar passes through the loops to provide reinforcement. A ferrule loop insert is manufactured by welding a wire loop to a steel ferrule that has been machined from bar stock. One end of the ferrule is closed, while the other end is drilled and tapped to accept a National Coarse threaded bolt. The ferrule loop insert is fastened to the forms using a mounting washer before concrete is placed. Sheet S5.1, Detail 15 shows similar ferrule installation for operable walls.

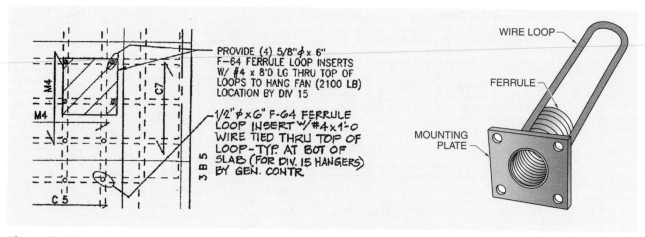

Figure 8-43. Ferrule loop inserts provide a connection to suspend heavy items from the ceiling.

Another note to the right of the Framing Plan on Sheet S2.4 indicates that the small circles shown along Rooms 315 and 316 and in Room 326 represent locations for ½" diameter by 6" long ferrule loop inserts. The ½" diameter ferrules are secured to #4 × 1'-0" long rebar using wire ties. Hangers are screwed into the ferrules to support fixtures hung from the ceiling below.

The exterior elevation of Room 326 is shown on Sheet A4.2, South Elevation. **See Figure 8-44.** Brick is the primary finish material for the exterior walls of Room 326. The windows for the room appear foreshortened on the elevation. Air intake louvers for the mechanical systems are to be installed at the office/administration level. Sheet A4.2, Detail 1, provides information regarding the construction around the louvers, including MDO plywood and a steel angle to secure the aluminum louver in position. The brick veneer is 3½" thick, with a 3" air space, ½" sheathing, 6" steel stud box header with insulation, and ⅝" gypsum wallboard interior finish. Sheet A11.2, Detail 2, and Sheet A11.3, Details 7 and 8 (not included with this set of prints), provide information regarding the masonry at the header, windowsill, and sides of the windows.

of the lamps, with leg "b" controlling one length of lamps, leg "a" controlling the first 8' of the other length of lamps, and leg "c" controlling the remaining 16' length of lamps. Type K luminaires are 8" diameter recessed incandescent luminaires fitted with one 150 W lamp. The incandescent luminaires are controlled by dimmer switches located at the entrances to the room, and are connected to circuits 18 and 20 on panelboard 2B. A Type Y luminaire is also to be installed above the chalkboard at the west end of Room 326. The Type Y luminaire is a 9" wide linear luminaire that is recessed in a gypsum wallboard soffit and is connected to circuits 7 and 9 on panelboard 4B.

As indicated on Sheet E3.3, 15 duplex electrical receptacles are to be installed in Room 326, including five isolated-ground receptacles. Seven of the duplex receptacles along the north wall of the room require 1½" shallow boxes. Seven data/power poles are to be installed at the south end of the prewired workbenches. A junction box for connection of a powered projection screen is to be installed on the west wall. All data/power poles and electrical receptacles are connected to panelboard 2C1, which is to be installed on the north wall of Room 326.

Figure 8-44. The South Elevation provides information regarding the exterior elevation of Room 326.

Data/power poles are vertical raceways that extend from the ceiling to the floor and enclose data and power conductors.

Mechanical and Electrical. In addition to the natural light provided through the windows, illumination is provided in Room 326 by eight rows of Type D luminaires and eight Type K luminaires. Type D luminaires are suspended indirect luminaires measuring 10" wide by 3⅝" high by 24' long and are connected to circuits 7 and 9 on panelboard 4B. The luminaires are suspended 18" below the ceiling line using aircraft cable. The fluorescent luminaires in Room 326 are to be provided with side-by-side switching

As shown on Sheet E4.3, clock outlets, mini-horns, and thermostats are to be installed on the north and west walls of Room 326. Four telecommunications outlets are also to be installed in the room, two on the north wall and one each on the south and west walls. An open-trough horizontal cable tray is to be installed above the ceiling in Room 326 with the top of the tray 11′-0″ above the finished floor. Sheet E5.2, Detail 2 (not included with this set of prints) provides information regarding the typical power pole connections to the cable trays. Three door-intrusion sensors are installed at the three doors leading into Room 326.

As shown on Sheet M2.3, three 3″ roof drains and drain lines and three 3″ roof drains and overflow drain lines are to be installed above the ceiling in Room 326 and connected to 4″ drain pipes extending north. A 3″ waste pipe extends up to the level above at two locations, with a cleanout at the west end and a 2″ vent at each end. Along the west wall, two 2″ waste pipes and one 3″ waste pipe extend up to the next level, with three 2″ individual vents. A 4″ waste pipe is installed along the north wall, extending toward the mechanical chase. As noted on Sheet M3.3, Room 326 is provided with a preaction fire protection system due to the large amount of electrical equipment housed in the room.

As indicated on Sheet M4.3, heat is provided to Room 326 by terminal units #29 and #30, which are identical units with a 1400 cfm maximum rating and 420 cfm minimum rating. Two 12″ diameter blower ducts extend from a 36″ × 16″ oval supply duct to the terminal units. As indicated on Sheet M5.3, hot water supply and hot water return pipes for both terminal units are ¾″ in diameter. An 18″ × 17″ supply air trunk extends from terminal unit #29. One 12″ × 12″ supply air branch and four 8″ diameter ducts extend from the 18″ × 17″ trunk. The 12″ × 12″ supply air duct then branches off to four 8″ diameter ducts. The 8″ diameter ducts terminate at eight 48″ long diffusers, each with a 175 cfm capacity.

A 26″ × 12″ supply air trunk extends from terminal unit #30. A 15″ × 10″, two 10″ × 10″, and one 8″ diameter supply air ducts branch off the 26″ × 12″ trunk. Four 8″ diameter ducts branch off the 15″ × 10″ supply duct and terminate at four 48″ long diffusers. Two 8″ diameter ducts branch off each of the 10″ × 10″ ducts and terminate at four 48″ long diffusers. The 8″ diameter supply duct terminates at a 48″ long diffuser. All air diffusers are rated at 175 cfm capacity.

Return air for the room is routed through 22″ square grilles and 22″ × 16″ and 22″ × 12″ return ducts. Sheet M4.3, Note 12, indicates that the damper is to be balanced to 3000 cfm at the maximum cooling level. A 45″ × 16″ return air duct extending up to the floor above is to be installed near the north wall of Room 326. A 36″ × 16″ supply air duct also extends up to the floor above, with floor penetration location coordinated with the other tradesworkers. Two thermostats are to be installed in Room 326 to control the temperature in the room.

Room 328—Custodial Closet

Room 328 (Custodial Closet) serves as a small vestibule for entry into the restrooms and as a workspace for the custodial staff. Minimal storage is provided in the room.

Architectural and Structural. As indicated on Sheet A2.6, the dimensions of Room 328 are 5′-2½″ × 6′-4¾″. The north, south, and west walls are Type 4 walls, which are constructed with 3½″ metal studs and covered with ⅝″ gypsum wallboard on both sides extending full height to the structure above, with sound-attenuation blankets between the studs. The east wall is a Type 9 wall consisting of a 3½″ metal stud wall with sound-attenuation blankets between the studs and ⅝″ gypsum wallboard on one side only extending to the structure above.

As described in the specifications, the floor finish is vinyl composition tile (VCT) with a rubber cove base. The walls are finished with a high-build glazed finish to repel water. Two doors are installed in Room 328. Door 328A is a 3′-0″ × 7′-0″ wood door with a 20 min fire rating. Door 328B is a 2′-0″ × 6′-0″ wood door with no fire rating and provides access to the plumbing chase between the restrooms.

Mechanical and Electrical. As indicated on Sheet E2.3, illumination in Room 328 and the restroom entryway is provided by one Type G and one Type H luminaire. A Type G luminaire is a chain-hung, surface-mounting, two-lamp, open-strip luminaire with a wire guard. A Type H luminaire is a round, recessed, compact fluorescent downlight with a clear reflector.

As shown on Sheet E3.3, one duplex electrical receptacle is to be installed on the west wall of Room 328. The duplex receptacle in Room 328 is on the same circuit as the GFI-protected duplex receptacles in the restrooms.

A mop basin is to be installed in the southeast corner of Room 328. A 2″ floor drain is to be installed near the mop basin. Sheet M7.2, Detail 1, provides additional information regarding the plumbing installation in this area.

As indicated on Sheet M4.3, heat is provided to Room 328 by terminal unit #15, which has an 840 cfm minimum and maximum rating. A 10″ diameter blower duct extends from the 35″ × 10″ supply duct to the terminal unit. As indicated on Sheet M5.3, hot water supply and hot water return pipes are ¾″ diameter. A 16″ × 15″ supply air branch extends from the terminal unit to a 6″ diameter branch connected to a 6″ diameter diffuser with a 90 cfm capacity.

Room 329—Telephone Closet

Room 329 is a small room located to the east of Room 313 and provides a secure area for the installation of telephone equipment. A small mechanical chase adjacent to Room 329 provides room for installation of mechanical equipment, pipes, and ducts.

Architectural and Structural. As indicated on Sheet A2.6, the dimensions of Room 329 are 8'-5⅝" × 9'-5⅛" with a 5'-1⅝" × 4'-8⅛" area provided in the southeast corner for the small mechanical chase. The north wall and east wall along Room 316 are Type 6 walls, which are 8" thick concrete walls with ⅞" furring channels and ⅝" gypsum wallboard on one side only. The south wall and portion of the east wall along the mechanical chase are Type 23 walls, which are shaft walls. The west wall of Room 329 is constructed of 3½" metal studs with ⅝" gypsum wallboard on both sides extending full height to the structure above.

As indicated in the specifications, the floor finish for Room 329 is exposed concrete. The door, which leads from Room 311 into the mechanical chase, is a 2'-6" × 7'-0" wood door with no fire rating.

As shown on Sheet S2.4, column J3 is located at the northeast corner of the mechanical chase. Column J3 is a 16" square concrete column that extends to an elevation of 1944'. Reinforcing steel for the classroom level of column J3 consists of four #8 rebar overlapping eight #11 rebar extending up from the floor below by 3'-9". Stirrups are fabricated from #3 rebar and are spaced 16" OC along the entire height of the classroom level. Sheet S3.2, Detail 29 (not included with this set of prints), provides additional

information about column J3. Sheet A2.7, Detail 4 (not included with this set of prints) provides information regarding the finish of column J3.

Mechanical and Electrical. As indicated on Sheet E2.3, illumination is provided in Room 329 by two Type G luminaires. Type G luminaires are chain-hung, surface-mounting, two-lamp, open-strip luminaires with a wire guard. A switch by the door controls power to the luminaires.

As shown on Sheet E3.3, five duplex isolated-ground electrical receptacles are to be installed in Room 329 and connected to circuits 28, 30, 32, 34, and 36 on panelboard 2X1.

As indicated on Sheet E4.3, a rate-of-rise heat detector is to be installed in Room 329. An open-trough horizontal cable tray is also to be installed around the perimeter of Room 329, with a vertical drop down to the floor below.

Additional information concerning plumbing in this area is provided on Sheet M7.2, Detail 2, especially related to the mechanical chase at the southeast corner of Room 329. Piping installed in the mechanical chase includes a 6" acid-resistant vent (ARV) stack, 1½" hot water riser, ½" hot water circulating (HWC) riser, 8" overflow drain stack and 8" roof drain stack, 4" waste pipe down, 6" vent stack, and 2" cold water riser. Placement of all the piping in this area is coordinated with the ductwork installation. Additional notes on Sheet M7.2, Detail 2, indicate that the slope of the overflow and roof drains is 1%. A 24" × 24" duct is installed to the upper floor in the mechanical chase.

Refer to the CD-ROM in the back of the book for Chapter 8 Quick Quiz® and related printreading and reference material.

Plans—SIRTI Building

TRADE COMPETENCY TEST

Name ___Paul Sullm___   Date _____

Refer to SIRTI project (Sheet A2.5). Cross-referencing between several sheets may be required to answer questions.

T   (F)   **1.** Special brick shapes "C" and "D" are to be laid at column G7.

_____   **2.** Section BB, which is shown on Sheet A5.1, is taken along the ___ wall.
                  A. north
                  B. south
                  C. east
                  D. west

_____   **3.** The operable wall receiver shown on Detail 3 is on the ___ side of column G5.

___ 1 ___   **4.** There is/are ___ Type 1 window(s) in the Student Lounge.

(T)   F   **5.** Door 327 is a double door.

_____   **6.** The abbreviation "DF" in the Elevator Lobby indicates a(n) ___.

X (T)   F   **7.** A 16" square column is to be formed at grid line intersection 2G.

_____   **8.** A total of ___ ceiling-mounted and down-pointing television cameras are specified on the classroom level.

_____   **9.** There are ___ curtains shown on Sheet A2.5 for covering exterior windows.

(T)   (F)   **10.** An interior elevation of the north wall of Room 323 is not provided in the set of prints.

_____   **11.** The elevation of the raised floor in Room 308 is ___ above the finish floor elevation of the classroom level.

_____   **12.** The fiberglass insulation on the east side of column K7 is ___" thick.

A   X   **13.** The abbreviation "PS" on the south wall of the Classroom Exhibit Room (Room 319) indicates a ___.
                  A. pull switch
                  B. projection screen
                  C. plaster sheet
                  D. pullout shoe

T   F   **14.** Doors 315A and 315B are left-handed swinging doors.

15. The heavy dashed line in Corridor 314 near the east wall indicates ___.
    A. a cable tray
    B. the column grid line offset
    C. hydronic piping
    D. an enlarged floor plan

(T)   F   **16.** The operable windows are based on an alternate bid.

Refer to SIRTI project (Sheet A2.6). Cross-referencing between several sheets may be required to answer questions.

3½"   **1.** The wall between Rooms 303 and 306 is framed with ___" wide steel studs.

(T)   F   **2.** A 2'-0" space is provided between the tile patterns in Corridor 314.

10¾"   **3.** The east wall of Stairway 327 is ___" thick.

T   (F)   **4.** The depth of the Men's Toilet and Women's Toilet are the same in the north-south direction.

7"   **5.** The center of the south wall of Room 305 is ___" from grid line 6.

(T)   F   **6.** There are no interior cast-in-place concrete walls on the classroom level.

1'-9"   **7.** The distance from the center of the column grid lines to the face of the masonry walls at column H2 is ___.

8. The center of the northernmost column on the classroom level is located ___' from grid line 8.

(T)   F   **9.** Stairway 322 is centered on grid line 5.

(T)   (F)   **10.** The west wall of the elevator shaft is to be covered with ⅝" thick gypsum wallboard.

T   F   **11.** All interior walls of Room 312 are framed with 3½" steel studs.

12. The centerline of Door 322 is ___ from grid line 5.

3'   **13.** Type 1 windows are ___ wide.

14. The center-to-center distance from the north wall of the Men's Toilet to the south wall of the Women's Toilet wall is ___'.

15. The angular portions of the south wall of Room 305 are offset at a(n) ___° angle.

1'-9"   **16.** The exterior face of the east wall projects ___ beyond grid line K.

T   F   **17.** The interior walls of all stairways on the classroom level are framed with 3½" steel studs and covered with ⅝" gypsum wallboard with no sound-attenuation blankets.

Refer to SIRTI project (Sheet A4.2). Cross-referencing between several sheets may be required to answer questions.

__1947.50 X__   **1.** The windowsills of classroom level windows are at elevation ___.

_____   **2.** The air intake louvers on the South Elevation are above Room ___.

_____   **3.** The abbreviation "C.I.P." shown on the West Elevation refers to ___.

__~~ooo~~326__   **4.** The window with the reference for Section 8 on Sheet A11.3 is located in Room ___.

_____   **5.** The flagpole indicated on the South Elevation is located on the ___ side of the building.

_____   **6.** Brick expansion joints are to be ___" wide.

__X 55'-6"__   **7.** The elevation of the eaves line for the office/administration level is ___'.

   T   F   **8.** A 2" × 1¼" × ⅛" steel angle frame is to be installed around the perimeter of each louver in the brick-veneer walls.

_____   **9.** The top of the concrete trellis is ___' above the roof deck level.

__1906.0__   **10.** The elevation at the bottom of the loading dock is ___'.

   T   F   **11.** A ¼" gap is provided at the top and the bottom of louvers for fitting, shimming, and caulking.

_____   **12.** The top of the guardrail must be at least ___ above the roof deck elevation.

Refer to SIRTI project (Sheet A5.1). Cross-referencing between several sheets may be required to answer questions.

__Ⓐ B__   **1.** The ceiling finish in Room 317 is ___.
          A. exposed concrete
          B. suspended
          C. 14'-0" above the finished floor
          D. none of the above

_____   **2.** A(n) ___ fireproofing material is to be applied to the structural steel beams.
          A. cementitious
          B. intumescent
          C. exothermic
          D. endothermic

_____   **3.** UBC standard ___ is referenced on the In-Wall Column Fireproofing detail.

__10'-0"__   **4.** The ceiling height in Room 307 is ___.

_____   **5.** The steel deck below the concrete floor at the Mechanical Pit is ___ thick.

_____ 6. The precast concrete planter indicated in Alternate #7 is on the ___ side of the SIRTI building.
        A. north
        B. south
        C. east
        D. west

T   (F)   **7.** The construction of the Elevator lobby on the second and third floors is identical.

_____ **8.** The duct enclosure in the Fan Room has a(n) ___ hr fire rating.

_____ **9.** Door # ___ is shown in Room 326 on Building Section BB.

T   (F)   **10.** The difference in elevation between the entry/plaza level and classroom level and the office/administration level and classroom level is equal.

T   F   **11.** A crawl space is provided below the handicapped ramp.

_____ **12.** The elevation at the top edge of the sloping roof at the south side of the Mechanical Pit is ___'.

(T)   F   **13.** The column along the concrete trellis is to be finished with brick.

Refer to SIRTI project (Sheet A6.1). Cross-referencing between several sheets may be required to answer questions.

_____38_____ **1.** The ceiling insulation above the building entry has an R-___ minimum rating.

_____ **2.** The typical exterior wall construction includes ___.
        A. ⅝" exterior gypsum sheathing
        B. 3½" steel studs
        C. 1" tinted insulated glass
        D. CMU facing

_____ **3.** Information regarding the attachment of a steel pipe guardrail to the parapet is provided in Detail ___ on Sheet ___.

(T)   (F)   **4.** The top of the parapet around the Mechanical Pit is finished with steel coping.

_____ **5.** The tops of the window openings on the classroom level are ___ above the finished floor.

_____ **6.** The wall at the classroom level shown on Wall Section C is between Rooms ___ and ___.

(T)   (F)   **7.** Elastomeric sheet roofing is utilized at the Entry, Mechanical Pit, and planter area.

_____ **8.** The phenolic board insulation applied to the exterior of the cast-in-place concrete floors is ___" thick.

_____ **9.** The perimeter insulation along the interior face of foundation walls has an R-___ rating.

galvined steel **10.** The gutter along the parapet is fabricated from ___.

_____ **11.** Exposed concrete columns are to be ___.
  A. sandblasted
  B. rubbed smooth
  C. waterproofed with pitch
  D. unfinished

T   F **12.** Wood blocking is to be used when installing the steel pipe guardrail along the east parapet.

T   (F) **13.** R-19 insulation is installed the full height of the west parapet.

Refer to SIRTI project (Sheet A10.5). Cross-referencing between several sheets may be required to answer questions.

T   (F) ✗ **1.** Elevation 3 indicates two different sizes of operable wall panels for Rooms 306 and 307.

T   F **2.** All tackboards and chalkboards are to be installed the same distance above the floor.

_____ **3.** Door ___ is shown on Elevation 5.

_____5'_____ **4.** The tops of fire extinguisher cabinets are to be positioned ___ above floor level.

_____ **5.** Column ___ is shown along the right side of Elevation 16.

T   F **6.** Cable tray soffits are placed below the suspended ceiling in the corridors.

(T)   F **7.** The east wall of Corridor 314 is framed with 3½" steel studs between the concrete columns.

Refer to SIRTI project (Sheets A11.1 and A11.2). Cross-referencing between several sheets may be required to answer questions.

(T)   F **1.** Gypsum wallboard is fastened to metal furring channels on the east wall of Room 311.

_____ **2.** Special brick shape "J" at column 3G.5 is at a(n) ___° angle.

_____ **3.** The anodized aluminum-framed windows are ___" thick.

_____327_____ **4.** The doorjamb shown on Sheet A11.1, Detail 11, is for Door ___.

T   F **5.** The wall described on Sheet A11.2, Section 1, is to the south of the elevator shaft.

_____ **6.** The abbreviation "CB" on Detail 7 indicates a(n) ___.

✗ 1½" **7.** Steel framing members for the south wall of the Elevator Lobby project ___" past the end of the cast-in-place concrete wall at the mechanical chase.

T   F **8.** Metal and wood furring support the gypsum wallboard at the southwest corner of Corridor 315.

_____5'_____   **1.** Horizontal concrete joists are to be formed ___' OC around each column in the waffle slab.

_____   **2.** Keyways formed at concrete construction joints in the waffle slab measure ___.

_____   **3.** Reinforcing steel installed at the top of the floor slab at concrete construction joints is ___" in diameter.
          A. ¼
          B. ⅜
          C. ½
          D. ¾

_____8"_____   **4.** The minimum width at the bottom of two-way joists is ___".

T   F   **5.** The flange width of typical dome forms varies with the height of the forms.

_____   **6.** The faces of perimeter beams project ___ beyond the grid lines.

_____1½"_____   **7.** The distance between the exterior face of the perimeter beams and the vertical cage reinforcing steel is ___".
          A. 1
          B. 1½
          C. 2
          D. 2½

T   F   **8.** Web shear reinforcing is set in the center of each concrete joist around columns.

T   F   **9.** The hidden line on Sheet S2.3A, Detail 28, represents the suspended ceiling.

_____6"_____   **10.** Horizontal bottom rebar for floor reinforcement overlaps ___" at middle grid lines.

_____   **11.** A ___ dome form is the largest dome form used to form the two-way joist system.
          A. type 2
          B. type 3
          C. type 4
          D. typical

_____   **12.** The steel beam to support the operable walls weighs ___ lb/ft.

_____Clear_____   **13.** The abbreviation "CLR" on Sheet S5.1, Detail 23, indicates ___.

T   F   **14.** A keyway is to be formed at the tops of concrete columns to provide a connection between the columns and floor slab.

_____   **15.** The column at the southwest corner of Room 305 is reinforced with ___ rebar stirrups spaced 16" OC.
          A. four #8 rebar with #3
          B. eight #11 rebar with #4
          C. eight #7 rebar with #4
          D. 12 #11 rebar with #4

_____7/8" 1/2"_____ **16.** The rebar used to reinforce the column on the west wall between Rooms 306 and 307 at the classroom level is ___" in diameter.

_____ **17.** The rebar used to reinforce the column at the northwest corner of Stairway 313 projects ___ above the classroom-level floor to allow for overlap.

_____ **18.** Sheet ___ contains information regarding footing sizes and depths.

Refer to SIRTI project (Sheet S2.4). Cross-referencing between several sheets may be required to answer questions.

_____12"_____ **1.** The concrete beam along the east side of the elevator shaft is ___" wide.

T   F **2.** Four type 2 domes are used to form the two-way joists along the southwest portion of Room 304.

_____ **3.** A construction joint is to be formed in the concrete floor below Room ___.
    A. 303
    B. 306
    C. 307
    D. 310

_____5/8"_____ **4.** The top reinforcing steel in exterior beam 3B6 is ___" in diameter.

_____ **5.** ___ type 3B4 beams are to be formed for the classroom level.
    A. Two
    B. Three
    C. Four
    D. Six

_____ **6.** The small circles shown on the Classroom Level Framing Plan in Corridor 315 represent ___.
    A. end views of rebar
    B. roof drains
    C. ferrules
    D. floor drains

T   (F) **7.** Typical floor slab reinforcing for the classroom level is comprised of ½" diameter rebar spaced 16" OC.

T   F **8.** Reinforcing in the bottom of the beam at the north of the elevator shaft projects 3" west of grid line H.

_____ **9.** The floor slab adjacent to Stairway 322 is to be ___" thick.

_____1/4"_____ **10.** The steel angle along Stairway 322 at the bottom of the floor slab is ___" thick.

T   F **11.** A ¼" camber is formed in the beam along the east side of Room 326.

_____ **12.** Stirrups used to reinforce beams on the classroom level are spaced ___" OC.

(T)   F **13.** The concrete floor slab is recessed 1½" in Rooms 324 and 325.

_____ **14.** The floor slab throughout most of the classroom level is ___" thick.

Refer to SIRTI project (Sheets M2.3, M7.1, and M7.2). Cross-referencing between several sheets may be required to answer questions.

(T) F   **1.** The roof drain line for Room 323 is 3" in diameter.

_____   **2.** A 2" vent pipe from Room 326 near the Women's Toilet is routed up to the vent stack on the ___ side of grid line 7.
     A. north
     B. south
     C. east
     D. west

T   F   **3.** Both remote chillers are installed above the suspended ceiling.

_____ B A X _____   **4.** ___ sleeves are to be installed in the concrete wall of the mechanical chase to route pipes to the Men's and Women's Toilets.
     A. Three
     B. Four
     C. Five
     D. Six

_____   **5.** Hose bibbs under the restroom lavatories are to be installed ___" above the finished floor.

_____   **6.** The roof drain above Room 308 is ___" in diameter.

(T) F   **7.** An 8" roof drain stack is installed in the mechanical chase near the Telephone Closet.

T   F   **8.** Roof drain lines and overflow drain lines are to be sloped 2% on the classroom floor level.

_____   **9.** Detailed diagrams of the plumbing risers between the Men's and Women's Toilets are included on Sheet ___.

gallon per minute   **10.** The abbreviation "GPM" indicates ___.

_____   **11.** Fixture P-5 is a(n) ___.

T   F   **12.** A 2" vent stack is installed in Corridor 314 to the west of door 313.

Refer to SIRTI project (Sheet M3.3). Cross-referencing between several sheets may be required to answer questions.

_____ 2" _____   **1.** A(n) ___" diameter drain pipe is installed for the fire protection sectional test valve.

_____   **2.** The abbreviation "WP" related to the fire protection system indicates ___.

T   F   **3.** The fire protection system for Room 308 is installed overhead and below the floor.

T   (F)   **4.** The main supply for the preaction fire protection system is installed in the plumbing chase between the Men's and Women's Toilets.

T   F   **5.** Test drain fire protection risers are installed in each stairway on the classroom level.

6. ___ areas of the classroom level are protected with a preaction fire protection system.
   - A. Two
   - B. Three
   - C. Four
   - D. Six

(T)   F   **7.** The Telephone Closet has a wet pipe fire protection system.

___   **8.** The city water main supplying the wet pipe fire protection system is ___" in diameter.

T   F   **9.** A 6" diameter air check valve is to be installed on the preaction, dry pipe system.

_12"_   **10.** The main drain valve to the floor drain for the wet pipe fire protection system is installed ___" above the floor.

Refer to SIRTI project (Sheet M4.3). Cross-referencing between several sheets may be required to answer questions.

_22"_   **1.** Return air grilles for Rooms 306 and 307 are ___" square.

T   F   **2.** The air supply duct along the south wall of Corridor 315 should be installed as close as possible to the bottom of the structure above.

T   F   **3.** The air supply ductwork is 16" high by 16" wide.

_C_   **4.** Type 21 diffusers have a ___ cfm capacity.
   - A. 4
   - B. 24
   - C. 150
   - D. none of the above

T   F   **5.** A thermostat is to be installed along the west side of door 300B.

___   **6.** The mechanical chase near the Telephone Closet contains a(n) ___" × ___" duct that extends upward.

T   (F)   **7.** Identical diffusers are to be installed in Rooms 317 and 318.

T   F   **8.** The three air conditioner diffusers in Room 308 are equal in diameter.

T   F   **9.** Diffusers and grilles are not installed in the stairways.

_48"_   **10.** The two air diffusers at the north end of Room 323 near Stairway 322 are ___" long.

___   **11.** ___ supply air diffusers are to be installed in Room 321.
   - A. Four
   - B. Six
   - C. Eight
   - D. Ten

_____ **12.** The return air duct between Room 312 and Corridor 315 measures ___″ × ___″.

T (F) **13.** Three thermostats are to be installed in Corridor 314.

Refer to SIRTI project (Sheet M5.3). Cross-referencing between several sheets may be required to answer questions.

Hot water heating Return — T F **1.** The abbreviation "HWR" on Sheet M5.3 indicates ___.

2. The water supply pipes for the hot water heating system are routed through the south mechanical chase.

_____ **3.** The hot water supply pipe above the ceiling in Room 320 is ___″ in diameter.

(T) F **4.** Refrigerant piping for the air conditioning unit near Room 308 extends up through the roof in the Elevator Lobby.

T F **5.** No terminal units are installed above the ceiling in Room 308.

T F **6.** All hot water return pipes in Corridor 314 are 1¼″ in diameter.

J5-H5 **7.** Hot water supply and return pipes are offset near columns ___ and ___.

T F **8.** Ceiling-mounted and down-pointing cameras are not shown on Sheet M5.3.

Refer to SIRTI project (Sheet E2.3). Cross-referencing between several sheets may be required to answer questions.

317 ✗ **1.** The low-voltage relay control cabinet is mounted on the north wall in Room ___.

_____ **2.** ___ Type B luminaires in the corridors and the Elevator Lobby are to be installed on the emergency lighting circuits.
  A. Four
  B. Seven
  C. Ten
  D. Twelve

_____ **3.** Panelboard ___ provides power to the fluorescent luminaires in Room 308.

5 **4.** ___ rows of suspended indirect fluorescent luminaires are shown in Room 317.
  A. Two
  B. Three
  C. Four
  D. Five

_____ **5.** Dimmer switches for the incandescent luminaires in Room 326 are installed near doors ___ and ___.

_____ **6.** There are ___ junction boxes in Room 319.

T ✗ (F) **7.** The master control panel in Room 308 controls luminaires in Room 311.

_____ **8.** Panel 2C is installed on the wall adjacent to door ___.

_____ **9.** All track lights on the classroom level are powered by a panelboard in Room ___.

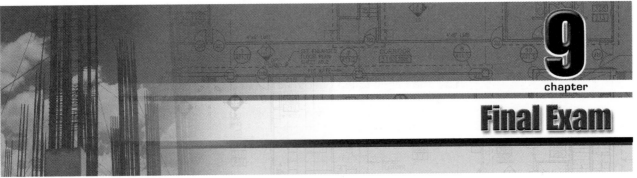

chapter 9

Final Exam

Name _____ Date _____

Refer to Riverpoint project (Sheets 2 and 47 to 58).

T   F   **1.** The two storm drain pipes that flow into catch basin 10 are equal in diameter.

T   F   **2.** The distance between the project northern building wall and the main storm drain line on the project northern side of the building is 8″.

_____ **3.** The closest grid line intersection to the second-floor drinking fountain is ___.

_____ **4.** The thickness of the roof insulation on the Second Floor Plan—South between grid lines G and H and south of grid line 8 is ___″.

T   F   **5.** The building lot generally slopes downward toward the north.

_____ **6.** The small squares throughout the vestibule area represent ___.
      A. square suspended ceiling tiles
      B. ceramic square tiles
      C. slate paving tiles
      D. construction joints

_____ **7.** The parapet cap on the circular classroom is ___.
      A. limestone
      B. precast concrete
      C. concrete masonry units
      D. shown on the Partial West Elevation/Section—Courtyard

T   F   **8.** The connection of the storm drains to the city sewer system is made at catch basin #1.

_____ **9.** The distance from the floor to the bottom of the tackboard in Room 207 is ___.

_____ **10.** There are ___ equal-size base cabinets on the east wall of the Teaching Computer Lab (Room 209).

_____ **11.** The height of the countertop in Room 209 at grid point K6 is ___.

T   F   **12.** The reinforced concrete column shown on the Partial West Elevation is outside of the doors and windows.

T   F   **13.** All windows on the north and west courtyard elevations have precast concrete sills.

T   F   **14.** There is no basement under the first-floor circular classroom.

_____ **15.** The height to the bottom of the soffit at the north wall of Room 229 is ___.

T   F   **16.** On the north elevation, the column at building line E is offset at the third floor elevation.

_____ **17.** The two squares with Xs to the north of Room 225 indicate ___.
     A. elevators
     B. air ducts
     C. service drain tubs
     D. electrical conduit chases

T   F   **18.** Each of the nine louvers on the North Elevation/Section—Courtyard is 8'-0" wide.

_____ **19.** There are ___ shelf standards on the east wall of Room 238.

_____ **20.** The shelves in Room 238 require ___ shelf brackets.

_____ **21.** The wall finish on the east wall of Room 238 is ___.
     A. exposed concrete
     B. brick veneer
     C. painted gypsum wallboard
     D. fabric-covered gypsum wallboard

_____ **22.** Column B6 in Room 238 is ___.
     A. left exposed
     B. wrapped with metal studs and gypsum wallboard
     C. covered with gypsum wallboard and fabric
     D. shown on wall detail 1A

_____ **23.** On Sheet A11.6, the door on Interior Elevation 76 leads to a(n) ___.
     A. office
     B. closet
     C. hallway
     D. stairwell

T   F   **24.** A construction joint is to be formed in exterior concrete walls at each column grid line.

T   F   **25.** The concrete snap tie locations are commonly 2'-0" OC.

_____ **26.** The wall between the Open Computer Lab (Room 207) and the Teaching Computer Lab (Room 209) on which the liquid marker board is mounted is constructed of ___.
     A. metal studs and gypsum wallboard
     B. materials that are only shown on detail 6A
     C. wood studs and gypsum wallboard
     D. demountable partition sections

T   F   **27.** There is no basement below the first floor at Wall Section I.

_____ **28.** As indicated on Sheet A11.6, interior elevation 76, cabinetry is to be finished with ___.
  A. varnished wood veneer
  B. painted particleboard
  C. painted wood veneer
  D. plastic laminate

_____ **29.** The irregular wall line on the right side wall of Interior Elevation 77 of Sheet A11.6 indicates ___.
  A. shelving
  B. recessed cabinetry
  C. a window
  D. a liquid marker board

T   F   **30.** On Wall Section J, the interior wall finish at elevation 1950' is exposed masonry.

_____ **31.** On Sheet A11.6, the concrete column on Interior Elevation 74 is located at the intersection of grid lines 6 and ___.

T   F   **32.** On Wall Section J, batt insulation is placed between the cast-in-place concrete wall sections and the interior brick wall finish.

T   F   **33.** On Wall Section J, all ceilings are exposed concrete.

_____ **34.** Detail information concerning the guardrails around the openings in the second-floor lobby is included on Sheet ___.

_____ **35.** Note 1A SB in Corridor 203 denotes ___.
  A. wall type 1A with a sound batten
  B. wall type 1A with solid brick
  C. receptacle type 1A with a safety barrier
  D. switch type 1A with a single bypass

_____ **36.** The liquid marker board in Room 210 is ___' wide.
  A. 3
  B. 4
  C. 5
  D. 6

T   F   **37.** Three waterproof receptacles are to be installed on the third floor outside wall along grid line K.

T   F   **38.** Isolated-ground receptacles on the third floor are connected to panelboard P2-3F7-1.

T   F   **39.** Details of the construction of the seismic joint are included on Sheet A12.1.

_____ **40.** The outside of the first-floor circular classroom is finished with ___.
  A. architectural concrete
  B. brick veneer
  C. stone veneer
  D. plaster

T   F   **41.** The South Elevation/Section—Courtyard indicates open web structural steel joist construction.

_____   **42.** Where the top-level roof slab joins the exterior cast-in-place concrete south and east walls, steel reinforcing bars are overlapped a distance of ___.

_____   **43.** The distance from the top of the parapet to the top of the third-floor projecting balcony roof on the south and east walls is ___.

T   F   **44.** Two isolated-ground receptacles are to be installed in Room 316.

T   F   **45.** The receptacles in the Chair's Office (Room 362) are on two separate circuits.

T   F   **46.** Five receptacles are to be installed in Corridor 304.

_____   **47.** The existing grade on Riverpoint Boulevard at the northern entrance drive is ___'.

_____   **48.** The radius of the curve shown on C2.1 for Trent Avenue is ___'.
  A. 190.52
  B. 584.25
  C. 650.00
  D. 782.6

T   F   **49.** Fluorescent luminaires in AV Storage Room 314 are recessed above the ceiling.

_____   **50.** The three concentric ovals with the numbers 1918, 1919, and 1920 on the southwest side of the building on C2.1 indicate ___.
  A. storm drains
  B. catch basins
  C. a depressed area in the grade
  D. a small mound

T   F   **51.** The luminaires in Room 328 have 3" deep 32-cell louvers.

T   F   **52.** The main circuit breaker on panelboard P2-3F7-1 has a higher circuit rating than the main circuit breaker on panelboard P2-3F7-2.

_____   **53.** The slab shown on the project east side of the building on Sheet C2.1 has a total downward slope at the north and south corners of ___".

_____   **54.** On the project south side of the building, the difference in elevation between the first-floor finish floor and the finish grade is ___".

T   F   **55.** The luminaires in Room 346 are controlled by three-way switches.

_____   **56.** Stormwater from the south side of the building flows into catch basin ___.
  A. 2A
  B. 4
  C. 7
  D. 10

_____   **57.** The area of each louver on the South Elevation/Section—Courtyard is ___ sq ft.

T   F   **58.** Ten luminaires are to be installed in the Faculty Lounge (Room 364).

T   F   **59.** Six exterior windows are to be installed in the Teaching Computer Lab (Room 209).

T   F   **60.** On Sheet A3.10, the elevation of the drain inlets for the small south roof sections is the same as the finish floor elevation.

_____ **61.** On Sheet S4.2, Section 1, the outside dimension of the tubular steel that reinforces the third-floor exterior metal-framed wall between the columns is ___″.

_____ **62.** The water supply pipe to the water closet in Room 344 is ___″ in diameter.

T   F   **63.** Office 232 is the only office on the Second Floor Plan—South with a closet.

_____ **64.** Brick-veneer walls are capped with ___.
     A. precast concrete
     B. stone
     C. a brick soldier course
     D. cast-in-place concrete

T   F   **65.** The walls between Offices 212/213 and 220/221 are identical.

T   F   **66.** Office 234 does not have an exterior window.

T   F   **67.** There is no concrete column at grid line intersection D5.

_____ **68.** The finish of the second floor at Wall Section I is ___.
     A. exposed concrete
     B. vinyl tile
     C. carpet
     D. ceramic tile

_____ **69.** On the South Elevation/Section—Courtyard, the distance from the rustication joint at elevation 1933.75′ to the top of the concrete wall supporting the second-story windows is ___.

T   F   **70.** The liquid marker board on the west wall of Room 209 is centered between door 209C and the adjacent south wall.

_____ **71.** All roof drain piping should slope ___%.

T   F   **72.** Door 209B has a glass panel at the west side.

T   F   **73.** A gypsum wallboard soffit is to be installed on the south wall of Room 229.

_____ **74.** On Sheet M2.9, the abbreviation "VTR" in Room 316 indicates ___.

T   F   **75.** The four liquid marker boards in Rooms 225, 226, 227, and 228 are equal in size.

_____ **76.** At Wall Section I, the first-floor ceiling is ___.
     A. exposed concrete
     B. gypsum board
     C. suspended acoustical tiles
     D. plaster

_____ **77.** The height of the wall-mounted fixture on the north wall of the southeast stairwell near grid line intersection K7 is ___′.

T   F   **78.** The tackboards in Rooms 227 and 240 are equal in size.

_____ **79.** The lower portion of the interior basement wall at Wall Section J is covered with ___.
       A. ½" gypsum wallboard
       B. ⅝" gypsum wallboard
       C. exposed masonry
       D. exposed concrete

_____ **80.** Detail information regarding the depth of the rustication joints is obtained on Sheet ___.

_____ **81.** The slope on the small roof deck shown on Sheet A3.10 between grid lines H and J and south of grid line 8 is ___" per foot.

_____ **82.** On the roof section on Sheet A3.10 between grid lines B and C and south of grid line 8, the difference in elevation between the overflow drain and the roof drain is ___".

_____ **83.** On the Partial West Elevation/Section—Courtyard, ___ window frames are specified for the nine small windows at the left of the elevation.
       A. brick
       B. precast concrete
       C. aluminum
       D. wood

T   F   **84.** The concrete columns shown on Interior Elevations 85 and 87 are actually the same column shown from two different views.

T   F   **85.** Hooked steel reinforcing bars that tie the third floor south concrete wall into the top-level roof concrete slab are held back from the face of the wall a minimum of ¾".

_____ **86.** The difference in height between a standard water closet and a handicapped-accessible water closet is ___".

_____ **87.** The abbreviation "COIW" in Room 342 on Sheet M2.9 indicates a(n) ___.

T   F   **88.** Brick veneer is tied to the concrete slabs at each floor level with a 4" × 3" × ¼" steel angle.

_____ **89.** The heating air supply to box 106 is ___ cfm maximum.

T   F   **90.** The metal framing angle braces above the third floor south windows are spaced 4'-0" OC.

T   F   **91.** There is no first-floor balcony on the east side of the building at grid line K.

_____ **92.** Rock excavation is required ___.
       A. in the basement area
       B. for catch basins
       C. for storm sewer installation
       D. all of the above

_____ **93.** The home run wiring for lighting from the junction box in Room 313 has ___ conductors.

T   F   **94.** For cast-in-place concrete stairway landings, the lower rebar is ½" in diameter and spaced 12" OC.

T   F   **95.** Where rebar is hooked to tie foundation walls and footings together, all hooks are turned toward the outside of the building.

T   F   **96.** In Hallway 304, the drinking fountain is supplied with a 1¼" cold water supply line.

_____ **97.** The maximum capacity of the terminal unit located in Room 112 is ___ cfm.

T   F   **98.** On the east wall at grid line K, brick veneer is anchored to the concrete slab at the second-floor slab level with 4" × 3" × ¼" angles fillet welded to 4" × 5" steel plate shims that are welded to Type C embedments.

T   F   **99.** The sink in Room 364 has a food waste disposer.

T   F   **100.** The water closet in Room 315 is floor mounted.

T   F   **101.** Plumbing connections for a refrigerator-installed icemaker are provided in Room 364.

_____ **102.** On Sheet M3.4, the abbreviation "AHU-4" in the circular classroom refers to ___.

T   F   **103.** The lavatory in Room 315 and hospitality sink in Room 316A have a common set of shutoff valves.

T   F   **104.** At Wall Section J, the parapet is constructed using metal framing members covered with brick veneer.

T   F   **105.** At the top of stairway landings, steel stringers are supported by a ³⁄₁₆" steel plate embedded into the landing.

_____ **106.** The heating capacity of the terminal unit in Room 106 is ___ cfm maximum.

_____ **107.** The ceiling luminaire in Room 310 receives ___ fluorescent light bulbs.

_____ **108.** On Section 4 of Sheet S4.2, the distance from the face of the brick veneer to the face of the raised floor slab is ___".

_____ **109.** On Sheet E2.6, the notation to the right of the circular classroom that reads "¾", 2#10, 1#10G" is interpreted as a(n) ___.

_____ **110.** The roof drain lines in the southeast stairway are ___" in diameter.

_____ **111.** The hospitality sink in Room 316A is ___.
      A. stainless steel
      B. vitreous china
      C. terrazzo with a stainless steel cap
      D. polished brass

_____ **112.** Circuit 1 of panelboard P2-3F7-2 is connected to ___ receptacle(s).

T   F   **113.** Corridor lighting and emergency lighting for the third floor south is provided on circuits 7, 9, and 11 on panelboard P4-3F7.

T   F   **114.** All roof drain lines are routed to drain down through the chase in the south-west stairway.

T   F   **115.** The lavatory in Room 315 is provided with a vandal-proof faucet.

_____ **116.** The floor drain in Room 310 is ___" in diameter.
   A. 3
   B. 4
   C. 6
   D. there is no floor drain in Room 310

_____ **117.** The 1" diameter pipe in Hallway 302 to the south of Room 357 is for ___.
   A. cold water
   B. hot water
   C. a waste vent
   D. a waste pipe

_____ **118.** The minimum throat dimension for concrete stairs on structural fill is ___".

_____ **119.** The waste pipes extending down from the water closet in Room 344 are ___" in diameter.

_____ **120.** The line along the west side of Room 349 with a notation of 2" represents a ___ pipe.
   A. cold water
   B. hot water
   C. waste vent
   D. waste

_____ **121.** On Sheet E2.6, the abbreviation "GFI" in Room 316A refers to a(n) ___.

_____ **122.** There are ___ smoke dampers on the third floor.

T   F   **123.** All water closets on the third floor are handicapped-accessible.

T   F   **124.** The east and west stairways along grid line 7 have no air returns at the first-floor level.

_____ **125.** The vent pipe for the lounge sink in Room 346 is ___.
   A. 1¼"
   B. 1½"
   C. 3"
   D. manufacturer model number 8714

_____ **126.** The air supply duct to Stairway 100A is ___" in diameter.

_____ **127.** The weld symbol on Sheet S1.4, Detail 25, is interpreted as ___.

_____ **128.** The distance from the exterior face of the east second-floor stairway to grid line J is ___.

_____ **129.** The piping in Hallway 102 indicates a hydronic system with a ___-pipe system.
   A. one
   B. two
   C. three
   D. four

T   F   **130.** The thermostat for Room 108 is located near the north entry door into the room.

T   F   **131.** Room 107 has no direct air supply duct.

_____ **132.** The second-floor concrete slab elevation at the intersection of grid lines E.4 and 6 is ___'.

_____ **133.** Air flowing into the 18″ × 18″ return air duct at the south end of Room 109 is circulated ___.
   A. to Room 108 through a 56″ × 2″ insulated duct
   B. to terminal unit 66
   C. upward through the chase at the north end of Room 113
   D. downward through the chase at the north end of Room 113

_____ **134.** There are ___ electrical receptacles in Storage Room 314.

_____ **135.** The receptacles on the north walls of Rooms 319 through 322 are connected to circuit ___ on panelboard P2-3F7-2.

_____ **136.** Supply air for Room 112 is circulated through the chase at the ___.
   A. south end of Hallway 103
   B. north end of Room 111
   C. north end of Room 113
   D. north end of Room 115

T   F   **137.** Terminal unit 64 is controlled by a thermostat in Room 121/123.

T   F   **138.** Terminal unit 103 in Room 116 has a 16″ × 16″ air inflow pipe and two air outflow pipes that are 10″ and 12″ in diameter.

_____ **139.** Each balcony section on the south wall of the Second Floor Framing Plan is ___ wide.

_____ **140.** The center-to-center spacing for the open web steel joists on the circular classroom roof is ___.

_____ **141.** Heating and cooling for Room 119 are supplied by terminal unit ___.

_____ **142.** On Sheet M3.4, the triangle with the number 3 in Stairway 103A refers to the type of ___.
   A. diffuser
   B. smoke damper
   C. fan
   D. thermostat

T   F   **143.** Luminaire type F3 is an incandescent fixture.

T   F   **144.** The 40″ × 4″ type 9 supply air diffuser for Room 123 is supplied by the 14″ × 12″ supply duct at the north side of Hallway 103.

_____ **145.** The three third-floor electrical panels are located in Room ___.

_____ **146.** The parapet along the south wall of the building is ___″ thick.

_____ **147.** Details regarding the second-floor slab reinforcing steel are included on Sheet ___.

_____ **148.** A ___–conductor cable with a grounding conductor is used for the motorized screen switches on the third floor.
    A. two
    B. three
    C. four
    D. none of the above

_____ **149.** Electrical power is provided to the third floor by conduit extending through the floor near the intersection of grid lines ___ and ___.
    A. C; 7
    B. F; 6
    C. F; 7
    D. G; 7

T   F **150.** The smoke damper in the ductwork of Corridor 303 is connected to circuit 40 on panelboard P2-3F7-2.

_____ **151.** On Sheet S4.2, Section 1, the projection of the cast-in-place concrete shade screen above the third floor balconies slopes ___″ at the top to drain water away from the building.

_____ **152.** The bus rating of panelboard P2-3F7-2 is ___ A.

_____ **153.** On Sheet E3.6, luminaires that are shaded half light and half dark on an angle denote ___.
    A. a switching configuration
    B. four lamps
    C. dimmer switches
    D. battery backup

T   F **154.** Room 319 has the same lighting installation as Room 309.

T   F **155.** Rooms 108 and 121 are both served by the same air handler.

_____ **156.** The most common luminaire installed in third-floor corridors is F___.

_____ **157.** Regarding catch basin #2A, the abbreviation "IE" indicates ___.

_____ **158.** The overall drop in elevation in the storm drain pipe between catch basins #2A and #2 is ___′.

_____ **159.** Panelboard P4-3F7 is located in Room ___.

T   F **160.** Incandescent luminaires are to be installed in Room 342.

T   F **161.** The vertical dashed line through the center of Wall Section J indicates concrete reinforcing.

T   F **162.** Steel stairway stringers are fastened to concrete slabs with ¾″ expansion bolts.

_____ **163.** In the Dean's Office (Room 311), the switching arrangement allows for ___.
    A. dimming of the fluorescent fixtures
    B. turning on half of the fixtures in banks of north or south from the south door
    C. turning on half of the fixtures in banks of north or south from the west door
    D. turning on half of the fixtures in banks of north or south from either the south or west doors

_____ **164.** The notation "17 LF 12" SD" between catch basins #7 and #8 is interpreted as ___.

_____ **165.** The largest amount of parking lot slope shown on Sheet C2.1 is ___%.

_____ **166.** Lowercase letters, such as "a", "b", and "c", are used on Sheet E3.6 to de-note ___.
    A. type of luminaires
    B. switching arrangements
    C. number of bulbs per luminaire
    D. type of connecting cables

_____ **167.** Based on Sheet C2.1, the elevation to the top of the curb at the southeast corner of the parking lot on the north side of the building is ___'.

T   F **168.** The open web steel joists supported by the masonry walls for the circular classroom are fastened to ¾" steel plates with ¼" fillet welds that are 4" long.

T   F **169.** The overall height of the open web steel joists for the penthouse is 12".

_____ **170.** The third-floor finish elevation is ___'.

_____ **171.** Where top tracks for metal-framed walls must be spliced, the length of the splice piece is ___".

T   F **172.** Where non-load-bearing stud walls are not constructed full height to the floor above and are braced to the structure above, braces are placed 4'-0" OC and made of 3" × 3" × ¼" steel angle.

_____ **173.** The horizontal distance between control joints on the Partial West Elevation/ Section—Courtyard is ___.

_____ **174.** Steel joists in the central portion of the circular classroom are supported on the north end by a ___.
    A. concrete beam
    B. steel girder
    C. masonry wall
    D. concrete slab

_____ **175.** The corrugations for the roof deck on the circular classroom are ___" deep.
    A. ½
    B. 1
    C. 1½
    D. 2

_____ **176.** The exterior handrail along the ramp on the north side of Room 105 is ___ high.

_____ **177.** The elevation of the highest point of the building walls on the North Elevation/ Section—Courtyard is ___'.

_____ **178.** The outside diameter of the circular classroom is ___.

T   F **179.** No. 6 wafer-head screws are used for metal stud framing attachments.

T   F **180.** The intermediate fasteners for the exterior ½" gypsum sheathing applied to structural studs on Wall Section J are typically spaced 16" apart.

_____ **181.** The concrete sunscreens on the windows on the South Elevation/Section—Courtyard at grid line E are ___" thick.

_____ **182.** From grid line 5 to grid line 6, the circular classroom roof has a total slope of ___".
       A. 11¼
       B. 12⅞
       C. 18
       D. 39½

T   F   **183.** At column C8, the notation "8#4 ×128H" indicates an 8' × 4' mat of high-strength steel rebar that is 128" long.

T   F   **184.** The parapet along grid line K is exposed cast-in-place concrete.

T   F   **185.** The third-floor exterior balcony wall on Sheet S4.2, Section 1, is secured with a 2½" × 2½" piece of welded tubular steel to a Type F embedment.

_____ **186.** The foundation wall at grid line K is ___" thick.

_____ **187.** At wall section I, the perimeter slab insulation extends in from the inside face of the foundation wall a minimum of ___.

_____ **188.** On the South Elevation/Section—Courtyard, the distance from the first floor windowsills to the bottom of the sunscreen is ___.

T   F   **189.** A seismic joint is located at grid line 5.

T   F   **190.** A vandal-proof faucet is to be installed with the P15 lab sink.

_____ **191.** The concrete walls at elevation 1935' on Wall Section J are ___" thick.

T   F   **192.** The second-floor windows on either side of the stairway at grid line intersection K and 7 are equal in width.

T   F   **193.** The notation "E3" on the Second Floor Framing Plan indicates that two 1" diameter rebar are placed in two joists in each strip noted.

_____ **194.** The steel gauge for the pans on steel stair bases is #___.

_____ **195.** Handrail posts are embedded in concrete curbs to a depth of ___".

_____ **196.** The typical projection from the face of a wall to the outside of a pipe handrail fastened to the wall is ___".

T   F   **197.** There are no domes placed in the second-floor slab adjacent to the concrete columns.

T   F   **198.** The second-floor slab surrounding columns B8 and J6 have similar rebar patterns.

_____ **199.** The steel lintel at a 5'-0" wide opening is ___ long.

_____ **200.** The gauge of the 6" wall studs shown on the Second Floor Framing Plan is #___.

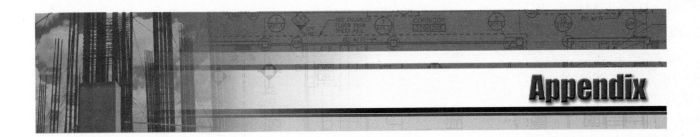

Appendix

ABBREVIATIONS . . .

Term	Abbreviation	Term	Abbreviation	Term	Abbreviation
A		beam	BM	ceramic-to-metal (seal)	CERMET
above	ABV	bearing	BRG	chamfer	CHAM or
above finished floor	AFF	bearing plate	BPL or		CHMFR
above surface of floor	ASF		BRG PL	channel	CHAN
access	ACS	bedroom	BR	check valve	CV
access panel	AP	below	BLW	chimney	CHM
acoustic	AC or ACST	bench mark	BM	chord	CHD
acoustical plaster ceiling	APC	beveled	BVL or BEV	cinder block	CINBL
acoustical tile	AT. or ACT.	beveled wood siding	BWS	circle	CIR
adhesive	ADH	bituminous	BIT	circuit	CKT
adjacent	ADJ	blocking	BLKG	circuit breaker	CB or
adjustable	ADJT or ADJ	board	BD		CIR BKR
aggregate	AGG or AGGR	board foot	BF or BD FT	circuit interrupter	CI
air circulating	ACIRC	boiler	BLR	circumference	CRCMF
air conditioner	AIR COND	bookcase	BC	cleanout	CO
air conditioning	A/C or	bookshelves	BK SH	cleanout door	CODR
	AIR COND	boulevard	BLVD	clear	CLR
alloy	ALY	boundary	BDRY	clear glass	CL GL
alloy steel	ALY STL	brass	BRS	closed circuit television	CCTV
alternate	ALTN	breaker	BRKR	closet	C, CL, CLO,
alternating current	AC	brick	BRK		or CLOS
aluminum	AL	British thermal unit	Btu	coarse	CRS
ambient	AMB	bronze	BRZ	coaxial	COAX.
American National	AMER NATL	broom closet	BC	cold air	CA
Standard	STD	building	BLDG or BL	cold-rolled	CR
American National		building line	BL	cold-rolled steel	CRS
Standards Institute	ANSI	built-in	BLTIN	cold water	CW
American Steel Wire		built-up roofing	BUR	collar beam	COL B
Gauge	ASWG			column	COL
American Wire Gauge	AWG	**C**		color code	CC
ampere	A or AMP	cabinet	CAB.	combination	COMB.
anchor	AHR	cable	CA	combustible	COMBL
anchor bolt	AB	canopy	CAN.	combustion	COMB.
appearance	APP	caulking	CK or CLKG	common	COM
apartment	APT.	cantilever	CANV	compacted	COMP
approximate	APX or APPROX	cased opening	CO	composition	COMP
architectural	ARCH.	casement	CSMT		or CMPSN
architecture	ARCH.	cast-in-place	CIP	compressed air	COMPA
area	A	casing	CSG	concrete	CONC
area drain	AD	cast iron	CI	concrete block	CCB or
as drawn	AD	cast iron pipe	CIP		CONC BLK
asphalt	ASPH	cast steel	CS	concrete floor	CCF,
asphalt roof shingles	ASPHRS	cast stone	CST or CS		CONC FLR,
asphalt tile	AT.	catch basin	CB		or
as required	AR	caulking	CK or CLKG		CONC FL
astragal	A	caulked joint	CLKJ	concrete masonry unit	CMU
automatic	AUTO.	cavity	CAV	concrete pipe	CP
automatic sprinkler	AS.	ceiling	CLG	concrete splash block	CSB
auxiliary	AUX	cellar	CEL	condenser	COND
avenue	AVE	Celsius	°C	conductor	CNDCT
azimuth	AZ	cement	CEM	conduit	CND
		cement floor	CF	construction	CONSTR
B		cement mortar	CEM MORT	construction joint	CJ or
barrier	BARR	center	CTR		CONSTR JT
barrier, moisture		centerline	CL	continuous	CONT
vapor-proof	BMVP	center matched	CM	contour	CTR
barrier, waterproof	BWP	center-to-center	C TO C	contract	CONTR or
basement	BSMT	central	CTL		CONT
bathroom	B	ceramic	CER	contraction joint	CJ or CLJ
bathtub	BT	ceramic tile	CT	contractor	CONTR
baten	BATT	ceramic-tile base	CTB	control joint	CLJ
		ceramic tile floor	CTF	conventional	CVNTL

... ABBREVIATIONS ...

Term	Abbreviation	Term	Abbreviation	Term	Abbreviation
copper	CU	double-pole switch	DP SW	finish two sides	F2S
corner	COR	double-strength glass	DSG	finished floor	FIN. FLR,
corner guard	CG	down	DN or D		FIN. FL, or
cornice	COR	downspout	DS		FNSH FL
corrugate	CORR	dozen	DOZ	firebrick	FBRK or FBCK
counter	CNTR	drain	D or DR	fire door	FDR
county	CO	drain tile	DT	fire extinguisher	FEX
cubic	CU	drawer	DWR	fire hydrant	FHY
cubic feet	CFT or CU FT	drawing	DWG	fireplace	FPL or FP
cubic foot per minute	CFM	dressed and matched	D & M	fireproof	FP or FPRF
cubic foot per second	CFS	drinking fountain	DF	fire resistant	FRES
cubic inch	CU IN.	drip cap	DC	fire wall	FW
cubic yard	CU YD	drop in pipe	D.I.P.	fixed transom	FTR
culvert	CULV	dryer	D	fixed window	FX WDW
current	CUR	drywall	DW	fixture	FIX. or FXTR
cutoff	CO	duplex	DX	flagstone	FLGSTN
cutoff valve	COV	duty cycle	DTYCY	flammable	FLMB
cut out	CO	dwelling	DWEL	flashing	FLG or FL
cutout valve	COV			floor	FLR or FL
cylinder lock	CYLL	**E**		floor drain	FD
		each	EA	flooring	FLR or FLG
D		east	E	fluorescent	FLUR or FLUOR
damper	DMPR	elbow	ELB	flush	FL
datum	DAT	electric or electrical	ELEC	foot or feet	FT
dead load	DL	electrical metallic tubing	EMT	footing	FTG
decibel	DB	electric operator	ELECT. OPR	foot per minute	FPM
deck	DK	electric panel	EP	foot per second	FPS
degree	DEG	electromechanical	ELMCH	foundation	FND or FDN
demolition	DML	electronic	ELEK	four-pole	4P
depth	DP	elevation	ELEV or EL	four-pole double-throw	
design	DSGN	elevator	ELEV	switch	4PDT SW
detail	DTL or DET	enamel	ENAM	four-pole single-throw	
diagonal	DIAG	end-to-end	E to E	switch	4PST SW
diagram	DIAG	entrance	ENTR or ENT	four-pole switch	4PSW
diameter	DIA or DIAM	equipment	EQPT	frame	FR
dimension	DIM.	equivalent	EQUIV	front view	FV
dimmer	DIM. or DMR	estimate	EST	frostproof hose bibb	FPHB
dining room	DR or	example	EX	full scale	FSC
	DNG RM	excavate	EXCA or EXC	full size	FS
direct current	DC	exchange	EXCH	furnace	FURN
direction	DIR	exhaust	EXH	furred ceiling	FC
discharge	DISCH	exhaust vent	EXHV	furring	FUR
disconnect	DISC.	existing	EXST	fuse	FU
disconnect switch	DS	expanded metal	EM	fuse block	FB
dishwasher	DW	expansion joint	EXP JT	fusebox	FUBX
distribution panel	DPNL	exterior	EXT	fuseholder	FUHLR
divided	DIV	exterior grade	EXT GR	fusible	FSBL
division	DIV				
door	DR	**F**		**G**	
door stop	DST	face brick	FB	gallon	GAL.
door switch	DSW	faceplate	FP	gallon per hour	GPH
dormer	DRM	Fahrenheit	F	gallon per minute	GPM
double-acting	DA or	fastener	FSTNR	galvanized	GV or GALV
	DBL ACT	fiberboard	FBRBD	galvanized iron	GI or
double-hung window	DHW	fiberboard, corrugated	FBDC		GALVI
double-pole double-throw	DPDT	fiberboard, double-wall	FDWL	galvanized steel	GS or
double-pole double-		fiberboard, solid	FBDS		GALVS
throw switch	DPDT SW	figure	FIG.	garage	GAR.
double-pole single-throw	DPST	finish	FIN. or FNSH	gas	G
double-pole single-		finish all over	FAO	gate valve	GTV
throw switch	DPST SW	finish grade	FG	gauge	GA
		finish one side	FIS	general contractor	GEN CONT

... ABBREVIATIONS ...

Term	Abbreviation	Term	Abbreviation	Term	Abbreviation
girder	G	inside diameter	ID	manufactured	MFD
glass	GL	install	INSTL	marble	MRB or MR
glass block	GLB or GL BL	insulation	INS or INSUL	masonry	MSNRY
glaze	GLZ	interior	INT	masonry opening	MO
glued laminated	GLULAM	iron	I	material	MTL or MATL
grade	GR	iron pipe	IP	maximum	MAX
grade line	GL			maximum working	
gravel	GVL	**J**		pressure	MWP
grille	G	jamb	JB or JMB	median	MDN
gross vehicle weight	GVW	joint	JT	medicine cabinet	MC
gross weight	GRWT	joist	J	medium	MDM
ground	GRD	junction	JCT	medium density overlay	MDO
ground (outlet)	G			meridian	MER
ground-fault circuit		**K**		metal	MET. or M
interrupter	GFCI	keyway	KWY	metal anchor	MA
ground-fault interrupter	GFI	kiln dried	KD	metal door	METD
gypsum	GYP	kitchen	K, KT, or KIT.	metal flashing	METF
gypsum board	GYP	knife switch	KN SW	metal lath and plaster	MLP
BDgypsum-plaster ceiling	GPC	knockout	KO	metal threshold	MT
gypsum-plaster wall	GPW			mezzanine	MEZZ
gypsum sheathing board	GSB	**L**		mile per gallon	MPG
gypsum wallboard	GWB	laminate	LAM	mile per hour	MPH
		laminated veneer lumber	LVL	mineral	MNRL
H		landing	LDG	minimum	MIN
hardboard	HBD	lateral	LATL	mirror	MIR
hardware	HDW	lath	LTH	miscellaneous	MISC
hazard	HAZ	laundry	LAU	miter	MIT
header	HDR	laundry tray	LT	mixture	MIX.
heat	HT	lavatory	LAV	modular	MOD
heated	HTD	leader	L	molding	MLD or MLDG
heater	HTR	left	L	monolithic	ML
heating	HTG	left hand	LH	mortar	MOR
heating, ventilating, and		length	L, LG, or	mullon	MULL.
air conditioning	HVAC		LGTH		
heavy-duty	HD	length overall	LOA	**N**	
height	HGT	level	LVL		
hertz	Hz	library	LIB	nameplate	NPL
hexagon	HEX.	living room	LR	National Electrical Code	NEC
high density overlay	HDO	light	LT	National Electrical	
high point	HPT	lighting	LTG	Safety Code	NESC
highway	HWY	light switch	LT SW	natural grade	NG
hinge	HNG	limestone	LMS or LS	negative	(–) or NEG
hollow-core	HC	linen closet	L CL	net weight	NTWT
hollow metal door	HMD	line	LN	noncombustible	NCOMBL
honeycomb	HNYCMB	lining	L	north	N
horizontal	HOR or HORZ	linoleum	LINO or LINOL	nosing	NOS
horsepower	HP	linoleum floor	LF or	not to scale	NTS
hose bibb	HB		LINO FLR	number	NO.
hot air	HA	lintel	LNTL		
hot water	HW	live load	LL		
hot water tank	HWT	living room	LR	**O**	
humidity	HMD	local	LCL	obscure glass	OBSC GL
hydraulic	HYDR	long	LG	octagon	OCT
		louver	LVR or LV	on center	OC
I		low point	LP	one-pole	OP
illuminate	ILLUM	lumber	LBR	opaque	OPA
incandescent	INCAND	luminaire	LUM	opening	OPG or OPNG
inch	IN.			open web joist	OJ, OW J, or
inch per second	IPS	**M**			OW JOIST
inch-pound	IN. LB	main	MN	opposite	OPP
infrared	IR	makeup	MKUP	optional	OPT

...ABBREVIATIONS...

Term	Abbreviation	Term	Abbreviation	Term	Abbreviation
ordinance	ORD	quarry tile floor	QTF	screen door	SCD
oriented strand board	OSB	quart	QT	screw	SCR
outlet	OUT.	quarter	QTR	scuttle	S
outside diameter	OD	quarter-round	¼RD	section	SEC or SECT.
out-to-out	O TO O	quick-acting	QA	select	SEL
overall	OA			self cleaning	SLFCLN
overcurrent	OC			self-closing	SELF CL
overcurrent relay	OCR	**R**		service	SERV or SVCE
overhead	OH or OVHD	radiator	RAD or RDTR		
		radius	R or RAD	sewer	SEW.
P		raised	RSD	sheathing	SHTH or SHTHG
paint	PNT	random	RDM		
panel	PNL	range	R	sheet	SHT or SH
pantry	PAN.	receptacle	RCPT	sheeting	SH
parallel	PRL	recessed	REC	sheet metal	SM
parallel strand lumber	PSL	rectangle	RECT	shelf and rod	SH&RD
partition	PTN	redwood	RWD	shelving	SH or SHELV
passage	PASS.	reference	REF	shingle	SHGL
peak-to-peak	P-P	reference line	REFL	shiplap	SHLP
penny (nails, etc.)	d	reflected	REFLD	shower	SH
perimeter	PERIM	refrigerator	REF or REFR	shower and toilet	SH & T
perpendicular	PERP	register	REG or RGTR	shower drain	SD
per square inch	PSI	reinforce or reinforcing	RE or REINF	shutter	SHTR
phase	PH	reinforced concrete	RC	siding	SDG
piling	PLG	reinforcing steel	RST	sidelight	SI LT
piping	PP	reinforcing steel bar	REBAR	sillcock	SC
pitch	P	required	REQD	single-phase	1PH
plaster	PLAS or PL	retaining	RETG	single-pole	SP
plastered opening	PO	reverse-acting	RACT	single-pole double-throw	SPDT
plastic laminate	PLAM	revision	REV	single-pole double throw switch	SPDT SW
plate	PL	revolution per minute	RPM	single-pole single-throw	SPST
plate glass	PG, PL GL, or PLGL	revolution per second	RPS	single-pole single-throw switch	SPST SW
		ribbed	RIB		
plate height	PL HT	right	R	single-pole switch	SP SW
platform	PLAT	right hand	RH	single-strength glass	SSG
plumbing	PLBG	rigid	RGD	single-throw	ST
plywood	PLYWD	riser	R	sink	SK or S
pneumatic	PNEU	road	RD	skylight	SLT
point	PT	roll roofing	RR	sliding door	SLD or SL DR
point of beginning	POB	roof	RF	slope	SLP
pole	P	roof drain	RD	soffit	SF
polyvinyl chloride	PVC	roofing	RFG	soil pipe	SP
porcelain	PORC	room	RM or R	soil stack	SSK
porch	P	rotor	RTR	solenoid	SOL
pound	LB	rough	RGH	solid core	SC
pound-foot	LB-FT	rough opening	RO or RGH OPNG	soundproof	SNDPRF
power	PWR			south	S
power supply	PWR SPLY	rough-sawn	RS	specific	SP
precast	PRCST	round	RND	specification	SPEC
prefabricated	PFB or PREFAB	rubber	RBR	splash block	SB
prefinished	PFN	rubber tile	RBT or R TILE	square	SQ
property	PROP	rustproof	RSTPF	square feet	SQ FT
property line	PL			square inch	SQ IN.
pull box	PB			square yard	SQ YD or SY
pull chain	PC	**S**		stack	STK
pull switch	PS	saddle	SDL or S	stained	STN
pump	PMP	safety	SAF	stainless steel	SST
		sanitary	S	stairs	ST
Q		S-beam	S	stairway	STWY
quadrant	QDRNT	scale	SC	standard	STD
quarry tile	QT	schedule	SCH or SCHED	steel	ST or STL
quarry tile base	QTB	screen	SCN, SCR, or SCRN		

. . . ABBREVIATIONS

Term	Abbreviation	Term	Abbreviation	Term	Abbreviation
steel sash	SS	triple-pole double-throw switch	3PDT SW	**W**	
stone	STN	triple-pole single-throw	3PST	wainscot	WSCT, WAIN., or WA
storage	STO or STG	triple-pole single-throw switch	3PST SW	walk-in closet	WIC
street	ST or STR	triple-pole switch	3P SW	wall	W
structural	STRL	truss	TR	wallboard	WLB
Structural Clay Products Research Foundation	SCR	two-phase	2PH	wall receptacle	WR
structural clay tile	SCT	two-pole	DP	warm air	WA
structural glass	SG	two-pole double-throw	DPDT	washing machine	WM
surfaced four sides	S4S	two-pole single throw	DPST	waste pipe	WP
surfaced one side	S1S	typical	TYP	waste stack	WS
supply	SPLY			water	WTR or W
survey	SURV			water closet	WC
suspended	SUSP	**U**		water heater	WH
switch	SW or S	underground	UGND	water line	WL
		unexcavated	UNEXC	water meter	WM
		unfinished	UNFIN or UNF	waterproof	WTRPRF
T		unit heater	UH	water-resistant	WR
telephone	TEL	unless otherwise specified	UOS	watt	W
television	TV	untreated	UTRTD	weatherproof	WTHPRF or WP
temperature	TEMP	utility	U or UTIL	weather-resistant	WR
tempered plate glass	TEM PL GL	utility room	UR or U RM	weatherstripping	WS
terra cotta	TC			weep hole	WH
terazzo	TZ or TER			welded wire reinforcement	WWR
thermostat	THERMO	**V**		west	W
thick	THK or T	vacuum	VAC	white pine	WP
three-phase	3PH	valley	VAL	wide	W
three-pole	3P	valve	V	wide-flange	W or WF
three-pole double-throw	3PDT	variance	VAR	window	WDO
three-pole single-throw	3PST	vent	V	wire glass	WG or W GL
three-way	3WAY	vent hole	VH	with	W/
three-wire	3W	ventilate	VEN	without	W/O
threshold	TH	ventilating equipment	VE	wood	WD
tile base	TB	vent pipe	VP	wood frame	WF
tile drain	TD	vent stack	VS	wrought iron	WI
tile floor	TF	vertical	VERT		
timber	TMBR	vestibule	VEST.	**Y**	
toilet	T	vinyl tile	VT or V TILE	yard	YD
tongue-and-groove	T & G	vitreous tile	VIT TILE	yellow pine	YP
top of curb	TC	void	VD		
total	TOT.	volt	V	**Z**	
township	T	voltage	V	zone	Z
tread	TR or T	voltage drop	VD		
triple-pole double-throw	3PDT				

ARCHITECTURAL SYMBOLS . . .

Material	Elevation	Plan View	Section
EARTH			
BRICK	WITH NOTE INDICATING TYPE OF BRICK (COMMON, FACE, ETC.)	COMMON OR FACE / FIREBRICK	SAME AS PLAN VIEWS
CONCRETE		LIGHTWEIGHT / STRUCTURAL	SAME AS PLAN VIEWS
CONCRETE MASONRY UNITS		OR	OR
STONE	CUT STONE / RUBBLE	CUT STONE / RUBBLE / CAST STONE (CONCRETE)	CUT STONE / CAST STONE CONCRETE / RUBBLE OR CUT STONE
WOOD	SIDING / PANEL	WOOD STUD / REMODELING / DISPLAY	ROUGH MEMBERS / FINISHED MEMBERS / PLYWOOD
PLASTER		WOOD STUD, LATH, AND PLASTER / METAL LATH AND PLASTER / SOLID PLASTER	LATH AND PLASTER
ROOFING	SHINGLES	SAME AS ELEVATION	
GLASS	OR / GLASS BLOCK	GLASS / GLASS BLOCK	SMALL SCALE / LARGE SCALE

. . . ARCHITECTURAL SYMBOLS

Material	Elevation	Plan View	Section
FACING TILE	CERAMIC TILE	FLOOR TILE	CERAMIC TILE LARGE SCALE / CERAMIC TILE SMALL SCALE
STRUCTURAL CLAY TILE			SAME AS PLAN VIEW
INSULATION		LOOSE FILL OR BATTS / RIGID / SPRAY FOAM	SAME AS PLAN VIEWS
SHEET METAL FLASHING		OCCASIONALLY INDICATED BY NOTE	
METALS OTHER THAN FLASHING	INDICATED BY NOTE OR DRAWN TO SCALE	SAME AS ELEVATION	SMALL SCALE / STEEL / CAST IRON / ALUMINUM / BRONZE OR BRASS
STRUCTURAL STEEL	INDICATED BY NOTE OR DRAWN TO SCALE	OR	REBARS / SMALL SCALE / LARGE SCALE / L-ANGLES, S-BEAMS, ETC.

PLOT PLAN SYMBOLS

NORTH	FIRE HYDRANT	WALK	ELECTRIC SERVICE
POINT OF BEGINNING (POB)	MAILBOX	IMPROVED ROAD	NATURAL GAS LINE
UTILITY METER OR VALVE	MANHOLE	UNIMPROVED ROAD	WATER LINE
POWER POLE AND GUY	TREE	BUILDING LINE	TELEPHONE LINE
LIGHT STANDARD	BUSH	PROPERTY LINE	NATURAL GRADE
TRAFFIC SIGNAL	HEDGE ROW	PROPERTY LINE	FINISH GRADE
STREET SIGN	FENCE	TOWNSHIP LINE	+ XX.00′ EXISTING ELEVATION

ELECTRICAL SYMBOLS . . .

Lighting Outlets

OUTLET BOX
AND INCANDESCENT
LIGHTING FIXTURE

CEILING   WALL

INCANDESCENT
TRACK LIGHTING

BLANKED OUTLET

DROP CORD

EXIT LIGHT AND OUTLET
BOX. SHADED AREAS
DENOTE FACES.

OUTDOOR POLE-MOUNTED
FIXTURES

JUNCTION BOX

LAMPHOLDER
WITH PULL SWITCH

MULTIPLE FLOODLIGHT
ASSEMBLY

EMERGENCY BATTERY PACK
WITH CHARGER

INDIVIDUAL
FLUORESCENT FIXTURE

OUTLET BOX AND
FLUORESCENT
LIGHTING TRACK FIXTURE

CONTINUOUS
FLOURESCENT
FIXTURE

SURFACE-MOUNTED
FLUORESCENT FIXTURE
UNDERFLOOR DUCT AND

Panelboards

FLUSH-MOUNTED
PANELBOARD
AND CABINET

SURFACE-MOUNTED
PANELBOARD
AND CABINET

Convenience Outlets

SINGLE RECEPTACLE OUTLET

DUPLEX RECEPTACLE OUTLET–
120 V

TRIPLEX RECEPTACLE OUTLET–
240 V

SPLIT-WIRED DUPLEX
RECEPTACLE OUTLET

SPLIT-WIRED TRIPLEX
RECEPTACLE OUTLET

SINGLE SPECIAL-PURPOSE
RECEPTACLE OUTLET

DUPLEX SPECIAL-PURPOSE
RECEPTACLE OUTLET

RANGE OUTLET   R

SPECIAL-PURPOSE
CONNECTION   DW

CLOSED-CIRCUIT
TELEVISION CAMERA

CLOCK HANGER RECEPTACLE   C

FAN HANGER RECEPTACLE   F

FLOOR SINGLE RECEPTACLE
OUTLET

FLOOR DUPLEX RECEPTACLE
OUTLET

FLOOR SPECIAL-PURPOSE
OUTLET

JUNCTION BOX FOR TRIPLE,
DOUBLE, OR SINGLE DUCT
SYSTEM AS INDICATED BY
NUMBER OF PARALLEL LINES

Busducts and Wireways

SERVICE, FEEDER,
OR PLUG-IN   B   B   B
BUSWAY

CABLE THROUGH
LADDER OR   C   C   C
CHANNEL

WIREWAY   W   W   W

Switch Outlets

SINGLE-POLE SWITCH S

DOUBLE-POLE SWITCH S_2

THREE-WAY SWITCH S_3

FOUR-WAY SWITCH S_4

AUTOMATIC DOOR SWITCH S_D

KEY-OPERATED SWITCH S_K

CIRCUIT BREAKER S_{CB}

WEATHERPROOF
CIRCUIT BREAKER S_{WCB}

DIMMER S_{DM}

REMOTE CONTROL SWITCH S_{RC}

WEATHERPROOF SWITCH S_{WP}

FUSED SWITCH S_F

WEATHERPROOF FUSED
SWITCH S_{WF}

TIME SWITCH S_T

CEILING PULL SWITCH S

SWITCH AND SINGLE
RECEPTACLE

SWITCH AND DOUBLE
RECEPTACLE

A STANDARD SYMBOL WITH $a.b$
AN ADDED LOWERCASE
SUBCRIPT LETTER IS $a.b$
USED TO DESIGNATE A
VARIATION IN STANDARD $S_{a.b}$
EQUIPMENT

. . . ELECTRICAL SYMBOLS

COMMERCIAL AND INDUSTRIAL SYSTEMS

PAGING SYSTEM DEVICE

FIRE ALARM SYSTEM DEVICE

COMPUTER DATA SYSTEM DEVICE

PRIVATE TELEPHONE SYSTEM DEVICE

SOUND SYSTEM

FIRE ALARM CONTROL PANEL — FACP

SIGNALING SYSTEM OUTLETS FOR RESIDENTIAL SYSTEMS

PUSHBUTTON

BUZZER

BELL

BELL AND BUZZER COMBINATION

COMPUTER DATA OUTLET

BELL RINGING TRANSFORMER — BT

ELECTRIC DOOR OPENER — D

CHIME — CH

TELEVISION OUTLET — TV

THERMOSTAT — T

UNDERGROUND ELECTRICAL DISTRIBUTION OR ELECTRICAL LIGHTING SYSTEMS

MANHOLE — M

HANDHOLE — H

TRANSFORMER-MANHOLE OR VAULT — TM

TRANSFORMER PAD — TP

UNDERGROUND DIRECT BURIAL CABLE

UNDERGROUND DUCT LINE

STREET LIGHT STANDARD FED FROM UNDERGROUND CIRCUIT

ABOVE-GROUND ELECTRICAL DISTRIBUTION OR LIGHTING SYSTEMS

POLE

STREET LIGHT AND BRACKET

PRIMARY CIRCUIT

SECONDARY CIRCUIT

DOWN GUY

HEAD GUY

SIDEWALK GUY

SERVICE WEATHERHEAD

PANEL CIRCUITS AND MISCELLANEOUS

LIGHTING PANEL

POWER PANEL

WIRING—CONCEALED IN CEILING OR WALL

WIRING—CONCEALED IN FLOOR

WIRING EXPOSED

HOMERUN TO PANELBOARD
Indicate number of circuits by number of arrows. Any circuit without such designation indicates a two-wire circuit. For a greater number of wires indicate as follows: —///— (3 wires) —////— (4 wires), etc.

FEEDERS
Use heavy lines and designate by number corresponding to listing in feeder schedule

WIRING TURNED UP

WIRING TURNED DOWN

GENERATOR — G

MOTOR — M

INSTRUMENT (SPECIFY) — I

TRANSFORMER — T

CONTROLLER

EXTERNALLY-OPERATED DISCONNECT SWITCH

PULL BOX

PLUMBING SYMBOLS . . .

Fixtures	Fixtures (continued)	Piping (continued)
STANDARD BATHTUB	LAUNDRY TRAY	CHILLED DRINKING WATER SUPPLY —— DWS ——
OVAL BATHTUB	BUILT-IN SINK	CHILLED DRINKING WATER RETURN —— DWR ——
WHIRLPOOL BATH	DOUBLE OR TRIPLE BUILT-IN SINK	HOT WATER
SHOWER STALL	COMMERCIAL KITCHEN SINK	HOT WATER RETURN
SHOWER HEAD	SERVICE SINK (SS)	SANITIZING HOT WATER SUPPLY (180°F)
TANK-TYPE WATER CLOSET	CLINIC SERVICE SINK	SANITIZING HOT WATER RETURN (180°F)
WALL-MOUNTED WATER CLOSET	FLOOR-MOUNTED SERVICE SINK	DRY STANDPIPE —— DSP ——
FLOOR-MOUNTED WATER CLOSET	DRINKING FOUNTAIN (DF)	COMBINATION STANDPIPE —— CSP ——
LOW-PROFILE WATER CLOSET	WATER COOLER	MAIN SUPPLIES SPRINKLER —— S ——
BIDET	HOT WATER TANK (HWT)	BRANCH AND HEAD SPRINKLER
WALL-MOUNTED URINAL	WATER HEATER (WH)	GAS—LOW PRESSURE —— G —— G
FLOOR-MOUNTED URINAL	METER (M)	GAS—MEDIUM PRESSURE —— MG ——
TROUGH-TYPE URINAL	HOSE BIBB (HB)	GAS—HIGH PRESSURE —— HG ——
WALL-MOUNTED LAVATORY	GAS OUTLET (G)	COMPRESSED AIR —— A ——
PEDESTAL LAVATORY	GREASE SEPARATOR (G)	OXYGEN —— O ——
BUILT-IN LAVATORY	GARAGE DRAIN	NITROGEN —— N ——
WHEELCHAIR LAVATORY	FLOOR DRAIN WITH BACKWATER VALVE	HYDROGEN —— H ——
CORNER LAVATORY	**Piping**	HELIUM —— HE ——
FLOOR DRAIN	SOIL, WASTE, OR LEADER—ABOVE-GRADE	ARGON —— AR ——
FLOOR SINK	SOIL, WASTE, OR LEADER—BELOW-GRADE	LIQUID PETROLEUM GAS —— LPG ——
	VENT	INDUSTRIAL WASTE —— INW ——
	COMBINATION WASTE AND VENT —— SV ——	CAST IRON —— CI ——
	STORM DRAIN —— SD ——	CULVERT PIPE —— CP ——
	COLD WATER	CLAY TILE —— CT ——
		DUCTILE IRON —— DI ——
		REINFORCED CONCRETE —— RCP ——
		DRAIN—OPEN TILE OR AGRICULTURAL TILE

... PLUMBING SYMBOLS

PIPE FITTING AND VALVE SYMBOLS

	FLANGED	SCREWED	BELL & SPIGOT		FLANGED	SCREWED	BELL & SPIGOT		FLANGED	SCREWED	BELL & SPIGOT
BUSHING				REDUCING FLANGE				AUTOMATIC BYPASS VALVE			
CAP				BULL PLUG				AUTOMATIC REDUCING VALVE			
REDUCING CROSS				PIPE PLUG				STRAIGHT CHECK VALVE			
STRAIGHT-SIZE CROSS				CONCENTRIC REDUCER				COCK			
CROSSOVER				ECCENTRIC REDUCER							
45° ELBOW				SLEEVE				DIAPHRAGM VALVE			
90° ELBOW				STRAIGHT-SIZE TEE				FLOAT VALVE			
ELBOW—TURNED DOWN				TEE—OUTLET UP				GATE VALVE			
ELBOW—TURNED UP				TEE—OUTLET DOWN				MOTOR-OPERATED GATE VALVE			
BASE ELBOW				DOUBLE-SWEEP TEE				GLOBE VALVE			
DOUBLE-BRANCH ELBOW				REDUCING TEE				MOTOR-OPERATED GLOBE VALVE			
LONG-RADIUS ELBOW				SINGLE-SWEEP TEE				ANGLE HOSE VALVE			
REDUCING ELBOW				SIDE OUTLET TEE—OUTLET DOWN				GATE HOSE VALVE			
SIDE OUTLET ELBOW—OUTLET DOWN				SIDE OUTLET TEE—OUTLET UP				GLOBE HOSE VALVE			
SIDE OUTLET ELBOW—OUTLET UP				UNION				LOCKSHIELD VALVE			
STREET ELBOW				ANGLE CHECK VALVE				QUICK-OPENING VALVE			
CONNECTING PIPE JOINT				ANGLE GATE VALVE—ELEVATION				SAFETY VALVE			
EXPANSION JOINT				ANGLE GATE VALVE—PLAN				GOVERNOR-OPERATED AUTOMATIC VALVE			
LATERAL				ANGLE GLOBE VALVE—ELEVATION							
ORIFICE FLANGE				ANGLE GLOBE VALVE—PLAN							

HVAC SYMBOLS

EQUIPMENT SYMBOLS	DUCTWORK	HEATING PIPING
EXPOSED RADIATOR	DUCT (1ST FIGURE, WIDTH; 2ND FIGURE DEPTH) — 12 X 20	HIGH-PRESSURE STEAM — HPS —
RECESSED RADIATOR	DIRECTION OF FLOW →	MEDIUM-PRESSURE STEAM — MPS —
FLUSH ENCLOSED RADIATOR	FLEXIBLE CONNECTION	LOW-PRESSURE STEAM — LPS —
PROJECTING ENCLOSED RADIATOR	DUCTWORK WITH ACOUSTICAL LINING	HIGH-PRESSURE RETURN — HPR —
UNIT HEATER (PROPELLER)—PLAN	FIRE DAMPER WITH ACCESS DOOR (FD | AD)	MEDIUM-PRESSURE RETURN — MPR —
UNIT HEATER (CENTRIFUGAL)—PLAN	MANUAL VOLUME DAMPER VD	LOW-PRESSURE RETURN — LPR —
UNIT VENTILATOR—PLAN	AUTOMATIC VOLUME DAMPER	BOILER BLOW OFF — BD —
STEAM	EXHAUST, RETURN OR OUTSIDE AIR DUCT—SECTION 20 X 12	CONDENSATE OR VACUUM PUMP DISCHARGE — VPD —
DUPLEX STRAINER	SUPPLY DUCT—SECTION 20 X 12	FEEDWATER PUMP DISCHARGE — PPD —
PRESSURE-REDUCING VALVE	CEILING DIFFUSER SUPPLY OUTLET 20" DIA CD 1000 CFM	MAKEUP WATER — MU —
AIR LINE VALVE	CEILING DIFFUSER SUPPLY OUTLET 20 X 12 CD 700 CFM	AIR RELIEF LINE — V —
STRAINER	LINEAR DIFFUSER 96 X 6-LD 400 CFM	FUEL OIL SUCTION — FOS —
THERMOMETER	FLOOR REGISTER 20 X 12 FR 700 CFM	FUEL OIL RETURN — FOR —
PRESSURE GAUGE AND COCK	TURNING VANES	FUEL OIL VENT — FOV —
RELIEF VALVE	FAN AND MOTOR WITH BELT GUARD	COMPRESSED AIR — A —
AUTOMATIC 3-WAY VALVE		HOT WATER HEATING SUPPLY — HW —
AUTOMATIC 2-WAY VALVE	LOUVER OPENING 20 X 12-L 700 CFM	HOT WATER HEATING RETURN — HWR —
SOLENOID VALVE		

AIR CONDITIONING PIPING

REFRIGERANT LIQUID	— RL —
REFRIGERANT DISCHARGE	— RD —
REFRIGERANT SUCTION	— RS —
CONDENSER WATER SUPPLY	— CWS —
CONDENSER WATER RETURN	— CWR —
CHILLED WATER SUPPLY	— CHWS —
CHILLED WATER RETURN	— CHWR —
MAKEUP WATER	— MU —
HUMIDIFICATION LINE	— H —
DRAIN	— D —

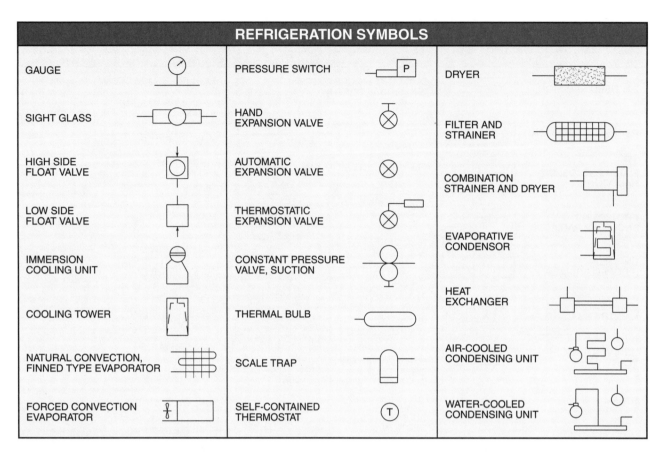

REFRIGERATION SYMBOLS

GAUGE	PRESSURE SWITCH	DRYER
SIGHT GLASS	HAND EXPANSION VALVE	FILTER AND STRAINER
HIGH SIDE FLOAT VALVE	AUTOMATIC EXPANSION VALVE	COMBINATION STRAINER AND DRYER
LOW SIDE FLOAT VALVE	THERMOSTATIC EXPANSION VALVE	EVAPORATIVE CONDENSOR
IMMERSION COOLING UNIT	CONSTANT PRESSURE VALVE, SUCTION	HEAT EXCHANGER
COOLING TOWER	THERMAL BULB	AIR-COOLED CONDENSING UNIT
NATURAL CONVECTION, FINNED TYPE EVAPORATOR	SCALE TRAP	WATER-COOLED CONDENSING UNIT
FORCED CONVECTION EVAPORATOR	SELF-CONTAINED THERMOSTAT	

WELD JOINTS AND POSITIONS

	BUTT	LAP	T	EDGE	CORNER
FLAT					
HORIZONTAL					
VERTICAL					
OVERHEAD					

JOINT TYPE	BUTT	LAP	T	EDGE	CORNER
FILLET	—			—	
SQUARE-GROOVE		—			
BEVEL-GROOVE					
V-GROOVE		—	—		
U-GROOVE		—	—		
J-GROOVE					
FLARE-BEVEL-GROOVE					
FLARE-V-GROOVE		—	—		
PLUG	—			—	
SLOT	—			—	
EDGE	—	—	—		—
FLANGED					
SPOT	—			—	
PROJECTION	—			—	
SEAM	—				
BRAZE				—	

WELD TYPES AND JOINTS

WELDING SYMBOL

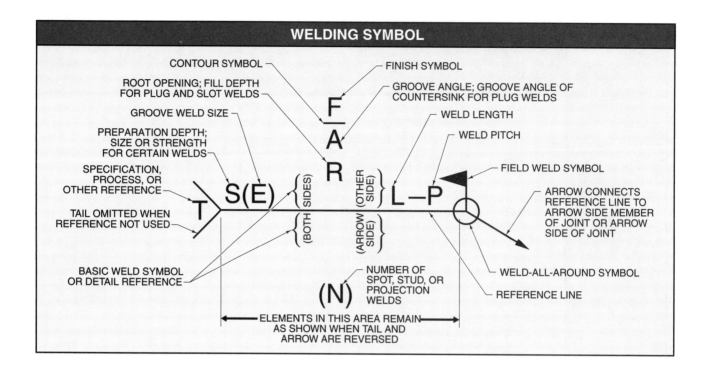

CONTOUR SYMBOL

FINISH SYMBOL

ROOT OPENING; FILL DEPTH
FOR PLUG AND SLOT WELDS

GROOVE ANGLE; GROOVE ANGLE OF
COUNTERSINK FOR PLUG WELDS

GROOVE WELD SIZE

WELD LENGTH

PREPARATION DEPTH;
SIZE OR STRENGTH
FOR CERTAIN WELDS

WELD PITCH

SPECIFICATION,
PROCESS, OR
OTHER REFERENCE

FIELD WELD SYMBOL

TAIL OMITTED WHEN
REFERENCE NOT USED

ARROW CONNECTS
REFERENCE LINE TO
ARROW SIDE MEMBER
OF JOINT OR ARROW
SIDE OF JOINT

BASIC WELD SYMBOL
OR DETAIL REFERENCE

NUMBER OF
SPOT, STUD, OR
PROJECTION
WELDS

WELD-ALL-AROUND SYMBOL

REFERENCE LINE

ELEMENTS IN THIS AREA REMAIN
AS SHOWN WHEN TAIL AND
ARROW ARE REVERSED

NONDESTRUCTIVE EXAMINATION SYMBOLS

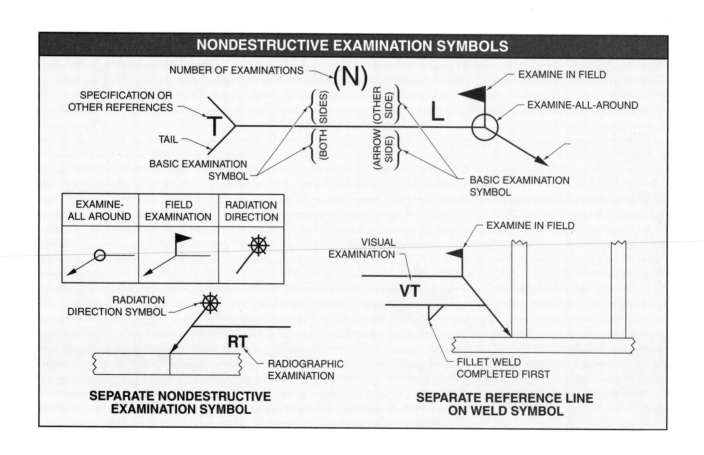

NUMBER OF EXAMINATIONS

EXAMINE IN FIELD

SPECIFICATION OR
OTHER REFERENCES

EXAMINE-ALL-AROUND

TAIL

BASIC EXAMINATION
SYMBOL

BASIC EXAMINATION
SYMBOL

| EXAMINE-
ALL AROUND | FIELD
EXAMINATION | RADIATION
DIRECTION |
|---|---|---|

RADIATION
DIRECTION SYMBOL

VISUAL
EXAMINATION

EXAMINE IN FIELD

VT

RT

RADIOGRAPHIC
EXAMINATION

FILLET WELD
COMPLETED FIRST

**SEPARATE NONDESTRUCTIVE
EXAMINATION SYMBOL**

**SEPARATE REFERENCE LINE
ON WELD SYMBOL**

ALPHABET OF LINES

Name and Use	Conventional Representation	Example	
Object Line Defines shape; outlines and details objects	—————————— THICK	OBJECT LINE	
Hidden Line Shows hidden features	⅛" (3 mm) THIN 1/32" (0.75 mm)	HIDDEN LINE	
Centerline Locates centerpoints of arcs and circles	1/16" (1.5 mm) THIN ⅛" (3 mm) ¾" (18 mm) TO 1½" (36 mm)	CENTERLINE CENTERPOINT	
Dimension Line Shows size or location **Extension Line** Defines size or location	DIMENSION LINE — DIMENSION THIN 2'–6" EXTENSION LINE	DIMENSION LINE 1¾ EXTENSION LINE	
Leader Calls out specific features	OPEN ARROWHEAD THIN X CLOSED ARROWHEAD 3X	1½ DRILL LEADER	
Cutting Plane Shows internal features	THICK ⅛" (3 mm) 1/16" (1.5 mm) A ▼ A ¾" (18 mm) TO 1½" (36 mm)	A A LETTER IDENTIFIES SECTION VIEW CUTTING PLANE LINE	
Section Line Identifies internal features	1/16" (1.5 mm) THIN	SECTION LINES	
Break Line Shows long breaks **Break Line** Shows short breaks	¾" (18 mm) TO 1½" (36 mm) THIN FREEHAND THICK	LONG BREAK LINE SHORT BREAK LINE	

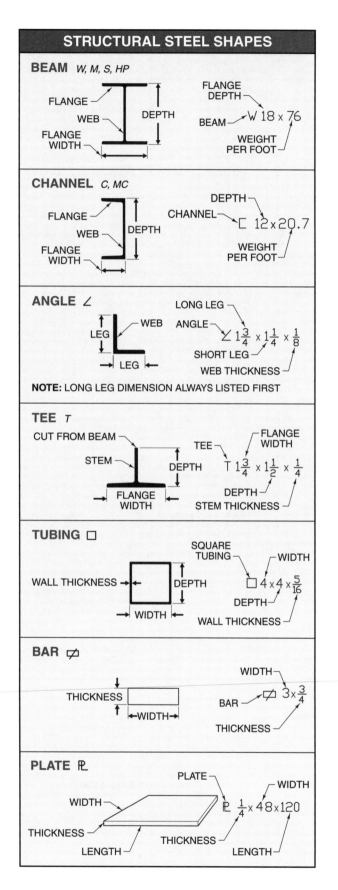

STRUCTURAL STEEL SHAPES

AMERICAN STANDARD CHANNELS

Designation	Depth d*	Flange		Web Thickness t_w*
		Width b_f*	Average Thickness t_f*	
C15 ×50	15	3¾	⅝	11/16
×40	15	3½	⅝	½
×33.9	15	3⅜	⅝	⅜
C12 ×30	12	3⅛	½	½
×25	12	3	½	⅜
×20.7	12	3	½	5/16
C10 ×30	10	3	7/16	11/16
×25	10	2⅞	7/16	½
×20	10	2¾	7/16	⅜
×15.3	10	2⅝	7/16	¼
C9 ×20	9	2⅝	7/16	7/16
×15	9	2½	7/16	5/16
×13.4	9	2⅜	7/16	¼
C8 ×18.75	8	2½	⅜	½
×13.75	8	2⅜	⅜	5/16
×11.5	8	2¼	⅜	¼
C7 ×14.75	7	2¼	⅜	7/16
×12.25	7	2¼	⅜	5/16
×9.8	7	2⅛	⅜	3/16
C6 ×12	6	2⅛	5/16	7/16
×10.5	6	2	5/16	5/16
×8.2	6	1⅞	5/16	3/16
C5 ×9	5	1⅞	5/16	5/16
×6.7	5	1¾	5/16	3/16
C4 ×7.25	4	1¾	5/16	5/16
×5.4	4	1⅝	5/16	3/16
C3 ×6	3	1⅝	¼	⅜
×5	3	1½	¼	¼
×4.1	3	1⅜	¼	3/16

* in in.

CHANNEL (C)

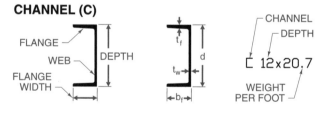

WIDE-FLANGE BEAM (W)

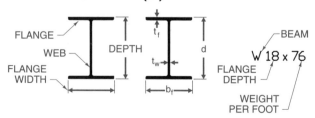

Note: See Wide-Flange Shapes—Dimensions for Detailing (next page).

WIDE-FLANGE SHAPES—DIMENSIONS FOR DETAILING

Designation	Depth d*	Flange Width b_f*	Flange Thickness t_f*	Web Thickness t_w*
W18 × 119	19	11 1/4	1 1/16	5/8
× 106	18 3/4	11 1/4	15/16	9/16
× 97	18 5/8	11 1/8	7/8	9/16
× 86	18 3/8	11 1/8	3/4	1/2
× 76	18 1/4	11	11/16	7/16
W18 × 71	18 1/2	7 5/8	13/16	1/2
× 65	18 3/8	7 5/8	3/4	7/16
× 60	18 1/4	7 1/2	11/16	7/16
× 55	18 1/8	7 1/2	5/8	3/8
× 50	18	7 1/2	9/16	3/8
W18 × 46	18	6	5/8	3/8
× 40	17 7/8	6	1/2	5/16
× 35	17 3/4	6	7/16	5/16
W16 × 100	17	10 3/8	1	9/16
× 89	16 3/4	10 3/8	7/8	1/2
× 77	16 1/2	10 1/4	3/4	7/16
× 67	16 3/8	10 1/4	11/16	3/8
W16 × 57	16 3/8	7 1/8	11/16	7/16
× 50	16 1/4	7 1/8	5/8	3/8
× 45	16 1/8	7	9/16	3/8
× 40	16	7	1/2	5/16
× 36	15 7/8	7	7/16	5/16
W16 × 31	15 7/8	5 1/2	7/16	1/4
× 26	15 3/4	5 1/2	3/8	1/4
W14 × 730	22 3/8	17 7/8	4 15/16	3 1/16
× 665	21 5/8	17 5/8	4 1/2	2 13/16
× 605	20 7/8	17 3/8	4 3/16	2 5/8
× 550	20 1/4	17 1/4	3 13/16	2 3/8
× 500	19 5/8	17	3 1/2	2 3/16
× 455	19	16 7/8	3 3/16	2
W14 × 426	18 5/8	16 3/4	3 1/16	1 7/8
× 398	18 1/4	16 5/8	2 7/8	1 3/4
× 370	17 7/8	16 1/2	2 11/16	1 5/8
× 342	17 1/2	16 3/8	2 1/2	1 9/16
× 311	17 1/8	16 1/4	2 1/4	1 7/16
× 283	16 3/4	16 1/8	2 1/16	1 5/16
× 257	16 3/8	16	1 7/8	1 3/16
× 233	16	15 7/8	1 3/4	1 1/16
× 211	15 3/4	15 3/4	1 9/16	1
× 193	15 1/2	15 3/4	1 7/16	7/8
× 176	15 1/4	15 5/8	1 5/16	13/16
× 159	15	15 5/8	1 3/16	3/4
× 145	14 3/4	15 1/2	1 1/16	11/16
W14 × 132	14 5/8	14 3/4	1	5/8
× 120	14 1/2	14 5/8	15/16	9/16
× 109	14 3/8	14 5/8	7/8	1/2
× 99	14 1/8	14 5/8	3/4	1/2
× 90	14	14 1/2	11/16	7/16
W14 × 82	14 1/4	10 1/8	7/8	1/2
× 74	14 1/8	10 1/8	13/16	7/16
× 68	14	10	3/4	7/16
× 61	13 7/8	10	5/8	3/8
W14 × 53	13 7/8	8	11/16	3/8
× 48	13 3/4	8	5/8	5/16
× 43	13 5/8	8	1/2	5/16
W14 × 38	14 1/8	6 3/4	1/2	5/16
× 34	14	6 3/4	7/16	5/16
× 30	13 7/8	6 3/4	3/8	1/4
W14 × 26	13 7/8	5	7/16	1/4
× 22	13 3/4	5	5/16	1/4
W12 × 336	16 7/8	13 3/8	2 15/16	1 3/4
× 305	16 3/8	13 1/4	2 11/16	1 5/8
× 279	15 7/8	13 1/8	2 1/2	1 1/2
× 252	15 3/8	13	2 1/4	1 3/8
× 230	15	12 7/8	2 1/16	1 5/16
× 210	14 3/4	12 3/4	1 7/8	1 3/16
W12 × 190	14 3/8	12 5/8	1 3/4	1 1/16
× 170	14	12 5/8	1 9/16	15/16
× 152	13 3/4	12 1/2	1 3/8	7/8
× 136	13 3/8	12 3/8	1 1/4	13/16
× 120	13 1/8	12 3/8	1 1/8	11/16
× 106	12 7/8	12 1/4	1	5/8
× 96	12 3/4	12 1/8	7/8	9/16
× 87	12 1/2	12 1/8	13/16	1/2
× 79	12 3/8	12 1/8	3/4	1/2
× 72	12 1/4	12	11/16	7/16
× 65	12 1/8	12	5/8	3/8
W12 × 58	12 1/4	10	5/8	3/8
× 53	12	10	9/16	3/8
W12 × 50	12 1/4	8 1/8	3/8	5/8
× 45	12	8	5/16	9/16
× 40	12	8	5/16	1/2
W12 × 35	12 1/2	6 1/2	5/16	1/2
× 30	12 3/8	6 1/2	1/4	7/16
× 26	12 1/4	6 1/2	1/4	3/8
W12 × 22	12 1/4	4	1/4	7/16
× 19	12 1/8	4	1/4	3/8
× 16	12	4	1/4	1/4
× 14	11 7/8	4	3/16	1/4
W10 × 112	11 3/8	10 3/8	1 1/4	3/4
× 100	11 1/8	10 3/8	1 1/8	11/16
× 88	10 7/8	10 1/4	1	5/8
× 77	10 5/8	10 1/4	7/8	1/2
× 68	10 3/8	10 1/8	3/4	1/2
× 60	10 1/4	10 1/8	11/16	7/16
× 54	10 1/8	10	5/8	3/8
× 49	10	10	9/16	5/16
W10 × 45	10 1/8	8	5/8	3/8
× 39	9 7/8	8	1/2	5/16
× 33	9 3/4	8	7/16	5/16
W10 × 30	10 1/2	5 3/4	1/2	5/16
× 26	10 3/8	5 3/4	7/16	1/4
× 22	10 1/8	5 3/4	3/8	1/4
W10 × 19	10 1/4	4	3/8	1/4
× 17	10 1/8	4	5/16	1/4
× 15	10	4	1/4	1/4
× 12	9 7/8	4	3/16	3/16
W8 × 67	9	8 1/4	15/16	9/16
× 58	8 3/4	8 1/4	13/16	1/2
× 48	8 1/2	8 1/8	11/16	3/8
× 40	8 1/4	8 1/8	9/16	3/8
× 35	8 1/8	8	1/2	5/16
× 31	8	8	7/16	5/16
W8 × 28	8	6 1/2	7/16	5/16
× 24	7 7/8	6 1/2	3/8	1/4
W8 × 21	8 1/4	5 1/4	3/8	1/4
× 18	8 1/8	5 1/4	5/16	1/4
W8 × 15	8 1/8	4	5/16	1/4
× 13	8	4	1/4	1/4
× 10	7 7/8	4	3/16	3/16

* in in.

ANGLES (EQUAL LEGS)— DIMENSIONS FOR DETAILING

Size and Thickness*	Size and Thickness*	Size and Thickness*
∠8 × 8 × 1⅛	∠4 × 4 × ¾	∠2 × 2 × ⅜
1	⅝	⁵⁄₁₆
⅞	½	¼
¾	⁷⁄₁₆	³⁄₁₆
⅝	⅜	⅛
⁹⁄₁₆	⁵⁄₁₆	∠1¾ × 1¾ × ¼
½	¼	³⁄₁₆
∠6 × 6 × 1	∠3½ × 3½ × ½	⅛
⅞	⁷⁄₁₆	∠1½ × 1½ × ¼
¾	⅜	³⁄₁₆
⅝	⁵⁄₁₆	⁵⁄₃₂
⁹⁄₁₆	¼	⅛
½	∠3 × 3 × ½	∠1¼ × 1¼ × ¼
⁷⁄₁₆	⁷⁄₁₆	³⁄₁₆
⅜	⅜	⅛
⁵⁄₁₆	⁵⁄₁₆	∠1 × 1 × ¼
∠5 × 5 × ⅞	¼	³⁄₁₆
¾	³⁄₁₆	⅛
⅝	∠2½ × 2½ × ½	
½	⅜	
⁷⁄₁₆	⁵⁄₁₆	
⅜	¼	
⁵⁄₁₆	³⁄₁₆	

* in in.

ANGLE (∠)

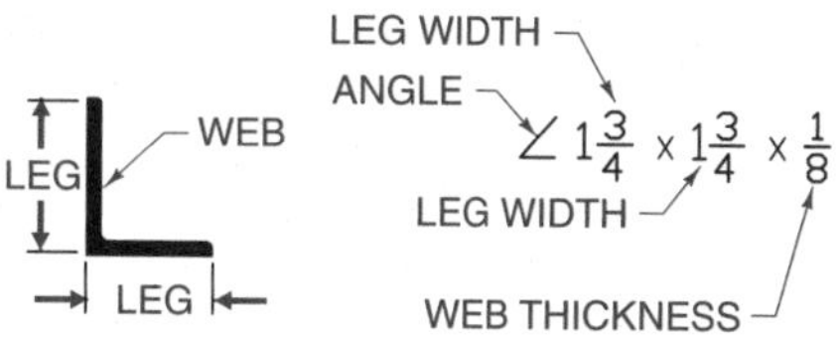

LEG — WEB — LEG

LEG WIDTH
ANGLE
LEG WIDTH
WEB THICKNESS

$$\angle\ 1\tfrac{3}{4} \times 1\tfrac{3}{4} \times \tfrac{1}{8}$$

ANGLES (UNEQUAL LEGS) — DIMENSIONS FOR DETAILING

Size and Thickness*	Size and Thickness*	Size and Thickness*
∠9 × 4 × 1	∠6 × 3½ × ½	∠3½ × 2½ × ½
⅞	⅜	⁷⁄₁₆
¾	⁵⁄₁₆	⅜
⅝	¼	⁵⁄₁₆
⁹⁄₁₆	∠5 × 3½ × ¾	¼
½	⅝	∠3 × 2½ × ½
∠8 × 6 × 1	½	⁷⁄₁₆
⅞	⁷⁄₁₆	⅜
¾	⅜	⁵⁄₁₆
⅝	⁵⁄₁₆	¼
⁹⁄₁₆	¼	³⁄₁₆
½	∠5 × 3 × ½	∠3 × 2 × ½
⁷⁄₁₆	⁷⁄₁₆	⁷⁄₁₆
∠8 × 4 × 1	⅜	⅜
⅞	⁵⁄₁₆	⁵⁄₁₆
¾	¼	¼
⅝	∠4 × 3½ × ⅝	³⁄₁₆
⁹⁄₁₆	½	∠2½ × 2 × ⅜
½	⁷⁄₁₆	⁵⁄₁₆
⁷⁄₁₆	⅜	¼
∠7 × 4 × ⅞	⁵⁄₁₆	³⁄₁₆
¾	¼	∠2½ × 1½ × ⁵⁄₁₆
⅝	∠4 × 3 × ⅝	¼
⁹⁄₁₆	½	³⁄₁₆
½	⁷⁄₁₆	∠2 × 1½ × ¼
⁷⁄₁₆	⅜	³⁄₁₆
⅜	⁵⁄₁₆	⅛
∠6 × 4 × ⅞	¼	∠2 × 1¼ × ¼
¾	∠3½ × 3 × ½	³⁄₁₆
⅝	⁷⁄₁₆	⅛
⁹⁄₁₆	⅜	∠1¾ × 1¼ × ¼
½	⁵⁄₁₆	³⁄₁₆
⁷⁄₁₆	¼	⅛
⅜		
⁵⁄₁₆		
¼		

* in in.

ANGLE (∠)

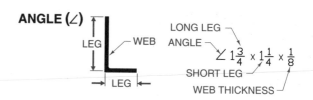

LEG — WEB — LEG

LONG LEG
ANGLE
SHORT LEG
WEB THICKNESS

$$\angle\ 1\tfrac{3}{4} \times 1\tfrac{1}{4} \times \tfrac{1}{8}$$

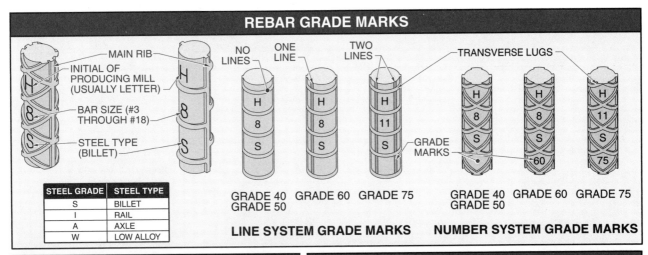

REBAR GRADE MARKS

MAIN RIB
INITIAL OF PRODUCING MILL (USUALLY LETTER)
BAR SIZE (#3 THROUGH #18)
STEEL TYPE (BILLET)

STEEL GRADE	STEEL TYPE
S	BILLET
I	RAIL
A	AXLE
W	LOW ALLOY

NO LINES — GRADE 40 / GRADE 50
ONE LINE — GRADE 60
TWO LINES — GRADE 75

LINE SYSTEM GRADE MARKS

TRANSVERSE LUGS
GRADE MARKS

GRADE 40 / GRADE 50
GRADE 60
GRADE 75

NUMBER SYSTEM GRADE MARKS

STANDARD REBAR SIZES

Bar Size Designation	Weight per Foot*	Diameter†	Cross-Sectional Area²†
#3	0.376	0.375	0.11
#4	0.668	0.500	0.20
#5	1.043	0.625	0.31
#6	1.502	0.750	0.44
#7	2.044	0.875	0.60
#8	2.670	1.000	0.79
#9	3.400	1.128	1.00
#10	4.303	1.270	1.27
#11	5.313	1.410	1.56
#14	7.650	1.693	2.25
#18	13.600	2.257	4.00

* in lb
† in in.

REINFORCING STEEL STRENGTH AND GRADE

Deformed Billet	Minimum Yield Strength*	Ultimate Strength*
ASTM A615		
Grade 40	40,000	70,000
Grade 50	50,000	90,000
Rail Steel ASTM A996		
Grade 40	50,000	80,000
Grade 50	60,000	90,000
Axle Steel ASTM A996		
Grade 40	40,000	70,000
Grade 50	50,000	90,000
Deformed Welded Wire Reinforcement ASTM A496	70,000	80,000
Plain Welded Wire Reinforcement ASTM A82		
Less than W1.2	56,000	70,000
Equal to or greater than W1.2	65,000	75,000

* in psi

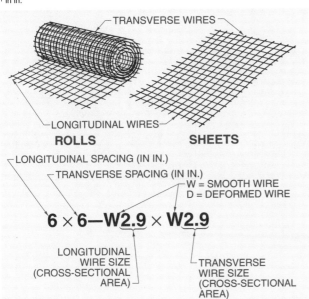

TRANSVERSE WIRES
LONGITUDINAL WIRES
ROLLS   **SHEETS**

LONGITUDINAL SPACING (IN IN.)
TRANSVERSE SPACING (IN IN.)
W = SMOOTH WIRE
D = DEFORMED WIRE

6 × 6—W2.9 × W2.9

LONGITUDINAL WIRE SIZE (CROSS-SECTIONAL AREA)
TRANSVERSE WIRE SIZE (CROSS-SECTIONAL AREA)

COMMON STOCK SIZES OF WELDED WIRE REINFORCEMENT

New Designation (W-Number)	Old Designation (Wire Gauge)	Diameter*	Weight†
6 × 6 — W1.4 × W1.4	6 × 6 — 10 × 10	⅛	21
6 × 6 — W2.0 × W2.0	6 × 6 — 8 × 8	5⁄32	29
6 × 6 — W2.9 × W2.9	6 × 6 — 6 × 6	3⁄16	42
6 × 6 — W4.0 × W4.0	6 × 6 — 4 × 4	¼	58
4 × 4 — W1.4 × W1.4	4 × 4 — 10 × 10	⅛	31
4 × 4 — W2.0 × W2.0	4 × 4 — 8 × 8	5⁄32	43
4 × 4 — W2.9 × W2.9	4 × 4 — 6 × 6	3⁄16	62
4 × 4 — W4.0 × W4.0	4 × 4 — 4 × 4	¼	85

* in in.
† in lb per 100 sq ft

				POWER FORMULAS—1ϕ, 3ϕ		
Phase	**To Find**	**Use Formula**	**Example**			
			Given	**Find**	**Solution**	
1ϕ	I	$I = \dfrac{VA}{V}$	32,000 VA, 240 V	I	$I = \dfrac{VA}{V}$ $I = \dfrac{32{,}000 \text{ VA}}{240 \text{ V}}$ $I = \textbf{133 A}$	
1ϕ	VA	$VA = I \times V$	100 A, 240 V	VA	$VA = I \times A$ $VA = 100 \text{ A} \times 240 \text{ V}$ $VA = \textbf{24,000 VA}$	
1ϕ	V	$V = \dfrac{VA}{I}$	42,000 VA, 350 A	V	$V = \dfrac{VA}{I}$ $V = \dfrac{42{,}000 \text{ VA}}{350 \text{ A}}$ $V = \textbf{120 V}$	
3ϕ	I	$I = \dfrac{VA}{V \times \sqrt{3}}$	72,000 VA, 208 V	I	$I = \dfrac{VA}{V \times \sqrt{3}}$ $I = \dfrac{72{,}000 \text{ VA}}{360 \text{ V}}$ $I = \textbf{200 A}$	
3ϕ	VA	$VA = I \times V \times \sqrt{3}$	2 A, 240 V	VA	$VA = I \times V \times \sqrt{3}$ $VA = 2 \times 416$ $VA = \textbf{832 VA}$	

				VOLTAGE DROP FORMULAS—1ϕ, 3ϕ		
Phase	**To Find**	**Use Formula**	**Example**			
			Given	**Find**	**Solution**	
1ϕ	VD	$VD = \dfrac{2 \times R \times L \times I}{1000}$	240 V, 40 A, 60′ L, 0.764 R	VD	$VD = \dfrac{2 \times R \times L \times I}{1000}$ $VD = \dfrac{2 \times 0.764 \times 60 \times 40}{1000}$ $VD = \textbf{3.67 V}$	
3ϕ	VD	$VD = \dfrac{2 \times R \times L \times I}{1000} \times 0.866$	208 V, 110 A, 75′ L, 0.194 R 0.866 multiplier	VD	$VD = \dfrac{2 \times R \times L \times I}{1000} \times 0.866$ $VD = \dfrac{2 \times 0.194 \times 75 \times 110}{1000} \times 0.866$ $VD = \textbf{2.77 V}$	

BRICK

Designation	Nominal Dimensions*			Joint Thickness*	Actual Dimensions*		
	t	h	l		t	h	l
STANDARD MODULAR	4	2⅔	8	⅜	3⅝	2¼	7⅝
				½	3½	2³⁄₁₆	7½
NORMAN	4	2⅔	12	⅜	3⅝	2¼	11⅝
				½	3½	2³⁄₁₆	11½
SCR	6	2⅔	12	⅜	5⅝	2¼	11⅝
				½	5½	2¼	11½
ENGINEER	4	3⅕	8	⅜	3⅝	2¹³⁄₁₆	7⅝
				½	3½	2¹¹⁄₁₆	7½
ECONOMY	4	4	8	⅜	3⅝	3⅝	7⅝
				½	3½	3½	7½

* in in.

CONCRETE MASONRY UNITS

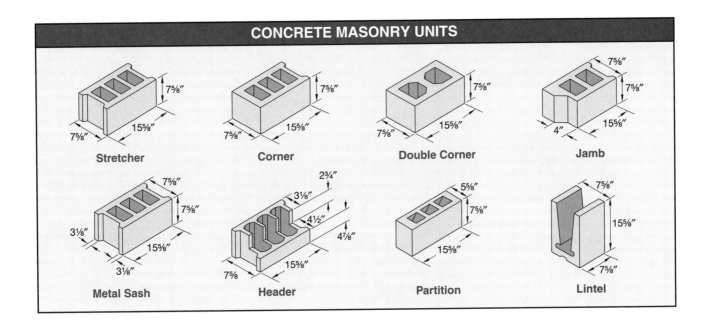

Stretcher   Corner   Double Corner   Jamb

Metal Sash   Header   Partition   Lintel

U.S. CUSTOMARY (ENGLISH) SYSTEM

LENGTH

Unit	Abbr	Equivalents
mile	mi	5280', 320 rd, 1760 yd
rod	rd	5.50 yd, 16.5'
yard	yd	3', 36"
foot	ft or '	12", 0.333 yd
inch	in. or "	0.083', 0.028 yd

AREA

$A = l \times w$

Unit	Abbr	Equivalents
square mile	sq mi or mi²	640 A, 102,400 sq rd
acre	A	4840 sq yd, 43,560 sq ft
square rod	sq rd or rd²	30.25 sq yd, 0.00625 A
square yard	sq yd or yd²	1296 sq in., 9 sq ft
square foot	sq ft or ft²	144 sq in., 0.111 sq yd
square inch	sq in. or in²	0.0069 sq ft, 0.00077 sq yd

VOLUME

$V = l \times w \times t$

Unit	Abbr	Equivalents
cubic yard	cu yd or yd³	27 cu ft, 46,656 cu in.
cubic foot	cu ft or ft³	1728 cu in., 0.0370 cu yd
cubic inch	cu in. or in³	0.00058 cu ft, 0.000021 cu yd

CAPACITY

WATER, FUEL, ETC.

VEGETABLES, GRAIN, ETC.

DRUGS

	Unit	Abbr	Equivalents
U.S. liquid measure	gallon	gal.	4 qt (231 cu in.)
	quart	qt	2 pt (57.75 cu in.)
	pint	pt	4 gi (28.875 cu in.)
	gill	gi	4 fl oz (7.219 cu in.)
	fluidounce	fl oz	8 fl dr (1.805 cu in.)
	fluidram	fl dr	60 min (0.226 cu in.)
	minim	min	⅛ fl dr (0.003760 cu in.)
U.S. dry measure	bushel	bu	4 pk (2150.42 cu in.)
	peck	pk	8 qt (537.605 cu in.)
	quart	qt	2 pt (67.201 cu in.)
	pint	pt	½ qt (33.600 cu in.)
British imperial liquid and dry measure	bushel	bu	4 pk (2219.36 cu in.)
	peck	pk	2 gal. (554.84 cu in.)
	gallon	gal.	4 qt (277.420 cu in.)
	quart	qt	2 pt (69.355 cu in.)
	pint	pt	4 gi (34.678 cu in.)
	gill	gi	5 fl oz (8.669 cu in.)
	fluidounce	fl oz	8 fl dr (1.7339 cu in.)
	fluidram	fl dr	60 min (0.216734 cu in.)
	minim	min	¹⁄₆₀ fl dr (0.003612 cu in.)

MASS AND WEIGHT

COAL, GRAIN, ETC.

GOLD, SILVER, ETC.

DRUGS

	Unit	Abbr	Equivalents
avoirdupois	ton		2000 lb
	short ton	t	2000 lb
	long ton		2240 lb
	pound	lb or #	16 oz, 7000 gr
	ounce	oz	16 dr, 437.5 gr
	dram	dr	27.344 gr, 0.0625 oz
	grain	gr	0.037 dr, 0.002286 oz
troy	pound	lb	12 oz, 240 dwt, 5760 gr
	ounce	oz	20 dwt, 480 gr
	pennyweight	dwt or pwt	24 gr, 0.05 oz
	grain	gr	0.042 dwt, 0.002083 oz
apothecaries'	pound	lb ap	12 oz, 5760 gr
	ounce	oz ap	8 dr ap, 480 gr
	dram	dr ap	3 s ap, 60 gr
	scruple	s ap	20 gr, .333 dr ap
	grain	gr	0.05 s, 0.002083 oz, 0.0166 dr ap

METRIC SYSTEM			
	Unit	**Abbr**	**Number of Base Units**
LENGTH	kilometer	km	1000
	hectometer	hm	100
	dekameter	dam	10
	*meter	m	1
	decimeter	dm	.1
	centimeter	cm	.01
	millimeter	mm	.001
AREA	square kilometer	sq km or km²	1,000,000
	hectare	ha	10,000
	are	a	100
	square centimeter	sq cm or cm²	.0001
VOLUME	cubic centimeter	cu cm, cm³, or cc	.000001
	cubic decimeter	dm³	.001
	*cubic meter	m³	1
CAPACITY	kiloliter	kL	1000
	hectoliter	hL	100
	dekaliter	daL	10
	*liter	L	1
	cubic decimeter	dm³	1
	deciliter	dL	.10
	centiliter	cL	.01
	milliliter	mL	.001
MASS AND WEIGHT	metric ton	t	1,000,000
	kilogram	kg	1000
	hectogram	hg	100
	dekagram	dag	10
	*gram	g	1
	decigram	dg	.10
	centigram	cg	.01
	milligram	mg	.001

* base units

METRIC TO ENGLISH EQUIVALENTS

	Unit	British Equivalents		
LENGTH	kilometer	0.62 mi		
	hectometer	109.36 yd		
	dekameter	32.81′		
	meter	39.37″		
	decimeter	3.94″		
	centimeter	0.39″		
	millimeter	0.039″		
AREA	square kilometer	0.3861 sq mi		
	hectacre	2.47 A		
	acre	119.60 sq yd		
	square centimeter	0.155 sq in.		
VOLUME	cubic centimeter	0.061 cu in.		
	cubic decimeter	61.023 cu in.		
	cubic meter	1.307 cu yd		
		cubic	**dry**	**liquid**
CAPACITY	kiloliter	1.31 cu yd		
	hectoliter	3.53 cu ft	2.84 bu	
	dekaliter	0.35 cu ft	1.14 pk	2.64 gal.
	liter	61.02 cu in.	0.908 qt	1.057 qt
	cubic decimeter	61.02 cu in.	0.908 qt	1.057 qt
	deciliter	6.1 cu in.	0.18 pt	0.21 pt
	centiliter	0.61 cu in.		338 fl oz
	milliliter	0.061 cu in.		0.27 fl dr
MASS AND WEIGHT	metric ton	1.102 t		
	kilogram	2.2046 lb		
	hectogram	3.527 oz		
	dekagram	0.353 oz		
	gram	0.035 oz		
	decigram	1.543 gr		
	centigram	0.154 gr		
	milligram	0.015 gr		

ENGLISH TO METRIC EQUIVALENTS

		Unit	Metric Equivalent
LENGTH		mile	1.609 km
		rod	5.029 m
		yard	0.9144 m
		foot	30.48 cm
		inch	2.54 cm
AREA		square mile	2.590 k²
		acre	0.405 hectacre, 4047 m²
		square rod	25.293 m²
		square yard	0.836 m²
		square foot	0.093 m²
		square inch	6.452 cm²
VOLUME		cubic yard	0.765 m³
		cubic foot	0.028 m³
		cubic inch	16.387 cm³
CAPACITY	U.S. liquid measure	gallon	3.785 L
		quart	0.946 L
		pint	0.473 L
		gill	118.294 mL
		fluidounce	29.573 mL
		fluidram	3.697 mL
		minim	0.061610 mL
	U.S. dry measure	bushel	35.239 L
		peck	8.810 L
		quart	1.101 L
		pint	0.551 L
	British imperial liquid and dry measure	bushel	0.036 m³
		peck	0.0091 m³
		gallon	4.546 L
		quart	1.136 L
		pint	568.26 cm³
		gill	142.066 cm³
		fluidounce	28.412 cm³
		fluidram	3.5516 cm³
		minim	0.059194 cm³
MASS AND WEIGHT	avoirdupois	short ton	0.907 t
		long ton	1.016 t
		pound	0.454 kg
		ounce	28.350 g
		dram	1.772 g
		grain	0.0648 g
	troy	pound	0.373 kg
		ounce	31.103 g
		pennyweight	1.555 g
		grain	0.0648 g
	apothecaries'	pound	0.373 kg
		ounce	31.103 g
		dram	3.888 g
		scruple	1.296 g
		grain	0.0648 g

DECIMAL EQUIVALENTS OF A FOOT

Inches	Decimal Foot Equivalent	Inches	Decimal Foot Equivalent	Inches	Decimal Foot Equivalent
1/16	0.0052	4 1/16	0.3385	8 1/16	0.6719
1/8	0.0104	4 1/8	0.3438	8 1/8	0.6771
3/16	0.0156	4 3/16	0.3490	8 3/16	0.6823
1/4	0.0208	4 1/4	0.3542	8 1/4	0.6875
5/16	0.0260	4 5/16	0.3594	8 5/16	0.6927
3/8	0.0313	4 3/8	0.3646	8 3/8	0.6979
7/16	0.0365	4 7/16	0.3698	8 7/16	0.7031
1/2	0.0417	4 1/2	0.3750	8 1/2	0.7083
9/16	0.0469	4 9/16	0.3802	8 9/16	0.7135
5/8	0.0521	4 5/8	0.3854	8 5/8	0.7188
11/16	0.0573	4 11/16	0.3906	8 11/16	0.7240
3/4	0.0625	4 3/4	0.3958	8 3/4	0.7292
13/16	0.0677	4 13/16	0.4010	8 13/16	0.7344
7/8	0.0729	4 7/8	0.4063	8 7/8	0.7396
15/16	0.0781	4 15/16	0.4115	8 15/16	0.7448
1	0.0833	5	0.4167	9	0.7500
1 1/16	0.0885	5 1/16	0.4219	9 1/16	0.7552
1 1/8	0.0938	5 1/8	0.4271	9 1/8	0.7604
1 3/16	0.0990	5 3/16	0.4323	9 3/16	0.7656
1 1/4	0.1042	5 1/4	0.4375	9 1/4	0.7708
1 5/16	0.1094	5 5/16	0.4427	9 5/16	0.7760
1 3/8	0.1146	5 3/8	0.4479	9 3/8	0.7813
1 7/16	0.1198	5 7/16	0.4531	9 7/16	0.7865
1 1/2	0.1250	5 1/2	0.4583	9 1/2	0.7917
1 9/16	0.1302	5 9/16	0.4635	9 9/16	0.7969
1 5/8	0.1354	5 5/8	0.4688	9 5/8	0.8021
1 11/16	0.1406	5 11/16	0.4740	9 11/16	0.8073
1 3/4	0.1458	5 3/4	0.4792	9 3/4	0.8125
1 13/16	0.1510	5 13/16	0.4844	9 13/16	0.8177
1 7/8	0.1563	5 7/8	0.4896	9 7/8	0.8229
1 15/16	0.1615	5 15/16	0.4948	9 15/16	0.8281
2	0.1667	6	0.5000	10	0.8333
2 1/16	0.1719	6 1/16	0.5052	10 1/16	0.8385
2 1/8	0.1771	6 1/8	0.5104	10 1/8	0.8438
2 3/16	0.1823	6 3/16	0.5156	10 3/16	0.8490
2 1/4	0.1875	6 1/4	0.5208	10 1/4	0.8542
2 5/16	0.1927	6 5/16	0.5260	10 5/16	0.8594
2 3/8	0.1979	6 3/8	0.5313	10 3/8	0.8646
2 7/16	0.2031	6 7/16	0.5365	10 7/16	0.8698
2 1/2	0.2083	6 1/2	0.5417	10 1/2	0.8750
2 9/16	0.2135	6 9/16	0.5469	10 9/16	0.8802
2 5/8	0.2188	6 5/8	0.5521	10 5/8	0.8854
2 11/16	0.2240	6 11/16	0.5573	10 11/16	0.8906
2 3/4	0.2292	6 3/4	0.5625	10 3/4	0.8958
2 13/16	0.2344	6 13/16	0.5677	10 13/16	0.9010
2 7/8	0.2396	6 7/8	0.5729	10 7/8	0.9063
2 15/16	0.2448	6 15/16	0.5781	10 15/16	0.9115
3	0.2500	7	0.5833	11	0.9167
3 1/16	0.2552	7 1/16	0.5885	11 1/16	0.9219
3 1/8	0.2604	7 1/8	0.5938	11 1/8	0.9271
3 3/16	0.2656	7 3/16	0.5990	11 3/16	0.9323
3 1/4	0.2708	7 1/4	0.6042	11 1/4	0.9375
3 5/16	0.2760	7 5/16	0.6094	11 5/16	0.9427
3 3/8	0.2813	7 3/8	0.6146	11 3/8	0.9479
3 7/16	0.2865	7 7/16	0.6198	11 7/16	0.9531
3 1/2	0.2917	7 1/2	0.6250	11 1/2	0.9583
3 9/16	0.2969	7 9/16	0.6302	11 9/16	0.9635
3 5/8	0.3021	7 5/8	0.6354	11 5/8	0.9688
3 11/16	0.3073	7 11/16	0.6406	11 11/16	0.9740
3 3/4	0.3125	7 3/4	0.6458	11 3/4	0.9792
3 13/16	0.3177	7 13/16	0.6510	11 13/16	0.9844
3 7/8	0.3229	7 7/8	0.6563	11 7/8	0.9896
3 15/16	0.3281	7 15/16	0.6615	11 15/16	0.9948
4	0.3333	7	0.6667	12	1.0000

DECIMAL EQUIVALENTS OF AN INCH

Fraction	Decimal	Fraction	Decimal	Fraction	Decimal	Fraction	Decimal
1/64	0.015625	17/64	0.265625	33/64	0.515625	49/64	0.765625
1/32	0.03125	9/32	0.28125	17/32	0.53125	25/32	0.78125
3/64	0.046875	19/64	0.296875	35/64	0.546875	51/64	0.796875
1/16	0.0625	5/16	0.3125	9/16	0.5625	13/16	0.8125
5/64	0.078125	21/64	0.328125	37/64	0.578125	53/64	0.828125
3/32	0.09375	11/32	0.34375	19/32	0.59375	27/32	0.84375
7/64	0.109375	23/64	0.359375	39/64	0.609375	55/64	0.859375
1/8	0.125	3/8	0.375	5/8	0.625	7/8	0.875
9/64	0.140625	25/64	0.390625	41/64	0.640625	57/64	0.890625
5/32	0.15625	13/32	0.40625	21/32	0.65625	29/32	0.90625
11/64	0.171875	27/64	0.421875	43/64	0.671875	59/64	0.921875
3/16	0.1875	7/16	0.4375	11/16	0.6875	15/16	0.9375
13/64	0.203125	29/64	0.453125	45/64	0.703125	61/64	0.953125
7/32	0.21875	15/32	0.46875	23/32	0.71875	31/32	0.96875
15/64	0.234375	31/64	0.484375	47/64	0.734375	63/64	0.984375
1/4	0.250	1/2	0.500	3/4	0.750	1	1.000

DECIMAL AND METRIC EQUIVALENTS

Fractions	Decimal Inches	Millimeters
1/16	0.0625	1.58
1/8	0.125	3.18
3/16	0.1875	4.76
1/4	0.250	6.35
5/16	0.3125	7.97
3/8	0.375	9.52
7/16	0.4375	11.11
1/2	0.500	12.70
9/16	0.5625	14.29
5/8	0.625	15.88
11/16	0.6875	17.46
3/4	0.750	19.05
13/16	0.8125	20.64
7/8	0.875	22.22
1	1.000	25.40

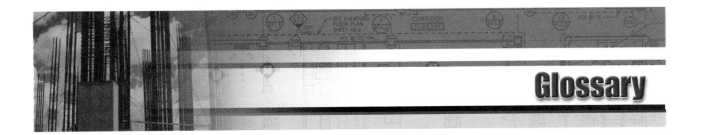

A

admixture: A substance other than water, aggregate, or portland cement that is added to concrete to affect its properties.

air diffuser: An air distribution outlet used to deflect and mix air; commonly fitted with vanes or louvers.

air entrainment: The occlusion of minute air bubbles in concrete during mixing.

air handling unit: A prefabricated air distribution assembly that uses fans, ductwork, heating and cooling coils, humidifiers, dehumidifiers, and controls to condition and distribute air throughout a building.

American Welding Society (AWS): A trade association that is devoted to promoting welding and related processes.

architect: A construction professional who designs and creates plans for a structure.

architectural concrete: Concrete that is exposed to view and requires care in selecting, forming, placing, and finishing of the surface to meet appearance standards.

B

bay: The space between the centers of adjacent columns along exterior walls.

beam and column construction: A structural steel construction method consisting of bays of framed structural steel that are repeated to create large structures.

blockout: A form that creates an open space within a concrete structure under construction in which fresh concrete is not to be placed.

bollard: A metal post, reinforced stone post, or concrete-filled metal post placed in a manner that inhibits vehicular traffic.

bolster: A continuous supporting device used to support reinforcing steel in the bottoms of slabs.

bond breaker: A blend of organic chemicals, inorganic chemicals, or polymers that prevents fresh concrete from adhering to previously set concrete.

borrowed light frame: A window opening in an interior partition between two interior areas.

building information modeling (BIM): An integrated, electronically managed system that aligns all working drawings, structural drawings, and shop drawings into a consistent system.

C

cable: A flexible assembly of two or more conductors with a protective outer sheathing.

cable tray: An assembly of sections and associated fittings that form a rigid structural system used to support cables and raceways.

caisson pile (caisson): A cast-in-place concrete pile made by driving a steel tube into the ground, excavating it, inserting reinforcement, and filling the cavity with concrete.

camber: A slight upward curvature in a horizontal structural member that is designed to compensate for deflection of the member under load.

cast-in-place concrete: Concrete formed using a system of wood, earth, fiberglass, or metal forming materials that are set in their specified locations and shapes and act as molds for the fresh concrete.

catch basin: A reservoir or tank in a surface water drainage system that is used to obstruct the flow of objects that will not readily pass through a sewer while allowing surface water to freely flow.

cathedral glass: A type of art glass available in a wide variety of colors and with many surface treatments.

C channel: A structural steel member in which the outside of the flanges and web are perpendicular to one another, forming a "C" shape.

chair: An individual supporting device used to support reinforcing steel in the proper position before and during concrete placement.

chase: An enclosure in a structure that allows for the placement of electrical, plumbing, or mechanical wiring and piping extending from floor to floor.

circuit breaker: An overcurrent protection device with a mechanism that may manually or automatically open a circuit when an overload condition or short circuit occurs.

column: The principle vertical load-bearing member in a steel structure.

combined sewer system: A drainage system that utilizes one set of waste piping for both stormwater and sanitary drainage systems.

commissioning agent: A project consultant who ensures that all building systems perform according to the design intent and the owner's operational needs.

compression test: A quality control test that is used to determine the compressive strength of concrete.

compressive strength: The measured maximum resistance of concrete to axial loading; expressed as force per cross-sectional area (typically pounds per square inch).

conductor: A wire used to control the flow of electrons in an electrical circuit.

conduit: Tube or pipe that supports and protects electrical conductors.

construction manager: An individual or company managing administrative and technical responsibilities of a construction project.

contraction joint: A groove made in a concrete surface to create a weakened plane and control the location of cracking.

curtain wall: A non-load-bearing, prefabricated or job-built, glass or metal panel supported by metal frame members or set with various clip systems that are attached to structural members.

dehumidification: The process of removing moisture from the air.

design-bid-build: A project delivery system where multiple contractors competitively bid against one another on a construction project.

design-build: A project delivery system where an owner negotiates the price of the construction project with a contractor.

detention pond: A low-lying area that is designed to temporarily hold stormwater while slowly draining to another location.

dome: A square prefabricated pan form used in two-way concrete joist systems.

door closer: A hardware device that closes a door and controls the speed and closing action of the door.

earthwork: Digging and excavating operations.

easement: A strip of land, commonly along the perimeter of property lines, that is used for placement and maintenance of utilities.

engineer: An individual educated in a specific discipline who is trained to analyze data and solve issues related to a construction project.

engineered fill: Material such as sand or crushed stone that meets certain physical requirements and placement methods.

estimator: A construction professional who determines overhead expenses, material quantities and costs, and labor needs and costs and assembles bids from others in the construction team to calculate project cost estimates.

expansion joint: A joint that separates sections of concrete to allow for movement caused by expansion and contraction of concrete.

exterior insulation and finish systems (EIFS): Exterior finish systems consisting of exterior sheathing, insulation board, reinforcing mesh, a base coat of acrylic copolymers, and a finish of acrylic resins.

falsework: Temporary shoring used to support work under construction.

fireproofing: The process of applying protective material to a structural or finish member to increase the fire resistance of the member and protect it from failure in the event of excessive heat or fire.

fire rating: The measure of resistance of a material or component to failure when exposed to fire; expressed as the number of hours a material or component will retain its integrity.

firestopping: The process of applying and installing a material or member that seals open construction to inhibit the spread of fire, smoke, and fumes in a structure.

float glass: Glass manufactured by floating liquid glass on a surface of liquid tin and slowly annealing it to produce a transparent, flat glass.

foreman: A skilled trades professional who manages a small trade-specific crew to perform work projects on a construction site.

form: A temporary structure or mold used to support concrete while it is setting and gaining adequate strength to be self-supporting.

formwork: The total system for supporting fresh concrete, including the sheathing that contacts the concrete, supporting members, hardware, and braces.

fuse: An overcurrent protection device that provides protection from short circuits and overloads.

gang form: A series of smaller prefabricated form panels fastened together to create a larger form.

gas metal arc welding (GMAW): An arc welding process that uses an arc between a continuous wire electrode and the weld pool.

gas shielded arc welding: A group of welding processes that includes gas metal arc welding and gas tungsten arc welding.

gas tungsten arc welding (GTAW): An arc welding process in which a shielding gas protects the arc between a nonconsumable tungsten electrode and the weld pool.

general contractor: An individual or company that agrees to fulfill an entire building agreement with various items or types of work to be completed such as carpentry and electrical and plumbing work.

geotextiles: Sheets or rolls of material that stabilize and retain soil or earth in position on slopes or in other unstable conditions.

geothermal heating system: A heating system that uses pipes installed below the surface of the ground to utilize the naturally occurring temperature of the earth to regulate fluid temperature.

girt: A type of purlin used as a horizontal stiffener between columns around the perimeter of a building.

glass block: An opaque or transparent hollow block made of glass that is used in non-load-bearing walls and partitions.

glued laminated (glulam) timber: An engineered wood product comprised of layers of wood members (lams) that are joined together with adhesives to form larger members.

gray water: Water that is collected at a building site and reused for irrigation or other nonpotable water applications.

grille: A decorative perforated or louvered cover installed at the inlet to the return air ductwork or in a ceiling or floor panel.

ground fault interrupter: An electrical device that automatically deenergizes a circuit or portion of a circuit when the grounded current exceeds a predetermined value that is less than the value required to operate the overcurrent protection device.

gusset: A piece of plate or sheet steel that is welded or bolted to all members at the connection point.

gypsum board: An interior surfacing material consisting of a fireproof gypsum core covered with heavy paper on both sides.

heat-absorbing glass: A type of float glass used to control heat and glare in large areas of glass.

heat exchanger: A device that transfers heat between two fluids that are physically separated and does not allow the fluids to mix.

heat-strengthened glass: Glass produced by reheating annealed float glass close to its softening point and rapidly cooling the glass using high-velocity air.

HP-shape beam: A structural steel member with wide parallel flanges that are joined by a perpendicular web.

humidification: The process of adding moisture to the air.

impervious paving: Paving that is watertight and does not allow for water to penetrate the surface of the paved area.

insulating glass: Glass made of two pieces of sheet glass that are separated by a sealed air space.

integrated project delivery (IPD): A project delivery method that provides an open information exchange for all those involved in the construction process.

invert elevation: The inside elevation at the bottom of the inside flow line of a pipe.

isolated-ground receptacle: A receptacle in which the grounding terminal is isolated from the device yoke or strap.

journeyman: A skilled trades professional who accomplishes the actual tasks of construction such as setting concrete forms, installing electrical conduit, or setting plumbing fixtures in place.

laminated glass: Specialty glass produced by placing a clear sheet of polyvinyl butyral (PVB) between two sheets of glass and subjecting the composite to intense heat and pressure.

lean construction: A project management system that utilizes various information and planning processes adopted from manufacturing to maximize value and minimize waste.

LEED Accredited Professional (AP): A consultant with extensive knowledge of green building practices and the Leadership in Energy and Environmental Design (LEED) rating system; employed by an owner, contractor, or construction manager to ensure that environmental processes are followed where specified.

long span construction: A structural steel construction method in which large girders and trusses constructed of large horizontal steel members are fastened together (built up) to span large areas.

low-emittance (low-E) coating: A metal or metallic oxide coating that reduces the passage of heat and ultraviolet rays through windows.

MasterFormat™: A uniform system of numbers and titles for organizing information about construction requirements, products, and activities into a standard sequence.

match line: An aligning mark on a print that is used when a drawing is too large to be contained on one sheet.

obscure glass: Glass used to obscure a view or create a design by sandblasting and/or etching one or both sides of the glass.

open web steel joist: A structural steel member constructed with steel angles and bars that are used as chords with steel angles or bars extending between the chords at an angle.

owner: The purchaser of a construction project.

oxyacetylene cutting: An oxygen cutting process in which heat is generated by an oxygen and acetylene flame to sever and remove the metal.

pan: A prefabricated form unit used in one-way concrete joist systems.

panelboard: A single frame or box, or a group of frames and boxes, with buses and overcurrent protection devices, such as circuit breakers, that may have switches to control light, heat, or power circuits.

panic bar: A door hardware device with a horizontal bar that releases a latch or bolt when pushed.

patterned glass: Glass that has one side finished with a fine grid or an unpolished surface so the glass is translucent.

pervious paving: A surface treatment that allows stormwater to run through the paved area into the ground below, thereby increasing the absorption of stormwater at the building site and minimizing stormwater runoff.

photovoltaic system: An electrical system that uses crystalline silicon wafers that are sensitive to sunlight to directly convert solar radiation into electricity.

pile: A slender concrete, steel, or timber structural member driven or otherwise embedded on end into the ground to support a load or compact the soil.

pile cap: A structural member placed on top of, and usually fastened to, the top of a pile or group of piles and used to transmit imposed loads to the pile or group of piles.

pipe pile: A steel cylinder with open ends that is driven into the ground with a pile-driving rig and then excavated and filled with concrete.

plasma arc cutting: An arc cutting process that uses a constricted arc to heat the metal; it removes the molten metal with a high-velocity jet of ionized gas.

plate glass: Sheet glass that is ground and polished after it is formed and cooled.

plate steel: Flat steel that is more than 3/16″ thick.

plat plan: A drawing of a parcel of land providing its legal description and showing existing and planned construction.

post-tensioning: A method of prestressing concrete in which steel tendons are tensioned after concrete is placed.

potable water: Water that is free from impurities that could cause disease or harmful physiological effects.

potable water supply system: A system of water service pipe, water distribution pipe, fittings, and valves inside or outside a building (but within the property lines) that supplies and distributes potable water to points of use within a building.

preaction fire protection system: A type of dry pipe system in which two separate events—smoke detection and heat development—must occur for the sprinkler system to activate.

precast concrete: Concrete that is formed, placed, and cured to a specific strength at a location other than its final location.

pre-engineered metal building construction: A structural steel construction method consisting of prefabricated structural steel members including beams, columns, girts, and trusses.

prestressed concrete: Concrete in which internal stresses are introduced to such a degree that tensile stresses resulting from service loads are counteracted to the desired degree.

pretensioning: A method of prestressing concrete in which steel tendons are tensioned before concrete is placed.

purlin: A horizontal support member that spans between beams, columns, or joists to carry intermediate loads, such as wall or roof decking materials.

R

rebar: Steel bars containing lugs (protrusions) that form an irregular surface to increase the ability of the bars to interlock and adhere to the surrounding concrete.

reflected ceiling plan: A plan view of a ceiling that indicates ceiling-mounted items such as air diffusers, exhaust fans, air intakes, and luminaires (lighting fixtures).

retention pond: Stormwater treatment system designed to hold a specific amount of water indefinitely.

sanitary drainage system: A system of sanitary drainage pipes and fittings that convey wastewater and waterborne waste from plumbing fixtures and appliances, such as water closets and sinks, to the sanitary sewer system.

security glass: Glass composed of multiple layers of polycarbonate plastic and/or glass bonded together under intense heat and pressure and coated with a PVB or polycarbonate film.

shaking out: The process of unloading steel members in a planned manner to minimize moving of members during erection.

sheet glass: Glass manufactured by drawing the glass vertically or horizontally and slowly annealing (cooling) it to produce a high-gloss surface.

sheet steel: Flat steel that is 3/16″ thick or less.

shielded metal arc welding (SMAW): An arc welding process in which the arc is shielded by the decomposition of the coating on an electrode that has a similar metallic composition to the steel being joined.

shop drawing: A detailed drawing or set of drawings, created by the contractor, subcontractor, supplier, fabricator, or manufacturer that shows how building elements will be fabricated or installed and contains other pertinent information such as construction materials, finishes, or dimensions.

silt fence: A barrier used to contain soil sediment that consists of geotextiles or straw bales secured in place with wood or metal stakes.

slump: The measure of consistency of freshly mixed concrete.

solar thermal heating system: A heating system that moves water through solar panels to preheat the water before it enters a boiler.

spandrel beam: A beam in the perimeter of a building that spans from column to column.

spandrel glass: Tinted glass or glass with a polyvinyl fluoride coating.

specifications: Written supplements to working drawings that provide additional construction information.

S-shape beam: A structural steel member with thickened parallel flanges that are joined with a perpendicular web.

steel angle: A structural steel member with an L-shaped cross section with equal- or unequal-width legs.

steel erection: The process of setting a structural framework in place.

step-down transformer: An electrical power regulating device with more windings in the primary winding than the secondary winding, resulting in a load voltage that is less than the initial applied voltage.

stirrup: Reinforcement used to resist shear and diagonal tension stresses in a structural member.

stormwater drainage system: A piping system used to convey rainwater or other surface water from landscaped areas, paved areas, and roofs to a storm sewer or other places of disposal or storage, such as catch basins, retention ponds, or detention ponds.

stormwater pollution prevention plan (SWPPP): A written plan detailing the pollution control measures that will be taken on a construction site to prevent stormwater and silt runoff.

structural steel joist: A lightweight beam spaced less than 4′ OC from adjacent joists.

subcontractor: An individual or company that performs work in one specific trade area on a construction project and works under an agreement with the general contractor.

superintendent: A construction professional who coordinates the many diverse operations on a construction site, such as managing the delivery and placement of project materials, working with various construction team members on labor allocations, and coordinating schedules for subcontractors.

supplier: A company that supplies materials, building components, or equipment to contractors and subcontractors.

tempered glass: Glass produced by heating a sheet of glass during manufacturing to near its softening point and then quickly cooling the glass under carefully controlled conditions.

terminal unit: A heating system component that transfers heat from hot water in a hot water system to the air in the building system.

theodolite: A precision surveying instrument that establishes and verifies vertical and horizontal angles through electronic distance measurement.

tie rod: A cylindrical steel member with threads on each end.

tilt-up concrete construction: A type of precast concrete construction in which concrete for large wall sections is placed on a flat slab adjacent to its final position.

total station instrument: A precision surveying instrument used to perform leveling, plumbing, and horizontal and vertical measurement operations.

transformer: An electrical device that contains no moving parts and is used to increase or decrease the voltage and current ratings of an alternating current (AC) circuit.

uninterruptable power supply (UPS): An electrical power system that provides continuous current to designated circuits in the event of a power outage.

valve: A device that controls the flow of fluids within pipes.

vent piping system: A system of vent pipes and fittings that provides circulation of air to or from a sanitary drainage system to protect trap seals from siphonage or back pressure.

wall bearing construction: A structural steel construction method in which horizontal steel beams and joists are supported by other construction materials such as masonry and reinforced concrete.

water-cement ratio: The ratio of the weight of water (in lb) to the weight of cement (in lb) per cubic yard of concrete.

water chiller: A device that cools water and is commonly used in large commercial buildings requiring large quantities of cooled water for cooling purposes.

welded wire reinforcement: Heavy-gauge wire joined in a grid and used to reinforce and increase the tensile strength of concrete.

wide-flange beam: A structural steel member with parallel flanges that are joined with a perpendicular web.

wire glass: Glass embedded with wire mesh to provide additional security.

Z channel: A structural steel member in which the flanges extend parallel from the web but in opposite directions.

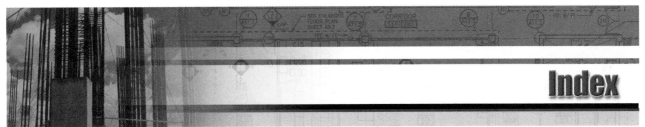

abbreviations, *16,* 16, *146*
above-grade slabs, 119–122, *122*
access control systems, 189
admixtures, *134,* 134
aircraft cables, *247*
air diffusers, *162,* 162
air entrainment, *134,* 134
air handling units, 160, *161*
American Standard beams, 85
Americans with Disabilities Act (ADA), *239,* 239, 253, *253,* 254
American Welding Society (AWS), 92
American Wire Gauge (AWG) numbers, 164
architects, 2
architectural (A) prints, 14, *110,* 110
architectural abbreviations, *214*
architectural concrete, 184
architectural notes, *194*
architectural symbols, 200, *201*
Architectural Woodwork Institute (AWI), *33,* 33
atmosphere factors, 6
AWG numbers, 164
AWS, 92

bar code cards, 189
bars, structural steel, 89–90
bays, 76
beam and column construction, *76,* 76
beam identification codes, *122*
beams, *78,* 85–86
  exposed, 192
  precast concrete, *132,* 132–133
bearing piles, *81*
bidding forms, 5
bidding information, 22, *23*

BIM, *5,* 5, *51,* 51
bituminous roofing, 190
blockout, 129
bollards, 62
bolster, 126, *127*
bolts, *90,* 90–91, *91*
bond breaker, 129
borrowed light frames, *196,* 196
boulders, 54
boundaries, 50
bowstring trusses, 87, *88*
brace inserts, 129, *130*
bracing, structural steel, 88–89, *89*
bricks, *248*
bridge construction, 12
bridge decking, 95
building codes, 8, *9*
building considerations, 1
Building Construction Index, 21
building decking, 93
building information modeling (BIM). *5,* 5, *51,* 51
building lines, 52
building materials, 6–8
building methods, 6–8
building planning, 1
  delivery systems, 3–5, *5*
  geographic requirements, 8–9, *9*
  participants, 1–3, *2*
  technological developments, *6,* 6–8, *7, 8*
butterfly valves, 150

cabinetry, *38,* 200, *201*
cables, 164
cable trays, *165,* 165, *229,* 229
CAD systems, 50
caisson piles, 111
camber, 224
carpeted floors, 196
casework, *38,* 200, *201, 223*

casting beds, 129
cast-in-place concrete, *11,* 11, 109–111, *110*
  above-grade slabs, 119–122, *122*
  column caps, 117, *118*
  columns, *116,* 116–117, *117*
  footings, 112, *113, 125*
  formwork, 109
  pile caps, *112,* 112
  piles, 111, *112,* 132
  prestressed concrete, 127–128
  reinforcing steel installation, 125–128
  reinforcing steel materials, 123–128
  slabs-on-grade, 113–116
  walls, 117–119, *119, 125*
catch basins, *57,* 57, *58*
cathedral glass, 186
caulk, 152
C channels, 80, *81,* 86
ceilings, 198–199
cementitious fireproofing, 152
chairs, 126, *127*
channels, 76
chases, 220
check valves, 150, *151*
chlorinated polyethylene (CPE), 190
circuit breakers, 166
civil (C) prints, 14
civil drawings. *See* site plans
civil engineering, 49–53, *50*
  legal information, 49–51
  site layout, 52–53
classroom/exhibit rooms, SIRTI building, 247–249, *249*
clay, 54
climate, 8
CMUs, 10
column caps, 117, *118*
columns, 76, *78*
  concrete, *116,* 116–117, *117*
  exposed, 192
  structural steel, 82–85, *83, 84*
combined sewer systems, 146
commercial construction prints, *16*
commissioning agents, 3
compactors, *56*

USING THE *PRINTREADING FOR HEAVY COMMERCIAL CONSTRUCTION* INTERACTIVE CD-ROM

Before removing the Interactive CD-ROM from the protective sleeve, please note that the book cannot be returned for refund or credit if the CD-ROM sleeve seal is broken.

System Requirements

To use this Windows®-compatible CD-ROM, your computer must meet the following minimum system requirements:
- Microsoft® Windows® 7, Windows Vista®, or Windows® XP operating system
- Intel® 1.3 GHz processor (or equivalent)
- 128MB of available RAM (256MB recommended)
- 335MB of available hard disc space
- 1024 × 768 monitor resolution
- CD-ROM drive (or equivalent optical drive)
- Sound output capability and speakers
- Microsoft® Internet Explorer® 6.0 or Firefox® 2.0 web browser
- Active Internet connection required for Internet links

Opening Files

Insert the Interactive CD-ROM into the computer CD-ROM drive. Within a few seconds, the home screen will be displayed allowing access to all features of the CD-ROM. Information about the usage of the CD-ROM can be accessed by clicking on Using This Interactive CD-ROM. The Quick Quizzes®, Illustrated Glossary, Flash Cards, Print Sets, Printreading Tests, Media Clips, and ATPeResources.com can be accessed by clicking on the appropriate button on the home screen. Clicking on the American Tech web site button (www.go2atp.com) accesses information on related educational products. Unauthorized reproduction of the material on this CD-ROM is strictly prohibited.